This book is to be returned on or before
the last date stamped below.

Worksho

Part I

An introductory cou

200

Workshop technology

Part I
An introductory course

SI UNITS

Dr. W. A. J. CHAPMAN

MSc (Eng), FIMechE, HonFIProdE

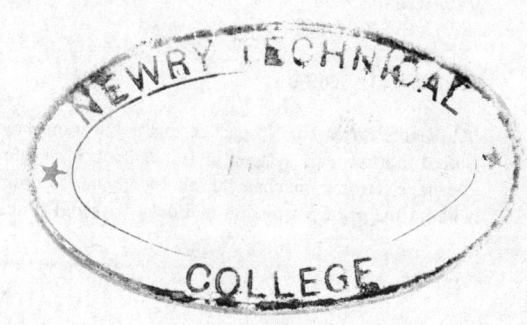

EDWARD ARNOLD

First Published 1943
by Edward Arnold (Publishers) Limited
25 Hill Street
London W1X 8LL

Reprinted 1944 (twice), 1945, 1947, 1950, 1952, 1953, 1954
Second Edition 1955
Reprinted 1956, 1957, 1958
Third Edition 1959
Reprinted 1960
Fourth Edition 1962
Reprinted 1963, 1964, 1965, 1968
Fifth Edition 1972
Reprinted 1973, 1974

ISBN 0 7131 3269 8

Printed in Great Britain by Butler & Tanner Ltd
Frome and London

Preface to first edition

The scheme of the present work was in my mind, and partly on paper, before the present catastrophe overtook us, and, with many other works of peaceful construction, was shelved. The trend of events, however, indicate that its publication should be of value to the war effort, so that I have persevered in completing this first portion.

My chief regret is that I was not qualified to write it fifteen years ago. At that time the status of our artisan workers was being allowed to fall relative to other members of the community. The artisan professions were neglected by educationalists, by writers and by public thought, a state of affairs which led to a dearth of suitably qualified prospective workers. What this policy has cost us cannot be assessed, but it is now obvious that it almost brought us to disaster.

That the artisan is worthy of a reasonable social position can be testified by those who have lived and worked with him. That the work he does calls for high skill and training should be evident from jobs such as that of turning the pair of rolls shown at Figs 39 and 41, an operation demanding a level of performance comparable with that in many so-called professions.

I have written for the men who do such work and for those who aspire to it. It is difficult to compromise between academic theory and everyday practice, but it is hoped that the fundamental principles underlying workshop processes have been clearly explained and that the reader will be able to understand more easily the processes described and to perform them in the workshop. The work covered in this volume is approximately that of the first two years of a senior part-time course in Technical Colleges and includes most of the material necessary for the City and Guilds Intermediate Examination in Machine Shop Engineering. It should also be useful to students for the preliminary work in Workshop Technology leading to the Higher National Certificate in Production Engineering.

In the preparation of the diagrams a great deal of help has been given by the kindness of various firms and institutions. Their names are appended to the illustrations concerned and I should like to offer them my sincere thanks. Finally, I feel that the Publishers in producing this book at such a price, and amid such difficulties, have performed a task worthy of commendation.

W.A.J.C.

Oakengates 1943

Preface to the fifth edition

The fifth edition of this book marks somewhat of a landmark in its career. To keep abreast of modern developments it is 'going metric' and this new edition has been revised to conform with the SI metric system of units and conventions. The Fourth edition, in British imperial units, will still be available for use in those countries where the SI system has not yet been adopted.

A further noteworthy aspect in the career of *Workshop Technology* is that, since it was first published in 1943, its circulation has exceeded $\frac{1}{4}$ million copies and it has found friends in all parts of the world.

The need to preserve high standards of quality in the workshop and a dedicated approach on the part of technicians and craftsmen is just as important now as it was at the time of the first publication. So, also, is the need to provide suitable and reliable information for those outside the workshop who seek the help of the book.

I have amended the text, where necessary, to cover developments in practice and equipment, and, on this account, I should like to record my appreciation to the firms and individuals who have given generous help and advice. The Publishers, as always, have produced a good job and I hope that the book will continue to further the aims I entertained when I originally produced it.

W.A.J.C.

Hatfield 1971

Contents

1 Introduction—materials— iron and steel

The wealth of a community is measured by the variety and quality of the articles it possesses for its use and consumption. All the material things we possess are made from substances which in the first place are won from the earth, or from Nature. Our prosperity depends upon our ability to convert these raw materials into useful articles of consumption, and to distribute these articles equitably amongst the various members of our community.

The production of our engineering workshops forms an important part of our general industrial scheme since a large proportion of our industries are of an engineering nature. Moreover, other industries, such as clothing, food, etc., depend to a large extent upon the help of workshop mechanics for making and maintaining the machinery they use. Our ability, therefore, to maintain a high standard of skill in our engineering workshops is an important factor in our general scheme of well-being, and the reader may be sure that the efforts he makes to acquire efficiency in workshop technique will react to the benefit of the community. From his own individualistic standpoint the fact of his being a master of his trade will add to his independence, increase his status and income, and ultimately enable him to enjoy a larger share of the commodities he is helping to make.

The knowledge that a skilled workshop engineer must possess takes many years of observant experience to acquire. The reader should note the term 'observant' experience, because unless he enters the workshops prepared to give thought and enquiry to every piece of work he will never acquire the sense and skill which go to make the thorough craftsman. One person may spend years in doing a certain job and learn less from his experience than another, who after a few weeks of studious application, has mastered the technique and is ready to advance further.

The amount of 'book' knowledge necessary to become a skilled craftsman is not great, and the reader need not be despondent of his chances if he is not very good at mathematics or English. What he must have, however, is an interest in mechanical things, and the ability to give patient application to a job until it is satisfactorily completed. An impatient worker, unable to concentrate his attention to the completion of what may be a rather tedious job, is unlikely to make a first-class artisan. The would-be workshop engineer should have a keen realisation of his responsibilities both towards his work and to his fellows. He should be capable of doing a fair day of good work without the need of constant supervision and should be scrupulously honest. The reader will find that if he adopts as his guiding principle the idea of giving service, without bothering overmuch about what profit he will get out of it, the financial and other rewards will follow also without his needing to worry about them.

Many of the aspects of workshop technique can only be acquired by experience, by teaching on the spot and by contacts with experienced craftsmen. There is much, however, which may be learned by reading, and the object of these two books is to provide the student with a preliminary insight into his work. He should endeavour to keep his knowledge up-to-date by reading technical journals, and it is hoped that the time he so spends may help him to bring a critical and scientific viewpoint to bear upon his work. Probably, when he has passed through the first stages of his experience, he will feel like studying more advanced books, and he can be assured that by his reading and experience he will find the workings of Nature to be as profoundly evident in the workshop as in any other branch of natural science.

Materials

Since the object of all manufacturing is to start with a raw material, and to add work to it until it is at some finished, or semi-finished state, it seems logical that we should commence by taking some note of the materials we use. The materials chiefly used in the shops are metals, and these are divided into those which come from iron and those which do not. The iron group which includes all irons and steels are called *ferrous* metals (ferrum = iron), while the others are specified as *non-ferrous*.

Iron and steel

There is scarcely a single aspect of our daily lives in which iron does not play an important part. Our food, our means of transport, the buildings we use, our fireside and its comforts—iron is necessary for them all. The magnetic property of iron alone, by making possible the electrical age in which we live, has conferred incalculable benefits on humanity.

Pure iron

Pure iron is a soft metal having a crystalline structure. A representation of its microstructure is shown in Fig. 1. This diagram, together with others which will be seen later, has been drawn to represent a photomicrograph, a method used extensively for the examination and study of metals. To obtain a photomicrograph a small piece of the metal is filed to a flat surface, and the surface is gradually brought to a high finish by successive polishings. The polished surface is then treated with dilute acid which etches the constituents and their boundaries, enabling them to be distinguished when seen through a microscope. By placing a camera at the microscope eyepiece, a photograph of the microstructure may be obtained. The microscope reveals the structure to an enlarged scale, and the magnification is usually stated underneath the photograph (e.g. $\times 200$ as shown in Fig. 1).

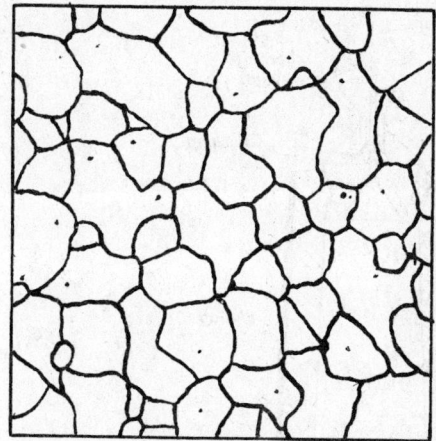

Fig. 1 Microstructure of pure iron $\times 200$

It will be noticed that the structure of pure iron looks like a map with fields separated by hedges. The shapes looking like the fields are the crystals, and the lines like hedges are where they join together. It may help us to visualise the structure of the metal if we imagine it as similar to bricks and mortar; the crystals being the bricks and their boundaries the mortar. A section through such a structure would appear somewhat similar to our photograph. If the reader measures the average size of the crystals on Fig. 1 and divides by the magnification factor, he will find that the crystal size of the specimen is about $\frac{4}{100}$ mm. The size of the crystals in iron and steel depends upon the treatment the metal has received, and in the steel used in the workshop the crystals may vary from $\frac{1}{100}$ to $\frac{20}{100}$ mm. These micrographs represent a portion of metal structure about as large as a pin point.

Pig iron

In its pure state, iron has very few practical uses, and if we tried to machine it we should find it so soft that it would tear badly and give a poor finish. Moreover, to obtain pure iron is a difficult process, because during smelting it is not easy to rid the metal entirely of certain elements for which it possesses a great affinity. We may regard iron in its pure state, therefore, as a metal mainly of theoretical interest, but one which may comprise 99% of

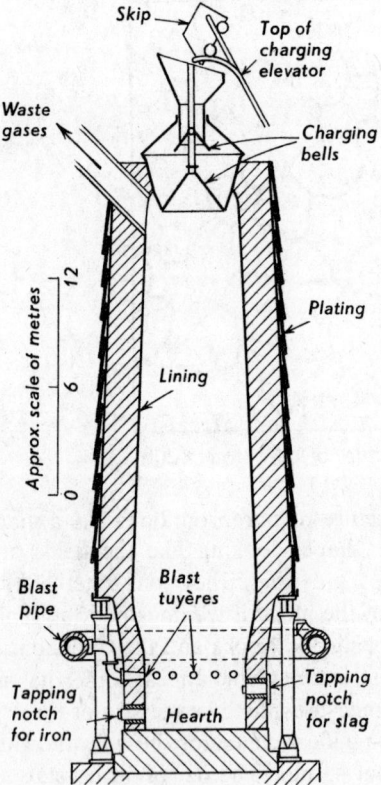

Fig. 2 Section through blast furnace

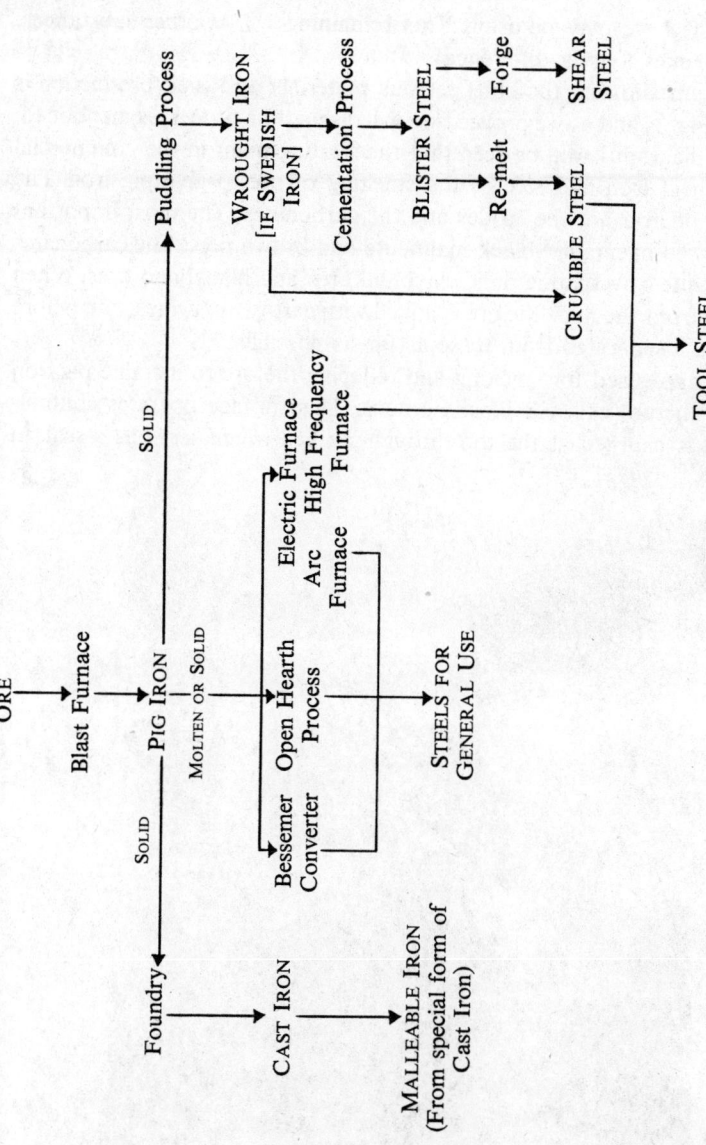

Fig. 3 Diagram showing production of the materials in the iron and steel group
(*The production of tool steel from the Cementation process is now mainly of historical interest only.*)

a certain steel we may be using. That remaining 1% of other substances, however, makes a great difference!

A diagram showing the chief ferrous materials and their production is shown at Fig. 3, and as we proceed we will discuss the processes mentioned. From the diagram it will be seen that the starting point in the commercial preparation of iron and steel is the smelting of ore to give pig iron. The chief ores of iron are the oxides and the carbonates. The most important oxides are red haematite, black magnetite and brown ores. The carbonates are represented by spathic ores, clayband ores and blackband ores. When it is taken from the mine the ore is mixed with earthy impurities, and before smelting it is separated from these as far as possible.

The furnace used for smelting and reducing the ore to metallic pig iron is a tall structure called a Blast Furnace. This furnace operates continuously and is charged at the top through the bell which serve as a seal. In

The United Steel Companies Ltd

Fig. 4 General view of blast furnace plant

addition to the ore, the furnace is charged with suitable quantities of coke and limestone. The coke provides fuel for maintaining the heat necessary to carry on the reducing action and at the same time some of the carbon (as carbon monoxide) from the coke combines with the iron oxides, reducing them to iron. The limestone serves as a flux and combines with the non-metallic portions of the ore to form a molten slag. Hot air is blown into the lower portion of the furnace via the tuyères, and as the action proceeds, the molten iron and slag fall to the bottom of the furnace, the slag floating on top of the iron. From time to time the slag is tapped off from the hole shown, and the molten iron from the hearth. Diagrams showing a blast-furnace section and a general view of an installation are shown at Figs. 2 and 4.

According to the type of plant the molten iron may be used in one or more of the following ways:

(a) Cast in pig beds.
(b) Cast in pig-casting machine.
(c) Transferred in hot metal ladles direct to an adjacent steelmaking process. This applies to composite iron and steel plants (Fig. 14).

Pig beds are sand moulds in the form of a grid prepared in the ground alongside the furnace, and the bars (pigs) cast in this way are about 1 metre long with a D-shaped cross-section measuring approximately 100 mm each way.

The modern method of casting pig iron which has largely superseded the older pig-bed method is to employ a pig-casting machine. This consists of a series of haematite iron or steel moulds, each about $\frac{2}{3}$ metre $\times$ 200 mm $\times$ 90 mm, fixed to an endless chain conveyor, the assembly being mounted on a suitable construction arranged on a slight slope. Molten iron from the blast furnace is conveyed in a ladle and poured into a channel feeding the moulds as they emerge from the underside of their travel at the foot of the slope of the conveyor. The moulds charged with iron move up the slope, the metal cooling off on the way up, and when the top is reached the turn-over of the moulds, as they reverse to the underside of the conveyor, causes the pigs to discharge into a wagon below. (The moulds are washed with lime or tar to prevent adhesion of the molten iron.) Generally, casting machines have two strands of pig moulds side by side with the pouring channel forked to feed both rows of moulds.

Nature of pig iron

Whilst it is in the blast furnace the iron absorbs varying amounts of carbon, silicon, sulphur, phosphorus and manganese, and these are present in the pig iron. The carbon present may be in its natural state as *graphite* and is normally finely dispersed throughout the metal in the form of small flakes (Fig. 5(a)), or it may be in the form of iron carbide, a chemical combination of carbon and iron called *cementite*.

Matrix of   Flakes of
pearlite   graphite

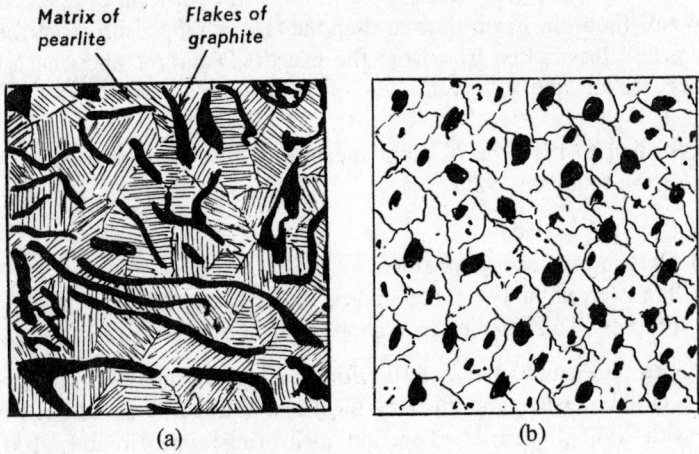

(a)   (b)

Fig. 5 Structure of cast iron
(*a*) Normally cooled with flaked graphite   (*b*) Spheroidal graphite (see p. 15)

Generally some of the carbon is free and some combined, the proportion of each depending mainly upon (*a*) the rate at which the iron cools down from its molten state and (*b*) upon the amount of silicon present in the iron. Slow cooling allows the carbon time to separate out and there is more free graphite. Quick cooling (called chilling) does not allow so much graphite to form, and a larger proportion of the carbon is in the combined form. When cooling is normal the fracture of iron has a greyish-black crystalline appearance and the graphilitic carbon may be easily discerned (Fig. 6(a)). The fracture of quickly cooled iron is whiter, showing that there is less free carbon, and when all, or nearly all, the carbon is combined the metal is called white iron (Fig. 6(b)).

The pigs, even with high silicon content, produced by machine casting generally have a white fracture since cooling is fairly rapid and the moulds are often cooled in a water bosh on the conveyor to speed the process.

The British Cast Iron Research Association

Fig. 6 Fractures of pig iron

(*a*) Grey iron   (*b*) White iron

Cementite is a very hard substance, so that the greater the amount of combined carbon the harder will be the iron. Cast iron is essentially an iron–carbon alloy modified by the presence of silicon, phosphorus, sulphur and manganese in varying amounts (see page 15).

Table 1. Approximate analysis of pig irons

Constituent		No. 1 Iron	No. 3 Iron	White Iron
Iron	%	92	92·5	94·2
Graphitic carbon . . .	%	3·5	3	Nil
Combined carbon . .	%	0·20	0·5	3·1
Silicon	%	2·8	2·5	1·0
Sulphur	%	0·05	0·05	0·3
Phosphorus	%	0·8	0·75	0·9
Manganese	%	0·65	0·70	0·5

Cast iron

The pig iron as tapped from the blast furnace is the crude form of raw material from which are prepared the various grades of iron and steel which we use. It is not suitable for making castings without some degree of refining, and this takes place in the foundry cupola. This is a small form of blast furnace, and a diagram of its essential details is shown at Fig. 7. The size of the cupola varies according to the nature of the work which has to be done; the diameters vary from 1 m to 2 m, with a height of from four to

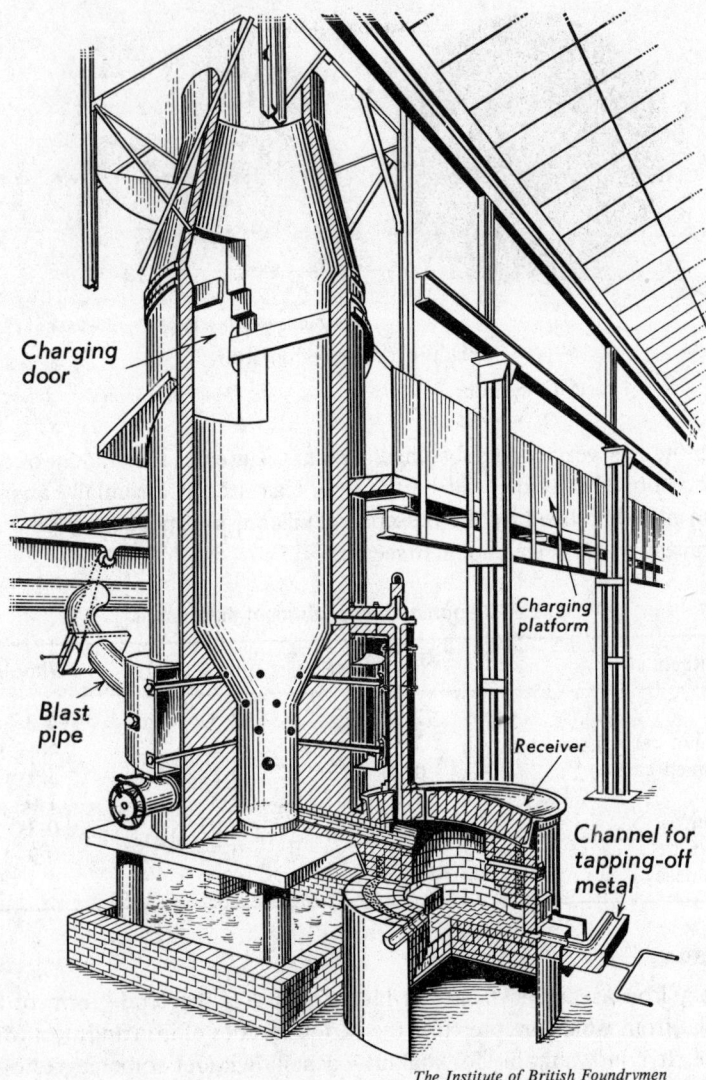

Charging door

Blast pipe

Charging platform

Receiver

Channel for tapping-off metal

The Institute of British Foundrymen

Fig. 7 Sectional view of cupola and hot metal receiver

five times the diameter. A 1 m diameter cupola 4 m high will melt about 2 tonnes of iron per hour.

Generally cupolas are not worked continuously as are blast furnaces, but are run only for such periods as may be required. To start the cupola a coke fire is first lit at the bottom and, when this is established, charging commences by adding alternate layers of coke and pig iron together with a little limestone. The pig iron is broken into pieces about $\frac{1}{3}$ m long before being fed in, and is generally mixed with an agreed proportion of iron and steel scrap, the proportion depending upon the desired quality of the melt. For example, adding low carbon scrap will reduce the percentage of carbon in the melt. An hour or two after charging, and when the charge has burnt up, the blast is gradually increased and the cupola closed up. The iron melts and sinks to the bottom of the furnace, and when sufficient has accumulated it is tapped into a ladle, or directly into the moulds. A receiver is attached to some cupolas into which the metal is directed when it falls down (see Fig. 7).

The amount of coke consumed is not large, as in the cupola the iron has only to be melted, there being no chemical changes to be produced as in the blast furnace. The amount of coke necessary varies from 50 kg to 125 kg per tonne of iron melted. The limestone is added to combine with the sand on the surface of the pig iron, and with the ash from the coke, to form a liquid slag which floats on top of the molten iron. This slag is tapped off separately. About 3 kg to 5 kg of limestone is necessary for each tonne of iron melted.

Properties and uses of cast iron. Grey iron is a very useful metal in engineering construction. The main advantages in favour of its use are as follows: (*a*) its cheapness, (*b*) its low melting temperature [1150 °C to 1200 °C] and fluidity when in the molten condition and (*c*) it is easily machined. A further good property of cast iron is that the free graphite in its structure seems to act as a lubricant, and when large machine slides are made of it a very free-working action is obtained. The reader may demonstrate this for himself by comparing the relative easiness with which scribing blocks with steel, and cast-iron bases may be slid about on a cast-iron surface plate. He will find that the movement of the cast-iron base is sweeter and it does not drag as much as the steel one.

The fluidity of iron when in the molten condition enables it to be used widely for making castings of parts having intricate shapes, such as the bodies of machines and other components. These are cast with metal from the cupola, in moulds prepared in sand, the moulds being made from wooden patterns. The process will be described in more detail in a later chapter.

The fracture of grey cast iron shows a crystalline or granular structure and a strong light will give a glistening effect due to reflection on the free graphite particles. The presence of this free graphite is also shown when filing or machining cast iron as it makes our hands black. Cast iron is brittle and may easily be broken if a heavy enough hammer is used. Small,

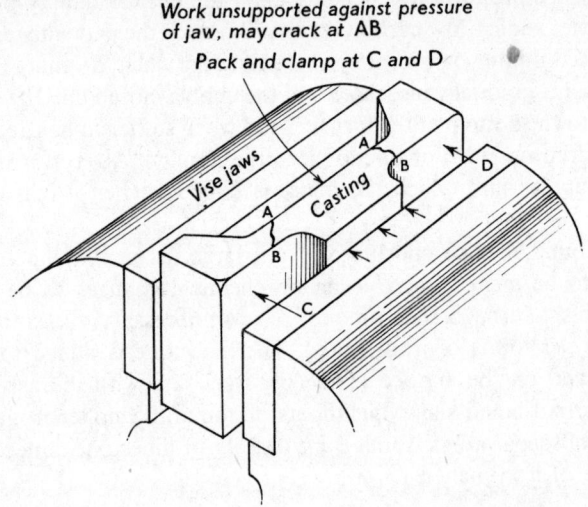

Fig. 8 Casting not supported under point of clamping

thin castings should not be dropped, because of the risk of their being too hard against unsupported sections of a casting, fracture may occur (Fig. 8).

The strength of iron is much greater in compression than in tension. It is usual to express the strength of a metal as the number of newtons required to fracture a section of the metal 1 mm^2 or 1 cm^2 in area. This load is called the Ultimate Strength of the material and it may be tensile (tension), compressive or shear. (Shear is when fracture is caused by forces which compel failure to occur across a plane parallel to the forces, Fig. 9.) The ultimate tensile strength of cast iron varies between approximately 120 and 300 newtons per square millimetre and depends on the composition of the iron.* In compression, iron will withstand about 600

* See 'Alloy Cast Irons', p. 15 giving recent developments in spheroidal iron.

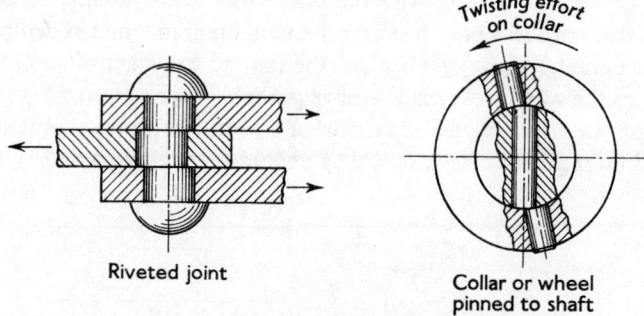

Riveted joint

Collar or wheel
pinned to shaft

Fig. 9 Examples of failures due to shearing

to 750 newtons per square millimetre before fracturing, whilst in shear its strength is approximately 150 to 225 newtons per square millimetre. Because cast iron is brittle and weak in tension it cannot be used where these deficiencies would be detrimental. Bolts and machine parts which are liable to tension, and which often require to have fine screw threads on them, could not be made of cast iron, because of its unreliability in tension, and because the cast iron would rapidly break and crumble away at the threads. Shafts which have to withstand bending loads cannot be made of brittle metal and therefore grey cast iron cannot be used for them. Parts subjected to such conditions of service are made of steel which can be hardened; grey cast iron cannot be hardened. There are many applications in engineering and machine tool practice, however, where an intricate shape is required, and where large tensile loads have not to be carried. Examples are the beds, slides and bases of machine tools, the bodies of electrical machines, the cylinders and beds of engines, and so on. For constructions of such types grey cast iron is an ideal material, because it can be cast to the general shape required, and machined only on such faces as are required to fit with mating components. The hard, outer skin of such castings gives a good appearance when cleaned of sand and painted, and is hard wearing against knocks or other damage.

Composition of cast iron. We have seen in our discussion of pig iron that the carbon may be in the free graphitic state or may be chemically combined as cementite. Also that cementite is hard and is caused by quick cooling, whilst slow cooling gives graphitic carbon and soft, easily machined castings. Quick cooling is generally called 'chilling' and the iron so produced is 'chilled iron'. All castings are chilled at their outer skin by contact of the molten iron with the cool sand in the mould, but on most castings this hardness only penetrates about 1 mm to 2 mm in depth, and if we put on enough cut to get beneath the skin we can machine it off without damage to the tool. Most readers have probably experienced what happens if they try to machine or file the skin of a casting without getting beneath the hard crust; if they have not, they should try it.

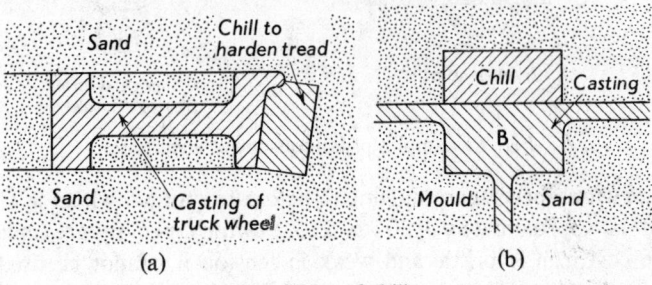

(a)   (b)

Fig. 10 Use of chills

Sometimes a casting is chilled intentionally and sometimes becomes chilled accidentally to a considerable depth. Intentional chilling is carried out by putting inserts of iron or steel (chills) into the mould, so that when the molten metal comes into contact with the chill its heat is rapidly conducted away and the formation of combined carbon is promoted. Chills are used on any faces of a casting which are required to be hard for the purpose of withstanding heavy wear and friction (Fig. 10(a)). Of course, chilling complicates the problem of machining, as ordinary tools will not stand up to their work and either specially hard tools or grinding must be employed. In some cases, chilling is employed to accelerate the cooling of a part of a casting and so avoid some portions cooling before others and thus giving rise to strains. In Fig. 10(b), the thin portions of the casting would normally cool before that at B and set up internal strains due to uneven contraction. If a chill is placed as shown the rate of cooling is equalised and the risk of distortion or cracking is minimised. Castings which have very thin sections of material cool quickly in the mould and are always liable to become chilled, and hence difficult to machine. To minimise this and promote the formation of graphite, an iron high in silicon content is often used.

We observed earlier that as well as carbon, cast iron generally contains small amounts of silicon, phosphorus, sulphur and manganese. It will be well for us to have some idea of the effect of these elements on the iron.

Silicon promotes the formation of free graphite and by so doing acts as a softener and gives an iron which is easily machinable. It has a high affinity for oxygen and thus helps to produce sound castings free from blowholes. The silicon content of the average cast iron varies up to about 3%.

Phosphorus, although not always wanted, gets into the iron as an impurity in the blast furnace and is very difficult to remove. In cast iron phosphorus makes for great fluidity and is used for castings which must be made to intricate and delicate shapes (e.g. thin switch boxes with cast lettering, etc.). Cast iron may contain up to $1\frac{1}{2}\%$ phosphorus.

Sulphur and Manganese. These two elements have the same general effect, viz. a tendency to harden the iron. Too much sulphur promotes unsound castings, and the sulphur content is generally kept as low as possible (0·1%). The amount of manganese present varies up to about $1\frac{1}{2}\%$, depending upon the type and use of the casting.

The following table shows the composition of cast irons for various uses:

Table 2. Constitution of various cast irons

Type	Iron	Total Carbon	Silicon	Manganese	Sulphur	Phosphorus
Light section machinery . .	93·65	3·2	2·2	0·5	0·1	0·35
Hydraulic cylinder .	94·45	3·2	0·9	1·0	0·1	0·35
Lorry cylinder . .	93·46	3·3	2·1	0·9	0·09	0·15
Switch boxes . .	91·6	3·5	2·8	0·8	0·1	1·2

Alloy cast irons

To overcome certain inherent deficiencies in ordinary cast iron and to give qualities more suitable for special purposes, whilst retaining the important casting advantage of this metal, a large number of alloy cast irons have been developed. Two recent examples are acicular and spheroidal cast irons. Acicular iron has nickel and molybdenum in its composition and is being used for cast crankshafts. In spheroidal iron the graphite content is converted from a flaky to a spheroidal form (see Fig. 5(b)) by the alloying of a small amount of magnesium or cerium, this change in the graphite form raising the tensile strength to about 600 to 750 newtons per square millimetre, and producing a tough metal which can be twisted and bent.

Wrought iron

Wrought iron was probably the first form of iron with which man was acquainted, and long ago, all the articles made from it would be shaped by the blacksmith, who was an important artisan. It is now one of a large selection of ferrous products, and of all the common types of iron and steel, wrought iron is the nearest approach to pure iron. The chemical analysis of the metal may show as much as 99·9% of iron. Even when strongly heated, wrought iron does not melt, but only becomes pasty, and in this form it may be forged to any shape, and separate parts joined by hand welding.

Puddling process. The chief point in the manufacture of wrought iron is the oxidation of nearly all the carbon and other elements from pig iron, and in this country the process is carried out in a puddling furnace. This is a coal-fired reverberatory furnace as shown in Fig. 11. The term reverberatory is applied to furnaces of this type because the charge is not in actual contact with the fire, but receives its heat by reflection from the shaped furnace roof.

In the process the furnace hearth is lined with iron oxide and grey pig iron and millscale (oxide) are fed on to it. The metal soon softens and melts, and as melting proceeds, the puddler, working with a long rake through the puddling door, hastens melting by drawing lumps of unmelted metal to the centre. When melting is complete the impurities form a slag with

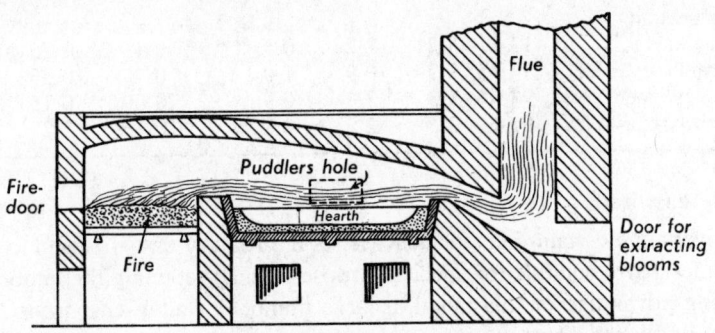

Fig. 11 Puddling furnace

some of the oxides present, and this slag floats on top of the bath, but from time to time the slag is stirred into the melt and more millscale is added. The addition of this oxide to the slag, already rich in oxide and being continuously stirred into the melt, brings about the oxidation of the carbon and the other elements originally present in the pig iron. Carbon monoxide bubbles from the mass of molten iron and burns at its surface, the mass for a time having the appearance of boiling, and at this stage some of the slag rises up and flows out from the slag notch. As the carbon becomes oxidised away, the bubbling subsides and the iron becomes stiffer and pasty in form. The temperature of the furnace is now raised to its highest possible point, the charge, meanwhile, being continuously puddled until at last it becomes a quiescent mass of pasty iron intermixed with slag. This is now taken from the furnace in the form of balls (or blooms) having a mass of about 50 kg, which, whilst white hot, are hammered to squeeze out a portion of the slag and then rolled into rough bars which rolls the remainder of the slag into long fibres. The rough bars are subsequently re-heated to a white, welding heat, and a number are bundled together and re-rolled, which welds the layers together and further elongates the slag.

The presence of slag gives the fracture of wrought iron a fibrous appearance and it is by this characteristic that it may easily be distinguished. When it is filed or machined to a surface parallel to the direction of rolling, the slag may be seen as long lines running along the surface. It is rather soft and tends to tear under such processes as screw-cutting.

Wrought iron is easily forged and welded at the forge. It is ductile (see p. 31), and easily bends when cold, in fact a medium-sized bar should withstand being doubled upon itself without the outer side cracking at the bend. The iron has the property of being able to withstand sudden and excessive shock loads without permanent injury, for which reason it is used for chains, crane hooks, railway couplings, etc. The fibrous nature of its composition gives visible warning on the surface of an impending fracture before complete breakdown takes place and arrangements can be made to replace the damaged part. Were the same conditions imposed on

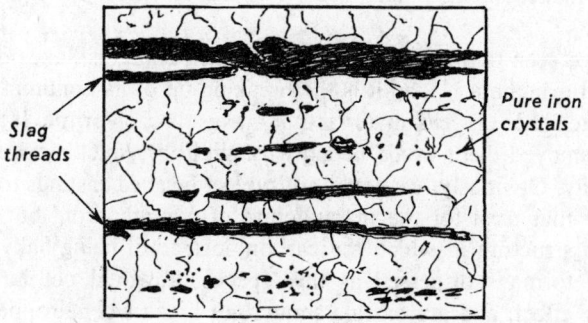

Fig. 12 Microstructure of wrought iron (in direction of rolling) × 100

any other metal it would probably show no signs until it fractured suddenly across its whole section.

Wrought iron has an ultimate tensile strength of about 350 newtons per square millimetre, and the approximate chemical composition of a Stafford-shire wrought iron is as follows:

Element	Carbon	Silicon	Sulphur	Phosphorus	Manganese	Slag	Iron
% Present	0·02	0·12	0·018	0·22	0·02	0·07 to 1	Remainder

The microstructure of wrought iron is shown at Fig. 12, in which the slag may be seen clearly.

Malleable cast iron

A malleable metal is one which may easily be caused to spread and flatten under pressure or hammering. Lead is the best example of a malleable metal we have in the workshop, and the reader may soon prove for himself that this metal is easily beaten out and flattened.

The application of the term 'malleable' to castings of that type is rather a misnomer because they are not very malleable when compared with the usual standards of malleability. When compared with grey iron castings, however, which are fairly brittle, malleable castings do possess a degree of toughness, and this is probably why they have been so named.

For many purposes it is necessary to produce small, thin parts having a reasonable degree of strength such as would not be obtained in a grey casting which, as we have seen, is brittle and would break under rough usage. Such parts could be built up or forged from steel, but it is much cheaper to produce castings, and the process of converting castings to the tougher form of 'malleable iron' has been evolved to meet the need for such products.

We have seen that the structure of cast iron contains flakes of graphite dispersed throughout it, and it is the breaking up of its continuity by these rather large flakes of weak material that makes the iron brittle. If the carbon were all removed there would be left wrought iron, a ductile material which bends easily. Obviously, we could not subject finished castings to a process similar to that used for the manufacture of wrought iron, but if we can produce a structure in which the carbon, instead of being flaky, or in the combined form, is dispersed as tiny specks, it would not have such a weakening effect, and our casting should not break when dropped. This, in general, is what takes place in the process of making malleable castings.

First, castings are made from an iron having all of its carbon in the combined form. Two methods are then used for malleabilising the castings: the Whiteheart, used in England, and the Blackheart, more common in America and for large castings in this country. The names refer to the colour of the fracture given by castings produced by each method.

In the Whiteheart process the castings, composed of cast iron with most of its carbon in the combined state, are packed in iron or steel boxes and surrounded with a mixture of used and new haematite ore. The boxes are heated to a temperature of 900 °C to 950 °C and maintained at that temperature for several days, during which time part of the carbon is oxidised out of the castings, and the remainder is dispersed in small specks throughout the structure. In thin sections the structure might be almost pure iron, but in thicker sections the outside approaches to nearly pure iron and the centre has the dispersed specks of carbon. The heating period is followed by a very slow cooling which occupies several more days and the result is a casting which is tough and which will stand hard treatment without fracture. The tensile strength of malleable castings is about 400 newtons per square millimetre.

Steel

The essential difference between cast iron and steel is in the amount of carbon contained in the constituency of the metal. Pure iron, as we have seen, is a soft metal having a structure made up of iron crystals or grains. In metallurgy, pure iron is called *ferrite*.

If now we make up a series of alloys of iron and carbon, with the carbon content increasing to about $1\frac{1}{2}\%$, we shall have a series of steels, the metal becoming harder and tougher as the carbon content increases. Up to a content of about $1\frac{1}{2}\%$, all the carbon is present in chemical combination with the iron, and none of it exists in its free graphitic state. If, however, we go on increasing the carbon above $1\frac{1}{2}\%$ a stage soon arrives when no more can be contained in the combined state and any excess must be present as free graphite. It is at this stage that the metal merges into the group termed cast irons, and we may go on increasing the carbon content up to about $4\frac{1}{2}\%$ whilst producing a range of cast irons.

Steel, then, is fundamentally an alloy of iron and carbon, with the carbon content varying up to $1\frac{1}{2}\%$, whilst cast iron is an alloy of these two elements with the carbon content ranging from about 2% to $4\frac{1}{2}\%$. For a material to be classed as steel there must be no free graphite in its composition; immediately free graphite occurs it passes into the category of cast iron. It is true that other elements are present in small quantities, and are put there to confer certain desired properties on the metal, but the carbon is by far the most important modifying element and, if he can appreciate this, the reader may study the effect of other elements later.

The plain steels are usually classified according to their carbon content, the commonest of the range being *mild steel* with a carbon content ranging from about 0·15% to 0·3%. When the carbon content is less than 0·15%, say 0·07 to 0·15, the steel is classed as a *dead mild steel*. *Medium carbon steels* include the range with the carbon content varying from 0·3% to 0·8%, whilst steels with a carbon content between 0·8% and 1·5% are classified as *high carbon steels*.

Application of steels. The following is an indication of the carbon content of steels for various purposes:

Table 3. Applications of carbon steels

Carbon %		Uses
Dead mild	0·1 to 0·125 .	Wire rod, thin sheets, solid drawn tubes, etc.
Mild	0·15 to 0·3 . .	Boiler plates, bridge work, structural sections, drop forgings, general workshop purposes.
Medium carbon	0·3 to 0·5 . .	Axles, drop forgings, high tensile tubes and wire, agricultural tools.
	0·5 to 0·7 . .	Springs, locomotive tyres, large forging dies, wire ropes, hammers and snaps for riveters.
	0·7 to 0·9 . .	Springs, small forging dies, shear blades, cold setts, wood chisels.
High carbon ·	0·9 to 1·1 . .	Cold chisels, press dies, punches, screwing dies, woodworking tools, axes, picks.
	1·1 to 1·4 . .	Razors, hand files, drills, gauges, metal-cutting tools.

As we shall see in a later chapter, the high carbon steels may be hardened by heating and quenching in water or oil. The low carbon steels, also, may be hardened on the surface by a process called case-hardening.

High carbon steels having a carbon content somewhat over 1% are often called 'Cast Steels', or 'Carbon Tool Steels'.

The manufacture of steel. The wide range of uses to which steel is applied and its consequent high consumption causes the supply of it to be an important problem. The consistency of its quality, also, is a factor upon which the success of our workshop production depends. The low and medium carbon steels are much more widely used than the high carbon and tool steels, these latter being used mainly for tools and for specialised work calling for high quality and performance. These considerations have caused a separation to be made between the methods employed in steel manufacture. We will deal with the low and medium carbon group first.

The open-hearth process. Most of the steel made in this country is produced by the Siemens Open-Hearth Process, so named after Siemens, its originator.

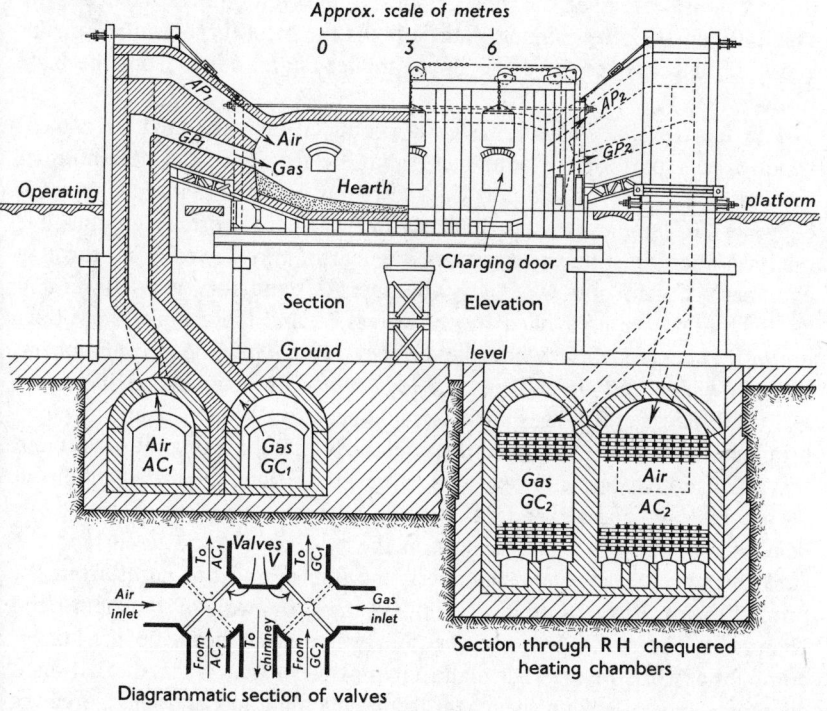

Proceedings of the Iron and Steel Institute

Fig. 13 An open-hearth furnace

The furnace employed is shown in section at Fig. 13 and furnace sizes range from those taking melts of 5 to 200 tonnes. In the larger sizes manipulative problems cause some difficulties, and the big furnaces are often made of the tilting type. Probably the most useful range for fixed furnaces is that between the 50 and 100-tonne sizes.

In the operation of the furnace the hearth is first prepared and well heated until its temperature is about 1500 °C. The charge generally consists of a mixture of selected pig iron and scrap in proportions varying according to the class of steel required. The ratio of pig iron to scrap may vary from 1 to 4, to 4 to 1, a good average being 3 of pig iron to 2 of scrap.

The pig iron is generally fed on to the hearth first, followed by the scrap, whilst in the larger furnaces charging is done at intervals to avoid too much cooling down of the furnace. The larger furnaces (over 30 tonnes) are generally charged by mechanical chargers, which pick up boxes containing the previously weighed-out charge, push them into the furnace and tip them. When the steel-making plant adjoins the blast furnaces the iron for charging is conveyed by ladle in its molten state, direct from the blast furnace (see Fig. 14).

The fuel for the furnace, which is producer gas generated by a plant which forms part of the equipment, may be fed on to the hearth either through chamber GC_1 and port GP_1, or it may be made to travel by way of chamber GC_2 and port GP_2. Similarly, the air, which forms a combustible mixture when meeting the gas over the hearth, may be fed either through chamber AC_1 and port AP_1, or by chamber AC_2 and port AP_2. The air and gas directions are controlled by the valves V, and the chambers are built up with chequered brickwork like a honeycomb. These heating chambers, through which the air and gas are fed, are a characteristic of the open-hearth furnace, and it was their development by William Siemens which brought about the success of this method of steel-making. It was found that intense heat was necessary to bring about the chemical reactions in the charge, and the air and gas heating chambers assist in obtaining this heat as follows: when the valve is in the position shown, the air and gas pass through the left-hand chambers, the flames and hot gases from the furnace being forced into the chimney stack by way of the right-hand chambers. The passage of the very hot waste gases heats up the honeycomb brickwork in the right-hand chambers, so that when the left-hand chambers have given up their heat to the incoming gas and air, the valves are moved over, causing the incoming products of combustion to enter through the newly heated right-hand chamber, and turning the furnace flames through the left-hand side. By changing the valves over at suitable intervals the incoming charges of gas and air are caused to pass through a network of very hot bricks which preheats them sufficiently to give the high temperature necessary when they burn at the hearth, and at the same time the hot waste gases are heating the opposite chambers ready for the next

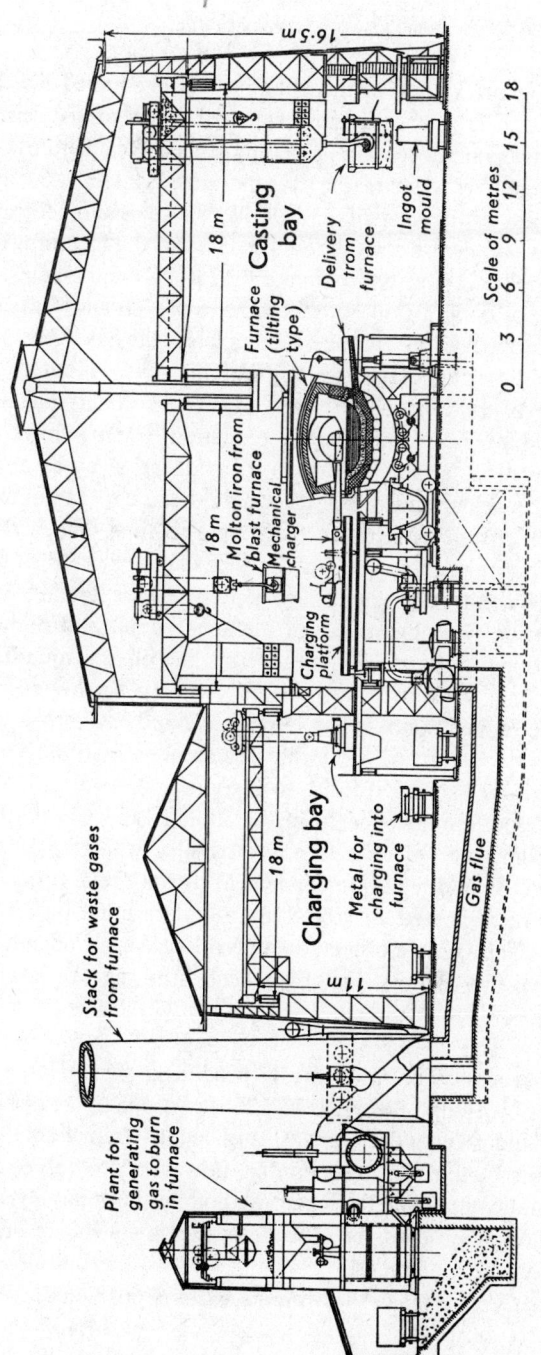

Fig. 14 Cross-section through open-hearth plant

Stack for waste gases from furnace

Plant for generating gas to burn in furnace

Charging bay

Metal for charging into furnace

Gas flue

11 m

18 m

Charging platform

Mechanical charger

Molton iron from blast furnace

Furnace (tilting type)

Casting bay

18 m

18 m

16.5 m

Delivery from furnace

Ingot mould

Scale of metres

0 3 6 9 12 15 18

change-over. The brickwork in the chambers is chequered like a honey-comb to provide a large area against which the gases must scrub during their passage. The chambers are generally called Regenerators, and the alternative change-over system, the Regenerative System. An impression of the effectiveness of the system, and of the prevailing temperatures in the furnace, may be obtained from the facts that when the incoming air leaves the heating chamber it is at approximately 1250 °C (white heat), and that the temperature of combustion above the hearth is about 1800 °C.

When the charging of the furnace is completed, the gas is put on and the melting completed as quickly as possible. This takes about 3 hours, by which time part of the carbon, and most of the silicon and maganese from the charge will have been removed by oxidation. Then follows the 'boil', during which iron ore is added to help in oxidising away as much of the carbon as is necessary, and during this period the carbon content slowly falls, samples of metal being taken from time to time. When the bath is nearing the requisite carbon content the top slag is allowed to thicken so as to slow up the action, and the furnace is prepared for tapping. Just before tapping, or during the time that the metal is running into the ladle, a deoxidiser is added to the steel to remove air and ensure soundness in the finished steel. Such deoxidisers may be ferro-silicon, ferro-manganese and aluminium. The tapped metal is run into a ladle and cast into ingot moulds to give slightly tapering and generally rectangular ingots $1\frac{1}{2}$ m to $1\frac{3}{4}$ m long having masses varying from 1 to 3 tonnes. After as short a time as possible, the ingots are stripped from the moulds and whilst still red hot, taken to the rolling mill for reduction into bars, or whatever other sections are being rolled. Should it not be convenient to roll them at once, the red-hot ingots can be re-heated, or stored in an underground furnace (soaking pit) until later. The complete cycle of operations for an open-hearth melt occupies from 6 to 12 hours. Fig. 14 indicates the scale of a steel-melting unit.

The Bessemer process. The discovery of producing steel in large quantities in a Converter, by Henry Bessemer in 1856, was one of the epoch-making events in our industrial development, and led to tremendous strides in engineering construction. In this country, however, due to certain advantages in cost and control, the Bessemer method has been largely superseded by the open-hearth process, but the Bessemer is still being employed in America and on the Continent.

The Bessemer converter is shown in its various positions at Fig. 15 and is constructed of an outer steel shell with ganister or basic lining, depending on whether the steel being made is acid or basic. The process starts by pouring molten pig iron into the converter until the level is just below the blast holes, when a blast of air at 1·5 to 2 bar* pressure is turned on. The converter is rotated into its upright position and that part of the reaction called the 'blow' begins. This is divided into three stages: (a) the preliminary, (b) the boil and (c) the finishing. These are all clearly indicated by variations in the smoke and flame issuing from the mouth of the converter, and the efficiency of the 'blower' in charge of operations is the ability with which he can interpret them.

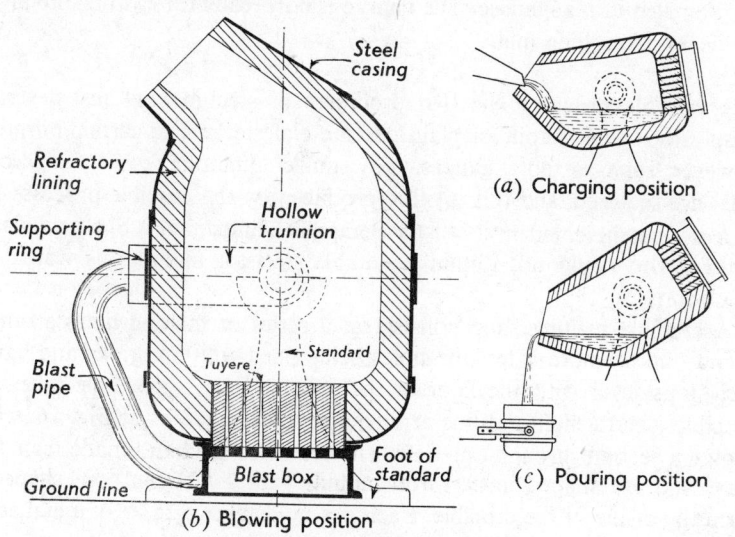

Fig. 15 Bessemer converter

During the preliminary stage of the blow almost the whole of the silicon and maganese are oxidised from the charge and brown fumes accompanied by sparks and a short flame issue from the converter mouth. The size and luminosity of this flame gradually increase until after about 5 minutes from the beginning of the blow the 'boil' commences. This stage is indicated by the ejection of molten particles of slag, and is accompanied by violent agitation of the contents of the converter caused by the escaping carbon monoxide gas which is formed by the chemical reaction of the air blast with the carbon in the melt. At this stage the blower estimates the heat

* 1 bar = 10 N/cm² (about 1 atmosphere).

of the charge and adds scrap or ferro-silicon according to whether it is too hot or too cold. When the carbon is almost completely oxidised away the boil is completed, the flame subsides and eventually drops altogether. The whole blow, which has taken about 20 minutes, is now completed and the converter is rotated to the horizontal position. Its contents consist of an almost carbonless iron, containing small percentages of silicon, sulphur and phosphorus. For most purposes the carbon content of this metal would be too low, and, as it would be impossible to stop the blow so as to leave a definite amount of carbon present, the requisite amount of carbon is now added in the form of spiegeleisen or ferro-manganese, which are varieties of pig iron rich in manganese. The charge is now ready for casting into ingots ready for the rolling mill.

The crucible process. The two processes of steel-making just described supply the main output of plain commercial steels. For certain purposes, however, such as tools, gauges, etc., smaller quantities of very high-class steel are required, and this is often produced by the crucible process. This process was invented in 1740 by Benjamin Huntsman, a Sheffield watch-maker, who could not obtain a suitable steel for making his watch and clock-springs.

A crucible melting shop consists of a series of melting holes arranged round one or more sides, the holes being lined with firebricks and having their tops level with the floor of the melting shop. Each hole takes two crucibles, and ashpit and flue arrangements are made as at Fig. 16, which shows a section through one hole. The crucibles are hand-made from fire-clay, and are shaped in cast-iron moulds with a wooden core shaped to form the inside of the crucible. Each pot holds about 25 kg of metal and is generally used for 3 melts before being discarded.

At the commencement of the process two pots with covers are put in the hole on stands or bricks and the coke fire made up. When the pots are white hot the charge is put in through a charger shaped like a funnel, the lid is adjusted and the melting hole is filled with coke. The charge consists of cut pieces of Swedish iron and blister bars followed by pig iron and any alloying elements that may be required. (Blister bars are bars of carbon steel made by heating pure iron in the presence of charcoal, which con-verts the pure iron into a carbon steel called 'blister steel'.)

After the charging the process consists of two stages—melting and killing. Two replenishments of the fire, which lasts about an hour, are usually sufficient to melt the contents of the pot, which will then have the

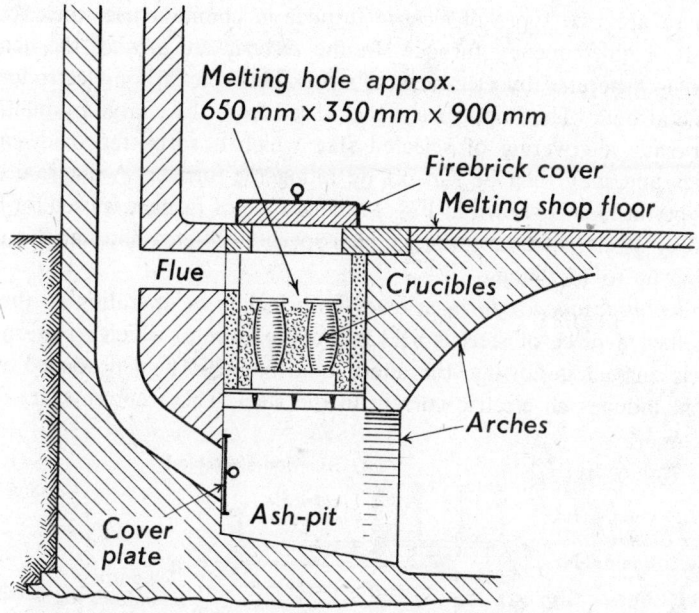

Melting hole approx.
650 mm x 350 mm x 900 mm

Firebrick cover
Melting shop floor

Flue

Crucibles

Arches

Cover plate

Ash-pit

Fig. 16 Section through a melting hole for crucible steel-making

appearance of a thick simmering liquid with the slag floating on the top. If an attempt were made to cast this into the ingot mould the gases in it would cause the metal to boil over the top of the mould and the ingot would be full of blowholes. The metal, therefore, must be 'killed' to rid it of these occluded gases. The last or 'killing fire' is made up and a small amount of manganese or aluminium added to the contents of the pot. This brings about chemical reactions in the metal and results in the boiling off of the gases. Finally, the crucible is pulled out of the hole with special tongs, and after the surface of the metal has been cleared of slag it is 'teemed' or poured into the previously prepared ingot mould.

Electric steel

The methods used in the crucible process have changed very little since Huntsman originated them. Within recent years the melting of steel by electric furnaces has developed rapidly, and although the crucible method yields a fine product, electric melting possesses advantages which indicate that this method is the one of the future for the preparation of high-class steels.

There are two types of electric furnace in common use: the arc type and the high-frequency furnace. In the *electric-arc furnace* the heat required is generated by electric arcs struck between carbon electrodes and the metal bath. The impurities are oxidised from the charge by melting it underneath a covering of selected slag which absorbs the oxidised impurities and may then be run off by tilting the furnace. A diagram of an arc-type furnace is shown in Fig. 17. This type of furnace is used for making alloy steels, such as stainless, high-speed steel, etc., and handles melts ranging up to 15 tonnes.

The *high-frequency furnace* owes its principle of operation to the fact that when a piece of steel is held in a coil of wire in which an alternating electric current is flowing, the alternating magnetic field produced by the current induces an electric current in the steel. These currents are called

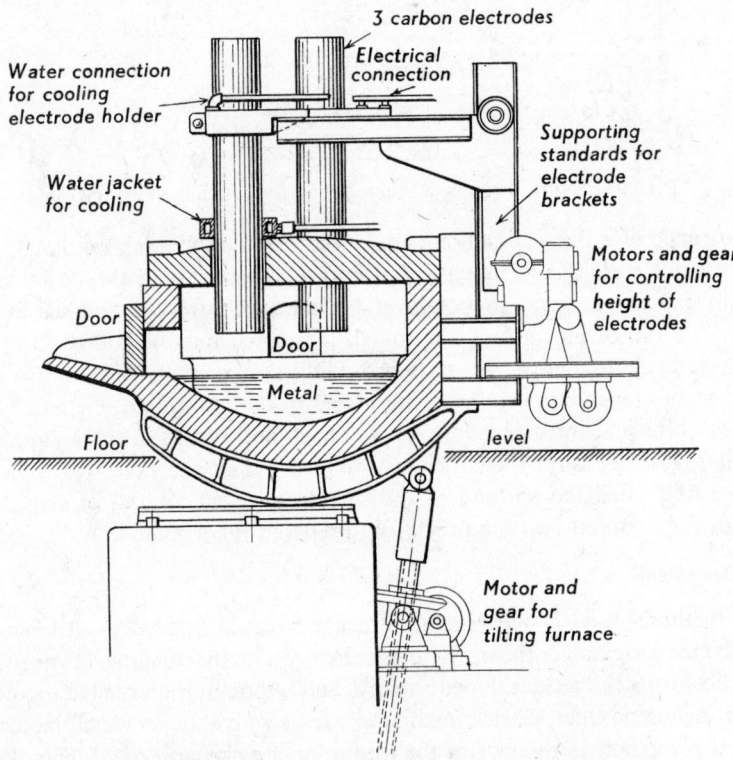

Fig. 17 Diagram of the Heroult electric-arc rocking furnace

eddy currents, and if they are powerful enough, their passage will cause the steel to be heated up.

The details of a high-frequency furnace are shown in Fig. 18. A is the crucible for containing the metal and is made of a suitable refractory material. B is the inductor coil which carries the alternating current and which is insulated and water-cooled. The outer ring, C, is made of special magnetic iron laminations and shields the casing from getting too hot as well as serving as a magnetic shield to increase the magnetic flux in the

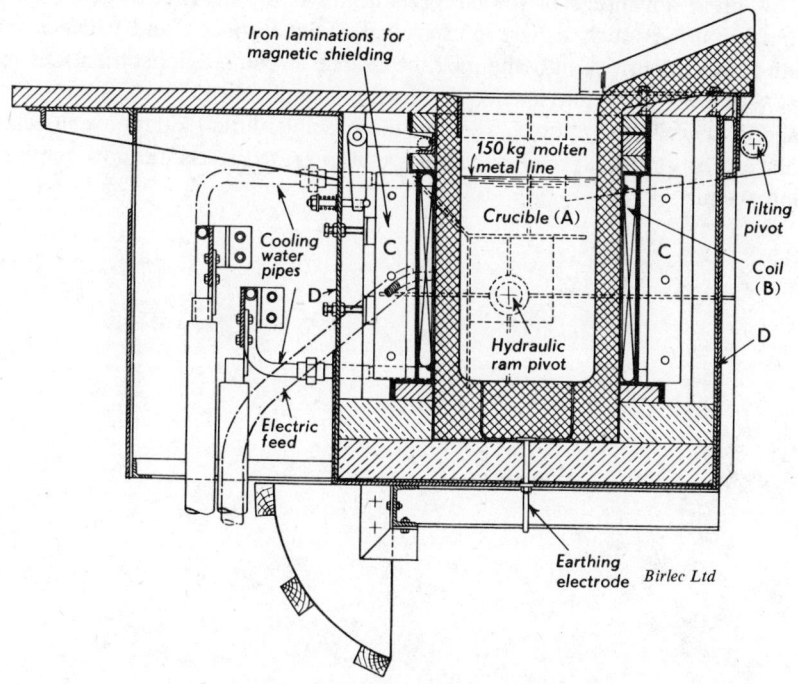

Fig. 18 High-frequency coreless induction furnace
(Tilting pattern, 150 kg capacity)

coil. D is the outer sheet steel casing. An auxiliary to the plant is the electrical apparatus necessary to produce the high-frequency current in coil B, the frequencies in use varying from 500 to 2000 hertz.

In the operation of this furnace the charge is introduced, together with the correct proportions of any alloying elements necessary and other materials as may be required to oxidise and flux the impurities from the melt. As the charge is brought to the molten condition the currents flowing in it cause a certain amount of agitation to take place which serves to promote efficient mixing of the constituents. When the refining and mixing are complete the metal is poured into ingot moulds by tilting the furnace on the pivots at lip-axis level.

High-frequency melting usually deals with charges up to about 500 kg and is used for high-purity steels such as are necessary for tool and die steels, ball-bearing steels, etc.

A great advantage of the electric furnace is the absence of gas, fumes and impurities such as are present in fuel-fed furnaces, and which may introduce impurities into the melt, or oxidise a required constituent out of it. Whatever is put into an electric furnace may be relied upon to stay there, and if a well-fitting top is kept on the crucible, little oxidation can take place. For certain special purposes vacuum-melted steels are now coming into vogue.

2 The properties and treatment of iron and steel

Mechanical properties of metals

Although we have mentioned certain properties already, it will be useful to study here the usual mechanical properties of metals. Later, when discussing treatment, we shall show how great is the influence of treatment on certain properties.

Brittleness is the property of breaking without much permanent distortion. It may be due to brittleness of the grain boundaries or of the crystals themselves. Cast iron is brittle because its structure is split up by flakes of graphite, which is a brittle material. Brittleness is often referred to as shortness. *Hot*, or *red-shortness* in steel, is when it is brittle in the red-hot state. It is caused by too much sulphur which is present as iron sulphide and forms a brittle membrane surrounding the steel crystals. *Cold-shortness* means that a metal is brittle when cold. In steel cold-shortness is produced when the phosphorus content is too high.

Ductility. A metal is ductile when it may be drawn out in tension without rupture. Wire drawing depends upon ductility for its successful operation. A ductile metal must be both strong and plastic, e.g. lead wire is difficult to draw because the strength of lead is low.

Elasticity. The elasticity of a metal is its power of returning to its original shape after deformation by force. A material may be stretched, compressed, or its volume changed by pressure on all sides (e.g. immersion in a liquid). Many materials behave to some extent like powerful elastic, and, within limits, will recover their shape when the load on them is removed. The *elastic limit* is the limit of the elasticity of a material, and is expressed in newtons per unit of area (mm^2, cm^2 or m^2). For example, if the elastic limit of a material were 23 kilonewtons per square centimetre, then, with a bar of the material 1 square centimetre in area, the material would return to its original length from a load of 23 kilonewtons. If this intensity of loading were exceeded the bar would take on a permanent stretch, often called *permanent set*.

Elongation. When a material is pulled in a testing machine for the purpose of finding its tensile strength, stretch takes place before the bar fractures. The elongation is the amount of this stretch and is generally expressed as a percentage of the original length. For example, if a test length of 50 mm stretched to 69 mm before fracturing, the elongation would be

$$\frac{69-50}{50} \times 100 = \frac{19}{50} \times 100 = 38\%$$

a good elongation indicates a ductile material.

Hardness. The hardness of a metal is a measure of its ability to withstand scratching, wear and abrasion, indentation by harder bodies, marking by a file, etc. The machinability and ability to cut are also hardness properties important in the workshop. A rough, but often reliable test for the hardness of a hardened tool, is to see if the edge of a fine file will touch it. The *Brinell hardness* of a metal is found by pressing a ball on to the surface of the metal, the hardness number being found by dividing the load on the ball by the surface area of the impression. For testing steel, the ball is 10 mm diameter and the load 3000 kilogrammes. Brinell machines are usually supplied with a chart which gives the hardness number when the diameter of the ball impression is known.

Malleability. This is the property of permanently extending in all directions without rupture by pressing, hammering, rolling, etc. It requires that the metal shall be plastic but is not so dependent on strength, e.g. lead is a very malleable metal.

Plasticity. This is a rather similar property to malleability, and involves permanent deformation without rupture. It is the extreme opposite to elasticity, as may be shown by comparing the behaviour of a piece of elastic rubber and a piece of plasticine under a straining force. Plasticity is necessary for forging, and metals may be rendered plastic by heating them, e.g. steel is plastic when at a bright red heat.

Strength. The strength of a metal is its ability to resist the application of force without rupture. In service a material may have to withstand tension, compression, or shear forces. The strength of a material is measured by loading it in a testing machine. The *ultimate strength* is the load necessary to fracture unit area (mm^2, cm^2 or m^2) of cross-section of the metal. The *tenacity* is the ultimate strength in tension. Ultimate strength and tenacity are always expressed in newtons per unit of area (usually in newton per square millimetre).

Toughness is the amount of energy a material can absorb before it fractures. A measure of the toughness of a metal may be obtained by nicking it, placing it in a vise and striking the end with a hammer. Certain woods are very tough and it is for this reason that hickory is a good material for sledge-hammer shafts.

Constitution of steel

Reverting back to our discussion of iron and steel, it will be advantageous for us now to examine more closely the constitution of the plain steels. The mechanical properties, which are of importance to us, are so closely connected with the structure, and the structure with the treatment, that we shall gain a more through insight to the subject if we have a clear conception of the nature of the metal.

The microstructure of pure iron (Fig. 1) showed the clear crystals of iron (ferrite) with their grain boundaries. This metal is very soft and ductile with an ultimate tensile strength of about 300 newtons per square millimetre. As soon as carbon is added to this iron a great change occurs in its structure and properties, and Fig. 19 shows the microstructure of a mild steel containing 0·25% to 0·3% carbon. The white constituent is the ferrite, whilst the dark patches represent that part of the structure which contains the carbon. It must be remembered that these dark patches are not actual carbon but contain it in a chemically combined form. A chemical combination of two elements can be formed in which the final result is totally unlike either of the elements of which the combination is composed. Thus, hydrogen and oxygen combine in a certain proportion to form water (H_2O), carbon and oxygen may form carbon monoxide (CO), or carbon

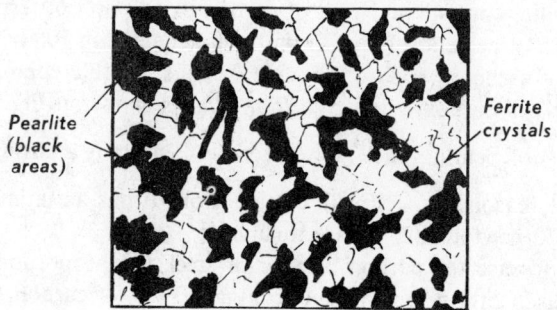

Fig. 19 Microstructure of mild steel (0·25 to 0·3% carbon × 100)

dioxide (CO_2), and so on. Now iron and carbon unite to form iron carbide (cementite) and they do so in the proportion by weight of 1 of carbon to 14 of iron. Iron carbide has the chemical formula Fe_3C and is a very hard, white and brittle substance, so that the more of it the steel contains the harder will it be.

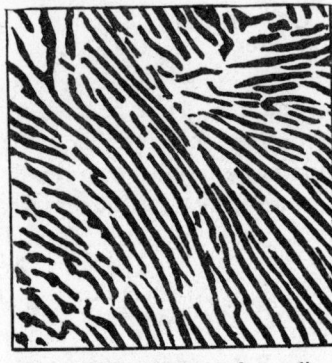

Fig. 20 Structure of pearlite
× 1000 approx.

Fig. 21 Microstructure of high
carbon steel

(Black areas of pearlite surrounded by a
membrane of cementite.)

If, now, we magnify further one of the dark portions of Fig. 19, it will show us that the dark, carbon-bearing constituent is in reality a substance built up of alternate light and dark plates, as shown at Fig. 20. These layers are alternately ferrite (iron) and cementite, and allowing for the great magnification it will be seen how thin the plates are. This substance is called *pearlite* and is made up of 87% ferrite and 13% cementite. We have, then, that 100 parts of pearlite contain 13 parts of cementite, and since cementite consists of 1 part of carbon to 14 parts of ferrite (i.e. 1 of carbon to 15 of cementite), the 13 parts of cementite in 100 of pearlite will contain $\frac{13}{15}$ of carbon = 0·87 or about 0·9. Thus, pearlite contains approximately 0·9% of carbon, and the 0·25% C steel shown at Fig. 19 contains $\frac{0·25}{0·9} = 28\%$ of pearlite and 72% of ferrite. Pearlite is a strong metal and may be cut reasonably well with cutting tools. It has an ultimate strength of about 770 newtons per square millimetre.

As we increase the carbon content of steel, the proportion of pearlite increases also, until when the steel contains 0·9% of carbon, its structure consists entirely of pearlite. If the carbon content is increased further still there will be some cementite left over and this will appear in the structure as a free constituent in the same way as free ferrite appears in the low carbon steels. This is shown in Fig. 21, which shows the microstructure of a 1.4% steel. Now, since ferrite is soft and not very strong, and cementite is hard and brittle but also without much strength, as the carbon (and the pearlite) is increased, the steel will get harder and stronger up to the point when it contains 0.9% of carbon. Beyond this, the cementite is increasing, but not the pearlite, so that its hardness will increase but its strength will decrease. This is illustrated in Fig. 22, which illustrates the effect of the carbon content on structure, strength, hardness and ductility.

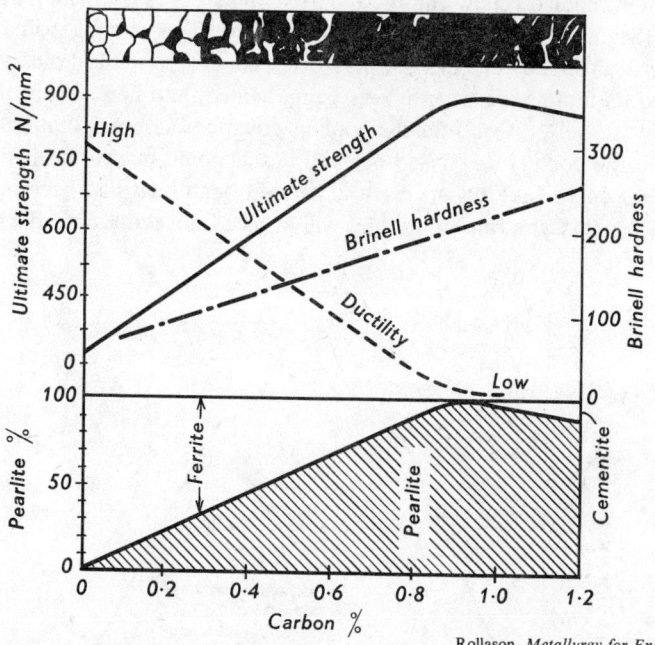

Rollason, *Metallurgy for Engineers*

Fig. 22 Relation between carbon content, structure, strength, hardness and ductility for steel

The behaviour of steel when heated

With the foregoing discussion fresh in our minds we will examine the changes which take place when steel is heated. These are of great importance, as they explain the reasons for, and effects of, heat treatments given to the metal, a subject upon which there is often much confusion.

If a piece of steel (say about 0·3% carbon) is put into a furnace, and gradually heated at a uniform rate, and the temperature of the steel observed at equal intervals of time, its temperature will at first rise uniformly, as would be expected. When, however, the temperature reaches 700 °C (a dull red heat) it will, for a time, remain stationary, and then rise at a somewhat slower rate until it reaches about 800 °C (a good red heat). After this, if the heating can be maintained, the temperature will continue to rise at approximately its initial rate.

Let us now assume that the piece of steel is heated to 900 °C in the furnace, then removed and observed in a darkened room. At 900 °C it be a bright reddish yellow colour, and as soon as it is taken from the furnace it will begin to cool and lose its brilliancy. The cooling will proceed normally until the temperature has dropped to about the point where it received its first check when it was being heated, and here, if it is observed carefully, it will be seen that the cooling down has stopped. Not only will a check in the cooling down be observed at this point, but the steel will probably be seen to take on an extra glow as though heat had been imparted to it. After this, the rate of cooling will proceed normally until the metal is cold.

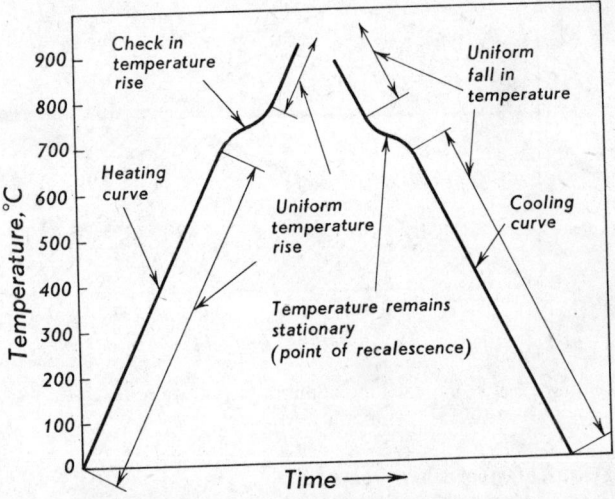

Fig. 23 Heating and cooling curves for steel

The heating up and cooling down may be represented on a graph of temperature-time. Fig. 23 shows the approximate form of such a graph.

Now when the steel was being supplied with heat at a constant rate in the furnace, had its structure remained in a stable and unchanged condition, its temperature would have shown a steady rise, and would not have remained stationary at 700 °C. Similarly, the slowing down, and actual evolution of heat at the same point on cooling, indicates that at this temperature a structural change takes place in the metal which absorbs heat when the steel is being heated and gives up heat when cooling. It has definitely been established that a structural change does take place at this point, and because the steel glows when cooling, it has been called the *Point of Recalescence.* We have seen that the structure of steel, less than 0·9% carbon, is made up of areas of ferrite (iron), surrounding areas of pearlite, which is a substance made up of plates of ferrite and cementite. When steel is heated, this structure remains stable until a temperature of about 700 °C is reached, when the carbon in the pearlite commences to dissolve in the iron. (The reader may wonder how this can be when the whole structure is solid, but we may have a *solid solution* of carbon in iron just as we may have a liquid solution of salt in water.) This change of state continues, until the whole structure of the metal consists of a solid solution of carbon in iron which is called *austenite*, and it was whilst this structure was building up that the steel in the furnace slowed up in its temperature rise, heat being absorbed to bring about the change. When the austenite is completely formed the temperature rise continues as at first. On cooling the reverse takes place, and at 700 °C the austenite changes back to pearlite

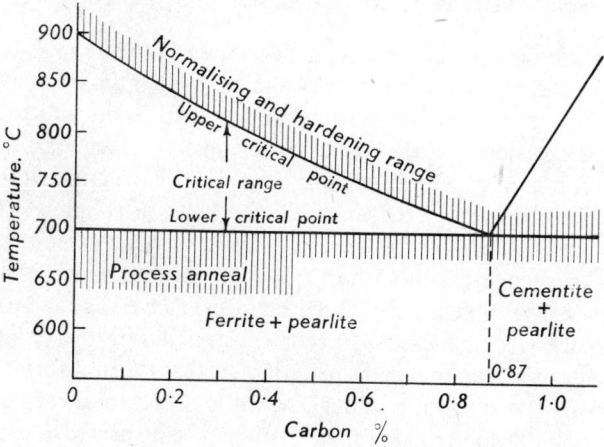

Fig. 24 Heat-treatment ranges of steels

again. This phenomenon is very important in the study of steel. The temperature points at which the change starts and ends are called the *critical points*, and the range including them the *critical range*. The temperature at which the change starts (lower critical point) is the same for all steels and is about 700 °C, but the finishing point of the transformation (the upper critical point) varies according to the steel carbon content. This is illustrated in Fig. 24, from which it will be seen that for a steel containing 0·87% carbon (wholly pearlite), there is only one critical point, the whole

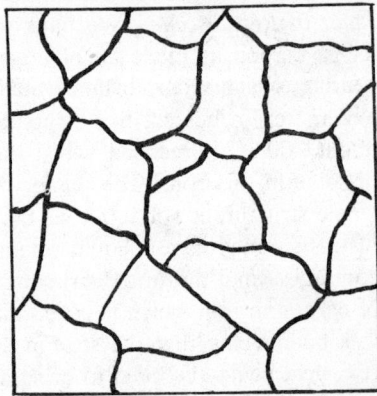

(a) Normal structure of ferrite and pearlite

(b) Austenite. Structure of the steel when the carbon has gone into solution

Fig. 25 Showing change in the structure of steel when heated to the critical range

transformation taking place at that temperature. Fig. 25 shows the change which occurs to the microstructure of the steel when the carbon goes into solution. Fig. 25(a) shows the normal structure of ferrite and pearlite and (b) shows the austenite at the end of the change. Austenite is a solid solution of carbon in iron, and all ordinary steels above the critical range are in this condition. Naturally, the amount of carbon in solution will depend upon how much is present in the steel (as we may have weak or strong salt solutions). Austenite can hold up to $1\frac{3}{4}\%$ of carbon, which is more than the amount ever found in steel. Other changes, as well as the formation of a solid solution of the carbon, occur to steel during the critical range: (1) Austenite is a non-magnetic material, so that when it forms, the steel loses its magnetic quality, a change which is useful to us for determining the upper point of the critical range. (2) When being heated, a considerable contraction occurs at the critical range, and when cooling there is a corresponding instantaneous expansion. (3) The metal becomes extremely plastic at this point.

To sum up: when steel is heated, no structural change takes place until the lower critical point is reached. At this point, the carbon in the steel commences to form a solid solution with the iron, and this change takes place through the critical range. The transformation is completed at the upper critical point and at temperatures above this the steel consists of a solid solution of carbon in iron called austenite, which is a hard, non-magnetic substance. When the steel is allowed to cool down normally, on passing through the critical range, the austenite is transformed back to pearlite, accompanied by ferrite in steels less than 0·9% C, and by cementite in steels above 0·9%C.

Heat-treatment of steel

For the proper heat-treatment of steel, some form of furnace is necessary together with an instrument for measuring the temperature inside the furnace. In the past a great amount of heat-treatment, particularly of tools, was done in the blacksmith's fire, and even to-day this method is still used in some places. At its best, however, this method of heating is not reliable, as for all except small tools the heating is not uniform, and, what is more important, the estimation of the correct temperature depends upon the skill and experience of the blacksmith. If the steel is made too hot it becomes burnt, and if the critical range is not attained the changes which are sought for in the treatment do not take place.

Furnaces are made in many shapes, sizes and varieties. The methods of heating are usually by coal, oil, gas or electricity, and furnaces may be obtained in capacities ranging from small ones with a chamber measuring about 150 mm wide × 100 mm high × 200 mm deep, suitable for small tools, up to huge structures many metres long for heat-treating large bars and forgings.

Salt-bath furnaces. For some purposes, particularly for the treatment of tools and special steels, furnaces are used which have a bath of molten salt as their method of heating. For example, sodium cyanide fuses and becomes molten at about 600 °C, and in its molten state may be heated up to about 900 °C. If, therefore, we wish to heat certain articles to temperatures between these limits an excellent method of doing it is to immerse them in a bath of molten cyanide until they have reached its

(a)   (b)   (c)   (d)

The Industrial Gas Information Bureau

Fig. 26 Corner of a heat-treatment shop

temperature. Whilst they are immersed in the liquid salt they are protected from the air and therefore do not oxidise and scale, and furthermore, they are being uniformly heated from all sides. Furnaces of this type are called salt-bath furnaces. The use of these furnaces involves taking certain precautions against the fumes given off, and care should be taken when quenching articles which may have a covering of molten salt because of the spitting which is liable to take place. To guard against this second risk, operators usually wear gloves and goggles.

A corner of an heat-treatment shop is shown at Fig. 26. The furnaces being (a) and (b) gas fired oven (or muffle) furnaces, (c) small tempering furnace with separate metal container and (d) gas heated salt-bath furnace (shown with doors closed).

Furnace temperatures. An important auxiliary to a furnace is some method of measuring its temperature, because the successful heat treatment of steel depends on close adherence to the correct temperature. There are many methods used for this, a simple one being to put in the furnace some substance which melts at the temperature it is desired to verify. The substances used for this are moulded in the form of cones from mixtures of Kaolin, lime, feldspar, magnesia, quartz and boric acid, with their melting temperatures arranged in steps from 600 °C to 2000 °C. When a furnace temperature is required, several of these cones, covering a range of melting temperatures within which the temperature of the furnace is judged to lie, are put in and observed. The temperature is then judged from which cones collapse, and which remain unaffected by the heat of the furnace. For example, to verify a temperature judged to be 810°–820 °C, cones having melting points of 790, 815 and 835 °C might be put in, the temperature then being estimated from their condition after sufficient time had elapsed for them to be affected. These cones are called *Seger Cones* or *Sentinels* and Fig. 27 shows how they appear after a test.

Fig. 27 Seger cones after being in furnace

Pyrometers. For modern heat-treatment furnaces the above method of measuring temperature is not very convenient because it is lengthy in operation and does not give a continuous reading of the temperature, as is often necessary. A more scientific and reliable method of measuring furnace temperatures is by an instrument called a *pyrometer*. There are various forms of pyrometers and two types in common use are: (1) The Thermo-Electric Pyrometer, (2) the Optical Pyrometer. The first type makes use of the principle that when two dissimilar wires are joined to form a complete electric circuit, and the two junctions maintained at different temperatures, an electric current flows in the circuit, the magnitude of the current depending upon the metals used, and the temperature difference of the junctions. The *hot junction*, which is placed in the furnace, is often made up of wires of platinum, and an alloy of platinum and rhodium welded together and is called a *thermocouple*. Leads from these wires are carried to a sensitive galvanometer which generally constitutes the cold junction, and which indicates the current flowing in the circuit due to the difference of temperature between the two junctions. The galvanometer is so calibrated, that instead of indicating electrical units, it reads in degrees of temperature. A diagram of this pyrometer is shown at Fig. 28(a).

The *optical pyrometer* compares the intensity of light being emitted from the furnace with that from some standard source. In the disappearing filament pyrometer the glow of a standard filament lamp is varied until it

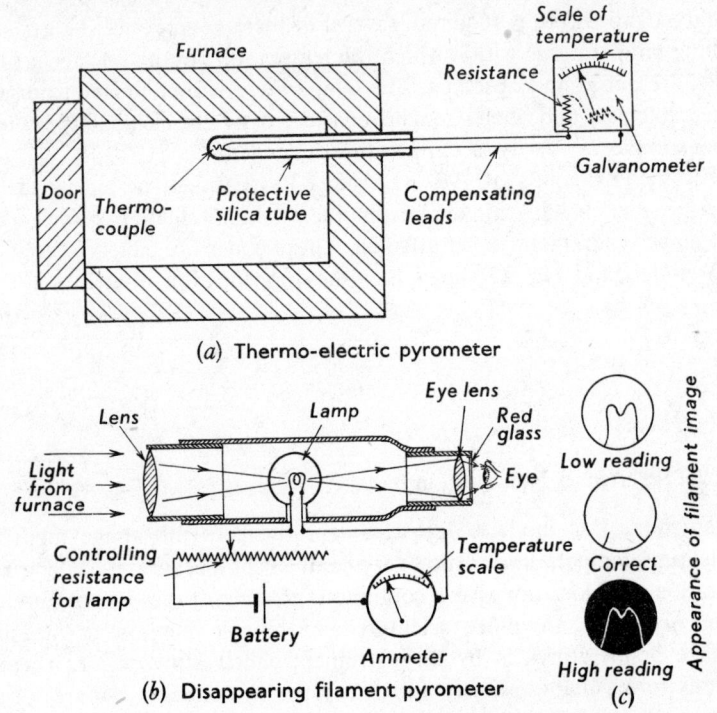

(a) Thermo-electric pyrometer

(b) Disappearing filament pyrometer   (c)

Fig. 28 Methods of measuring furnace temperatures

matches the light from the furnace and disappears when viewed through the telescope. The instrument, which is shown diagramatically at Fig. 28(b), is set up in front of the furnace and the light from the furnace is viewed through the eyepiece. The current through the lamp is varied by a resistance, and when a colour match is obtained the lamp filament disappears from sight, the lamp current required to cause this being indicated on an ammeter. This instrument may be calibrated so that it reads in degrees of temperature, instead of in units of electric current. The appearance of the filament, as seen through the telescope, is shown alongside the diagram Fig. 28(c).

Heat colours. The luminous colours corresponding to different furnace and metal temperatures are as follows:

Colour	Temperature °C	Process for which suitable
Black red (viewed in dull light or darkness)	426 ⎫ 482 ⎪ 538 ⎬ 593 ⎭	Toughening carbon tool steel after quenching. Tempering high-speed steel.
	648	
Dark red	704	
	760 ⎫	Hardening and annealing
Cherry red	815 ⎤⎬	carbon tool steel.
Light cherry red	871 ⎥	
	926 ⎬	Hardening alloy tool steel.
Orange red	981 ⎥	
	1036 ⎦	
Yellow	1093	
	1149	
	1204	
Yellow white	1259 ⎫⎬	Hardening high-speed steel.
	1315 ⎭	
White	1371	Welding.

Reasons for heat treatment. Metal is heat-treated to give it certain desired properties. Some of the properties which may be required and the treatments necessary are as follows:

To soften the metal—*Annealing.*

To harden it to resist wear, or to enable it to cut other metals—*Hardening.*

To remove some of the extreme brittleness caused by hardening—*Tempering.*

To refine the structure after it has been distorted by hammering or working when in the cold state—*Normalising.*

In addition there are various other treatments such as toughening the metal to better withstand shock, toughening soft steel so that it machines without tearing, treating special steels to increase their strength and so on.

Annealing

The purposes of annealing are (1) to soften the steel so that it may be more easily machined, (2) to relieve internal stresses which may have been caused by working the metal or by unequal contraction in casting. The process involves (a) heating slowly to the required temperature, (b) holding at that temperature for long enough to enable the internal changes to take place, (c) cooling slowly. We have seen that above the critical range steel consists of austenite. This is true whatever may have been the structural condition of the steel before heating. Furthermore, when austenite is cooled normally through the critical range, it changes to pearlite, mixed with ferrite or cementite, depending on the carbon content of the steel. This change occurs only if the cooling is slow. The true transformation to pearlite, therefore, is dependent on having true austenite to start with, and then allowing sufficient time for the metal to cool through the critical range for the soft pearlite to form. The temperature for annealing must be 30° to 50° above the higher critical point for steels up to 0·9% carbon, and about the same amount above the lower critical point for high carbon and tool steel* (above 0·9% C). The reader may obtain these from the graph (Fig. 24), and they are approximately as follows:

Carbon Content %	Annealing Temperature °C
less than 0·12 (Dead mild)	875 to 925
0·12 to 0·25 (Mild)	840 to 970
0·3 to 0·5	815 to 840
0·5 to 0·9	780 to 810
0·9 to 1·3 (Tool steels)	760 to 780

The time taken by the metal to reach the temperature of the furnace, and the period for which it should be 'soaked' at the annealing temperature vary according to the shape and dimensions of the article, but the process should not be hurried.† The best method of cooling is to turn off

---

* The reader may wonder, in view of the foregoing remarks, why tool steel should not be heated above the upper critical point for annealing. It is found that this may result in an unsatisfactory structure and cause cracking when the tools are hardened. Heating is therefore stopped when all the carbon in the pearlite has gone into solution but not all the free cementite.

† If metal is placed in a furnace at, say, 800 °C, at least 20 min per 25 mm of thickness should be allowed for the heating-up period.

the furnace and leave the work inside, allowing the whole to cool slowly together. If this is not possible the metal should be taken out and buried in a non-conducting material such as sand, lime or ashes.

Normalising

The object of normalising is to refine the structure of steel and remove strains which may have been caused by cold working. When steel is cold worked (hammered, rolled, bent, etc., in the cold condition) the crystal structure is distorted, and the metal may be brittle and unreliable. Also, when steel is kept heated for a considerable period well above the higher critical point (as may often be necessary for prolonged forging), a growth in the grain size takes place, and when cool the metal may have lost its toughness. If the steel is slowly heated to its annealing temperature, the structure is in the most refined state, and normalising consists of cooling it in air from this point.

Hardening and tempering of carbon steels

We now know that the critical range is the interval between two temperatures the lower of which is about 700 °C, and that when steel cools normally through this range, it is transformed from austenite to pearlite, plus a free constituent. If we can by some means lower the temperature of the transformation of austenite so that instead of taking place at 700° it takes place at about 300°, it will not decompose into pearlite, but will be transformed into a constituent called *martensite*. We can lower this temperature of transformation by cooling the metal suddenly, so that the change does not have time to take place at the normal point, but is forced to occur at some lower temperature. The sudden cooling is usually made to take place by quenching the steel in water or some other liquid, and the efficiency with which the quenching occurs determines how much of the austenite is transformed to martensite, and the hardness of the steel.

The exact constituency of martensite is not clearly known but it serves our purpose to know that it is a very hard substance capable of resisting extreme wear and of cutting other metals. It has a needle-like structure as shown in Fig. 29. It should be clear that martensite cannot be formed (and the steel hardened) by quenching until the steel is in the austenitic condition, i.e. above the lower critical temperature, the carbon then being in solution. For this reason, the term *hardening carbon* has been given to the form of the carbon when the steel is above the lower critical temperature, and *cement carbon* when below this point (i.e. in the pearlite). When the carbon is in the cement (softening) form, the steel cannot be hardened by quenching.

Before passing on to the practical consideration of hardening and tempering it will be necessary to qualify our remarks somewhat for low carbon steels.

Fig. 29 Representing the needle-like structure of martensite in quenched carbon-steel. The white background is austenite × 1000 approx

When steel contains less than 0·3% carbon, and this includes all the mild steels, the solution of carbon in iron which forms the austenite is naturally a much weaker solution, and contains more iron than for the higher carbon steels. When steel of this class is quenched, some of this extra iron is set free in the structure and this, together with the fact that the smaller amount of carbon results in a smaller amount of martensite, makes it impossible to harden mild steel in the manner just described. A mild steel will certainly be somewhat harder and of a more uniform texture as a result of quenching, but it will not be hard in the generally accepted meaning.

Table 4. Hardness of quenched steel

Carbon %	0·1	0·3	0·5	0·7	0·9	1·2
Brinell hardness	150	450	650	700	680	690

Tempering

Martensite, being an extremely hard and brittle substance, renders a dead hard tool made of it liable to cracking and chipping. The steel is therefore heated up again to a temperature below the lower critical temperature which causes a partial transformation of the martensite back to pearlite again, thereby taking away some of the hardness, but making the steel tougher.

The temperature at which tempering should be carried out depends upon the purpose for which the article or tool is to be used, and the table below gives the temperatures for some of the usual applications of high carbon steel.

When the article has been brought to the tempering temperature it may be quenched or allowed to cool off naturally. The temperature for this operation is often judged by the colour of the oxide film which appears on a freshly polished surface of the article, and these colours are given in Table 5. The reader should experiment with a few pieces of steel in order to familiarise himself with the order in which the colours appear.

Table 5. Tempering temperatures

Tool or Article	Temperature °C	Temper Colour
Turning tools.	230	Pale straw
Drills and milling cutters . . .	240	Dark straw
Taps and shear blades	250	Brown
Punches, reamers, twist drills, rivet snaps	260	Brownish purple
Press tools, axes	270	Purple
Cold chisels, setts for steel . . .	280	Dark purple
Springs	300	Blue
For toughening generally without undue hardness	450–600	

To sum up: carbon steels may be hardened by quenching from the annealing temperature (page 44), when they may be too brittle for their purpose. To toughen them they must be tempered by reheating to a suitable temperature.

Quenching

We have seen that to transform the austenite to martensite efficiently, the cooling must be so rapid that the temperature of transformation is lowered from about 750 °C to 300 °C. This involves very rapid cooling and brings troubles with distortion and cracking. There are two factors which tend to cause the metal to warp and crack:

(1) When metal is cooled it undergoes a general contraction which is not uniform, but occurs first at the outside surfaces, and in the thin sections of the article.
(2) When steel cools through the critical range an *expansion* takes place.

Now, if we could arrange to cool the metal so that its *whole volume* could be suddenly cooled at the same instant, we should not experience much trouble with volume changes, but unfortunately this is not possible. Let us examine what happens: after carefully heating the steel to the annealing temperature we quickly take it from the furnace and plunge it into water. The outer portion of the metal bring in contact with the water is immediately cooled and undergoes its critical range expansion, followed by its cooling contraction, becoming a hard, rigid skin of metal. The inner portion of metal, however, has not yet felt the quenching effect and is still red hot. An instant later, the quenching effect is transferred to this portion which, as it passes through the critical range, must expand. It can easily be imagined what is likely to happen to the hard and brittle outer layer of cold metal when this inner core undergoes its critical expansion: it is a matter of very good fortune if it does not crack, and most of us must have had that mortifying experience.

The general contraction which takes place may be to some extent allowed for by the direction in which we quench the tool. For example, if a long tool such as a tap is quenched on its side, the whole length of one side meets the cold water at once and contracts. This sudden shortening of one side causes the tap to warp, and we must remember, that to try and straighten it afterwards will most surely break it. If the tap is plunged vertically into the water, the warping effect is greatly minimised. The reader should study every job he has to harden and endeavour to quench it in such a manner that warping effects are minimised as much as possible. When an article has been quenched it should be moved about in the water to expedite the cooling as much as possible.

After what has been said regarding the slower cooling down of the inner metal of large and thick jobs, it can be imagined that the section of such an article will not be of the same hardness throughout, and the very centre may almost have had time to transform to the pearlitic condition. This may not be a disadvantage as a rather softer core to a tool helps to give it strength and we are not likely to require to use the metal near the very centre for cutting. When extreme hardness can be sacrificed in favour of extra toughness, oil may be used for cooling instead of water, with less risk of cracking during quenching. As the specific heat capacity of oil is less than that of water its cooling effect is less and the quenching effect is not so intense. This means that less martensite will be formed, and the tool will be softer and tougher.

It is often not necessary to harden the whole of a tool, but only that portion which is required to do the cutting and only the part required hard should be heated and quenched. This leaves us with a hard cutting portion merging on to a tough shank or body. Tools such as chisels, punches, drifts, etc., must on no account be hard where they are struck with the hammer or the metal will chip and may fly into the face or hands.

The heat required for tempering is generally obtained by placing the article on to a piece of plate which has been heated to redness. The portion of the tool to be tempered having been previously polished with emery cloth is carefully watched, and as soon as the correct tempering colour appears it may be cooled off. Round articles such as taps may be held inside a piece of red-hot tubing so that they are uniformly heated from all sides.*

For some tools such as chisels, punches, etc., time, and heat, may be saved by hardening and tempering in the same operation. The tool should be heated up to the hardening temperature for about half its length and then the cutting end quenched for a length of 30 to 60 mm. When it is quite certain that quenching is complete the tool should be removed from the water and the cutting edge quickly polished with emery cloth. The heat from the unquenched portion will soon travel, by conduction, to the end, and the tempering colours will show up. When the required colour appears the whole tool should be quenched. This method gives a good effect as the tool consists of the hardened and tempered cutting edge, with the metal gradually and uniformly decreasing in hardness towards the soft shank.

The working of steel

Practically the whole of the energy of the workshop engineer is aimed towards some aspect of changing the shape of metal and we shall now discuss the general question of hot and cold working. *Cold working* means the subjecting of metal to forces which cause it to undergo plastic deformation when it is below the lower critical temperature (pearlite condition). *Hot working* is the same thing when the metal is above the lower critical temperature. The processes included are forging, bending, rolling, drawing out, hammering, etc., and it will be realised that since cast iron is brittle and does not lend itself to any form of plastic flow, it is excluded from the

---

* When large quantities of work have to be tempered, an efficient way of heating is by immersion in a bath of molten metal or salt. A useful range of temperatures is provided by the tin-lead alloys which are molten between 183 °C (37% lead) and 327 °C (100% lead).

following remarks which refer to wrought iron and steel, and in a general way to the non-ferrous metals which can be worked.

Cold working. It may at first seem strange why working steel at 600 °C, which is a dull red heat, should be included in the category of cold work. When it is remembered, however, that the pearlitic structure remains unchanged up to the lower critical (just over 700 °C), it will be realised that the effect, on the pearlite, of plastic working, will be the same at any temperature below this.

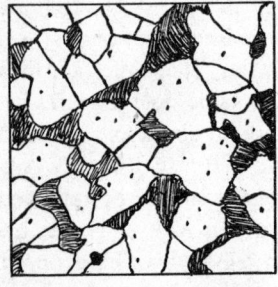 

(*a*) Normal structure of a mild steel   (*b*) Same steel after a considerable
(0·17% C)   reduction in wire drawing

Fig. 30 Effect of cold drawing on the structure of metal

Cold working distorts the crystal structure of metal and renders it harder and more brittle. Unless, however, the metal has been worked so much that it cracks, its structure may be brought back to its original state by annealing. Whatever distortion may be in the structure of the steel below the critical range, when it is brought to the upper critical temperature it recrystallises into austenite and then, if cooled normally, transforms back to pearlite bearing little trace of the original ill-treatment. The reader may imagine that when a metal has been bent and straightened again, or given a certain amount of permanent stretch, or has received some other such treatment, it has lost its 'nature' and is rendered worthless. This is not the case, as provided the metal has not cracked the treatment will only have made it a little harder, stronger, and more brittle, and it may be brought back to its original condition by annealing or normalising. If, of course, the treatment has made the metal so brittle that it has cracked, then we may say it has lost its 'nature' or any other word we like to use, for the only cure is the scrap-heap and re-melting. As we shall discuss later, cold rolled, drawn into wire. drawn into cups, cold headed, bent cold. etc. The distortion to the structure of a mild steel caused by being drawn into wire is shown at Fig. 30.

Hot working. This consists of changing the shape of steel when its temperature is above the upper critical and is used in hot rolling, forging, upsetting, etc. Hot work refines the structure of steel by smashing up large grain formations and closing up any cavities which may be present. The shaping of hot, plastic steel may be by blows such as from a hammer or drop stamp, or the metal may be caused to flow by a slower and more even movement as by the pressure of a hydraulic press. The hammering processes tend to give more thorough effects than the pressing, which is more in the nature of kneading. Where the thickness of metal is large the effects of hammering may not penetrate right through, and the surface metal will be better worked than that below. If forging is carried out with the strength requirements of the finished article in mind, a much improved and more reliable product may be obtained. This is because steel to some degree exhibits directional qualities in its grains, and by suitable forging, the flow of the metal may be so controlled that the direction of the grain adds strength. Fig. 31 is of a gear, solid with a shaft. At (a) is shown how the grain would run if the whole were turned from a solid bar, whilst at (b) the grain glow caused by forging up the end of a bar is shown. The metal in gear teeth cut on the gear at (b) will be more durable than at (a). Fig. 32 shows the grain direction in a forged crankshaft, where again it can be seen that the directional qualities of the grain add toughness to the webs, and where they join the round parts of the shaft.

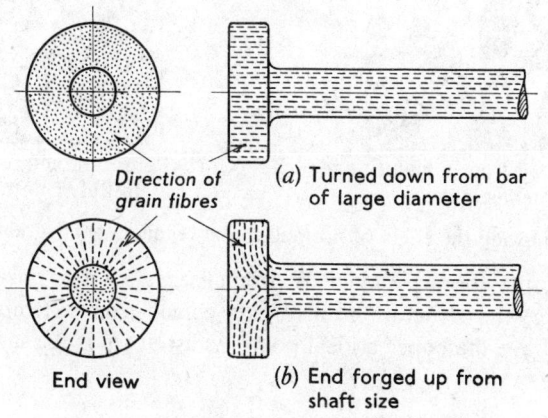

Direction of grain fibres

(a) Turned down from bar of large diameter

End view

(b) End forged up from shaft size

Fig. 31 Effect of upsetting end of shaft

Finishing temperature. The temperature at which the hammering of a forging is left off has an important influence on the properties of the forging. When steel is heated to, and soaked at, a temperature considerably above its upper critical point, its grain size increases, and if cooled from this point

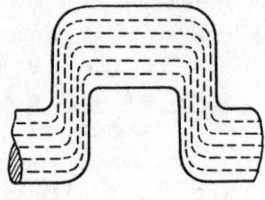

Fig. 32 Direction of grain in a forged crankshaft

its structure may be weak and without toughness. When a considerable amount of forging has to be done the metal must be made hot or it will cool too quickly and time will be wasted by continual re-heating. Whilst the metal is being hammered the grain growth is compensated for by the refining effect of the hammering, but if the hammering is discontinued at a high temperature, grain growth will occur as the forging cools. On the other hand, if the metal is hammered when it is below 700 °C it will be cold worked and may be given small hair cracks. The best time to leave off hammering is when the steel is just above the upper critical temperature and it will then have the best possible structure when it is cold. If there is

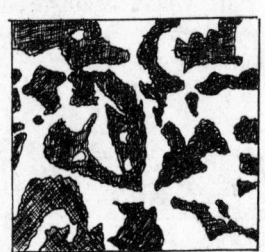

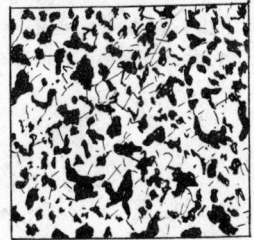

(a) As cooled from a high forging temperature

(b) Refined structure of the same steel after normalising

Fig. 33 Showing the effect of normalising in refining coarse-grain structures

any doubt, the forging should be normalised after it is finished. Fig. 33 shows at (a) the structure of a forging cooled from a temperature considerably above the upper critical point, whilst at (b) is the same steel after normalising.

Overheated steel. Sometimes, by accident, steel is heated to a very high temperature—almost melting. This causes the structure to be rendered brittle, and is brought about by a film of oxide getting into the grain boundaries. When in this state the metal is often referred to as 'burnt' and nothing except re-melting will bring it back to a usable condition.

Toughening mild steel. When the ferrite in the microstructure of mild steel, instead of being well dispersed amongst the pearlite is collected together in large patches, the steel gives unsatisfactory results in machining and the surface tears, especially for such operations as screw-cutting. By quenching this steel from the critical range it is often possible to improve matters. The quenching breaks up the segregations of ferrite and confers enough additional toughness to avoid the tendency to tear.

Case-hardening. Case-hardening is a method of hardening the surface of mild steel to a depth varying up to about 2 mm. We have seen that heating and quenching has no appreciable effect on low carbon steel, but if we can convert the surface of mild steel into a high carbon steel we may harden it by heating and quenching in the usual way. This is what actually takes place in case-hardening and the operation is divided into two main parts: (1) converting the outer skin (case) to a high carbon steel (carburising) and (2) hardening this case and refining the core. To increase the carbon content of the case, the steel is packed in cast-iron or steel boxes together with a substance rich in carbon such as charcoal granules, hoof clippings, bone dust, etc., and the boxes are charged into a furnace at about 900 °C to 950 °C. At this temperature the carbon infuses into the surface of the metal, converting it to a high carbon steel, and the depth to which this takes place depends upon the time of treatment, but generally 3 or 4 hours are sufficient. At the end of this time the box is allowed to cool slowly and when removed the steel parts consist of a soft mild steel core with a case of high carbon steel. Due to the prolonged heating at a high temperature the grain structure of the core will be relatively coarse, so that as well as hardening the case, a treatment will be necessary to refine and toughen the core. The core may first be refined by heating to about 900 °C and quenching in oil, after which the case is hardened by heating to about 770 °C and quenched in water. This leaves an article having a soft, tough core and a glass-hard case which should have its brittleness relieved by tempering at about 200 °C or over, depending on the final hardness required. Fig. 34 shows the fracture of a case-hardened shaft.

On some jobs it is necessary for certain portions to be left in the soft condition. For example, if a thread is hardened it will chip in service owing

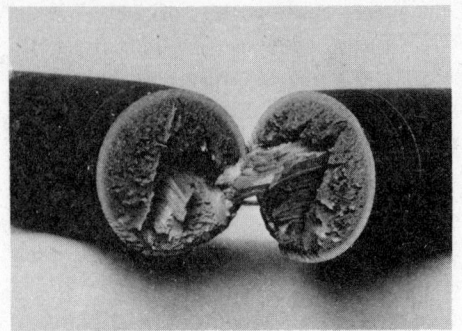

W. T. Flather Ltd

Fig. 34 Fracture of a case-
hardened bar showing the hard
'case' and tough 'core'

to its brittleness. There are two ways of leaving any portion of a case-
hardened part soft: (1) by preventing infusion of carbon, (2) by cutting
away the high carbon steel case before it is quenched. In the first method
the part to be left soft is covered with clay, or copper plated. This retains
it in the low carbon condition by preventing infusion of carbon. In the
second method extra metal is left on the portions that must be left soft.
After carburising this is machined off, which exposes the mild steel underneath and when quenching takes place this is left unhardened.

Case-hardening is applied to parts such as spindles, pins and similar
components which require to be tough but hard, and wear-resisting on the
surface. The steel used should be of low carbon content (0·08 to 0·25%).
Some tools, such as hand-reamers, which are only subjected to light working conditions, may be made satisfactorily from case-hardened steel.

Superficial hardening. For the rapid surface hardening of small parts such
as screws, clamps, washers, etc., a form of surface hardening may be used..
This uses cyanide, or prussiate of potassium, and may be carried out either
by placing the article in the fused salt, holding it at the quenching temperature for a few minutes and then quenching out, or by sprinkling the red-hot
article with the powdered salt, heating up again and quenching. This
method is often spoken of as case-hardening, but is not a true case-hardening process.

Tests for iron and steel

It is sometimes necessary to make sample tests on the metal being used.
The test may be to differentiate one metal from another, to test the hardness

or toughness of a metal and so on. Whilst the reader may not have access to elaborate testing machines and metallurgical apparatus he can, with the workshop tools at his disposal, gain a useful insight into many of the properties of metals. With the help of a simple lens he can follow the course of hardening cracks and use the information to help him to do better next time. The colour and quantity of the sparks given off by grinding various steels will tell him about the steels. The appearance of a fracture will give an indication as to whether he is forging at too high a temperature and getting grain growth and so on.

The reader should purposely submit steel to certain treatments, e.g. hammering hot, hammering cold with and without normalising, prolonged heating, severe overstraining (of a wire), quenching different ways and in various liquids and so on. After such experiments he should put the metal to test and examination in order to study the effect of what he has done, and compare it with a normal sample of the same steel. In this way he will develop an instinct which, in later life, will serve him in good stead. We may learn a great deal from books and very soon forget what we have learned, but the knowledge we gain by doing and experiencing things is rarely, if ever forgotten.

The following tests may all be carried out with the appliances and tools in the workshop. At (a) are given tests for properties. These will not give actual numerical results, but will be useful for comparing the degree of each property possessed by different metals. The distinguishing tests under (b) will help the reader to recognise the various grades of iron and steel. Distinguishing tests for high-speed steel have also been given. This is a special alloy steel used for making cutting tools, and its properties and treatment will be discussed later.

Table 6

(a) Tests for Properties

Hardness. (a) *Unhardened metals.* Use some form of pressure (e.g. vise or arbor press) to force a 10 mm ball on to the polished surface of the metal. Exert the same pressure each time and compare the diameters of the impressions made.

(b) *Hardened metals.* Try to file the metal with the edge of a small, fine file. Grind smooth, and scratch with a needle point. Use scratch test to try for hardening at the *centre* of a piece of quenched steel (after previously nicking and breaking the bar).

Strength. Take bars about 12 mm to 20 mm diameter and 1 metre long and support at each end. Load centre with weights, and compare sag at centre. Or, clamp one end of $\frac{1}{2}$ m to $\frac{3}{4}$ m long bar in vise and load the other end. Strong metals give least deflection. Keep loading within the elastic limit of the material (i.e. no permanent set when load removed).

Toughness. Saw a 12 mm bar about $\frac{1}{3}$ through, hold in vise and hit with 1 kg hammer until severed. Force, and number of blows necessary is an indication of toughness. Test on quenched carbon steels which have had varying degrees of temper up to 600 °C.

Brittleness. Brittle metals break easily under the above test for toughness.

Ductility
(1) Take a piece of the material 6 mm to 9 mm diameter, hold it in the vise with about 50 mm protruding. Bend the end backwards and forwards and count the number of times it may be bent before it fractures. A ductile material will stand a greater number of bends than one which is not ductile.
(2) Take a bar or strip and double it over on itself like a hairpin. Examine the outside of the bend for cracks, and if the material is ductile it should be free from them.
(3) By some means, pull wires of the material until they fracture. Compare the lengths under tension before and after fracture. Wires which have stretched the most before fracture are the most ductile.

Malleability. Hammer a round bar until it has been flattened. Examine the edges of the flattened portion for cracks. Materials which are most easily flattened, and which show least signs of cracking are the most ductile. This test also gives a measure of the *plasticity* of a material.

Table 6. Workshop tests for iron and steel
(b) *Distinguishing Tests*

Nature of Test	Cast Iron	Wrought Iron	Mild Steel	Medium Carbon Steel	Cast Steel	High Speed Steel
Appearance of bar	Grey and sandy. Shows line of casting.	Red and scaly.	Smooth finish. Bluish sheen.	Bluish black sheen. Smooth.	Bright black. Very smooth. Sharp corners on square bars.	Not as smooth as cast steel. Often painted.
File with rough file	Skin very hard. Black powder from filed metal. Resistance low.	File drags and tends to clog. Whitish filings. Slag visible.	Not as much drag as W.I. White finish and filings.	Increasing difficulty in making file bite into metal. Surface becomes more glazed as carbon increases.		Not as hard as cast steel.
Turn in lathe	Cuts easily. Black crumbly chips. Black powder when surface wiped.	White curly turnings. Poor finish. Slag lines visible.	Turns easily. White curly turnings.	Increasing hardness under tool. Turnings break into short pieces and may be brown or blue. Rather glazed finish.		Turns fairly easily. Long chips. Distinctive smell from scale.
Hammer at full red heat	Crumbles under hammer.	Flattens very easily.	Flattens fairly easily.	Increasing resistance to flattening as carbon increases.		Considerable resistance to flattening.

Table 6 (*continued*)

Nature of Test	Cast Iron	Wrought Iron	Mild Steel	Medium Carbon Steel	Cast Steel	High Speed Steel
Quench from full red	No appreciable change. May crack.	No appreciable change.	May harden slightly but not much.	Hard when tested with file.	Hard when tested with file.	Moderately hard.
Grind on emery wheel	Small stream of dull red sparks with an occasional bright burst.	Lighter sparks than C.I. and greater quantity	Stream of long white sparks.	As carbon content increases spark stream becomes more bushy with secondary 'bursts'		Dull red sparks rather like C.I.
Drop a 12 mm to 20 mm dia. bar, ¼ m long, on to ground.	Very dull sound.	Dull, but metallic sound.	Medium metallic sound.	Higher note than mild steel.	High, ringing sound.	Lower ring, more like mild steel.
Saw about ⅓ through a 12 mm bar, 25 mm from end. Hold in vise and break off with 1 kg hammer	Snaps easily.	Bends well over.	Bends over, then breaks.	Bends a little before breaking.	Good resistance to blow. Then breaks off.	
Examine fracture of last test	Large crystals. Bright specks of free carbon.	Earthy, fibrous fracture.	Medium crystalline fracture.	Finer fracture than mild steel.	Very fine crystalline fracture.	Fine, velvety fracture.

3 Materials (cont.)—non-ferrous metals and alloys used in the workshop—alloy steels—'rare' metals —ceramic tools—preparation of metals

Although iron and steel are the most common and important of the materials with which we have to deal, there are other metals used in the workshop about which it will be necessary for us to have some knowledge. The chief pure metals in the non-ferrous group with which the reader will have to deal are aluminium, copper, lead, tin and zinc. These may be mixed to give *alloys*, many of which are of great importance. The reader, if he likes, may regard the above metals in the light of primary colours which may be mixed in different proportions to give a wide range of varying shades.

Aluminium

Aluminium is a white metal produced by electrical processes from the oxide (alumina), which is prepared from a clayey mineral called bauxite. Bauxite is found in large quantities in various parts of the world and the successful extraction of the metal depends upon the supply of large amounts of cheap electricity. Owing to its light weight aluminium is used largely for aircraft and automobile components where the saving of weight is an advantage. The relative density of aluminium is about $2 \cdot 68$ as compared with $7 \cdot 8$ for steel. In its pure state the metal would be too weak and soft for most purposes, but when mixed with small amounts of other alloys it becomes hard and rigid. Aluminium is very ductile and malleable. It can be rolled into leaf $\frac{2}{100}$ mm thick and drawn into wire $\frac{1}{10}$ mm diameter. It can be given a high finish by burnishing and polishing. Aluminium melts easily and may be formed into parts by casting. In its natural state (about

99% pure), it has a tensile strength varying from 90 to 150 newtons per square millimetre depending upon the amount of mechanical treatment it has received. Its good electrical conductivity is an important property and aluminium is used for overhead cables on the Grid system pylons. To give the strength necessary to carry the large spans, the cables are made of aluminium with a thin core of high tensile steel wire. The high resistance of aluminium to corrosion makes it a useful metal for cooking pans. It owes this resistance to a thin film of oxide which covers its surface and protects it. Aluminium foil is used for wrapping chocolates and cigarettes and for sealing milk bottles, whilst the powdered metal is used as the base for aluminium paint. Weight for weight, aluminium is only exceeded in tensile strength by the best cast steel.

Aluminium alloys. It is when alloyed with small amounts of other metals that aluminium finds its widest uses. The addition of small quantities of other elements converts this soft, weak metal into hard, strong metals with a wide range of applications. For the casting of crank cases and for general engineering use, aluminium is alloyed with small amounts of copper and zinc. The alloy containing $12\frac{1}{2}$ to $14\frac{1}{2}\%$ of zinc with $2\frac{1}{2}$ to 3% of copper (BS 1490)* has a minimum tensile strength of 170 newtons per square millimetre and is used for castings. For use in bar form BS 1476 specifies an alloy containing 2 to 4% copper, 4 to 8% zinc and not more than 1% of iron and silicon. These last two metals are nearly always present in aluminium as impurities. For castings which must be tough to withstand shocks and severe stresses, an aluminium-copper alloy is used containing 12% of copper. An important series of casting and forging alloys having high strength have recently been developed for use in aeroplane construction. These contain copper, nickel, magnesium and zinc together with small amounts of other substances. One example of such alloys is as follows: zinc 5%, magnesium 3%, copper 2·2%, nickel up to 1%, aluminium the remainder. To bring this alloy to its maximum strength and hardness it must be heat-treated, and when this has been carried out an ultimate tensile strength of over 500 newtons per square millimetre may be obtained. Another alloy containing copper, nickel and magnesium, and which may be cast or wrought is known as *Y-alloy*. The wrought alloys are specified in BS 1470 and BS 1476 and contain $3\frac{1}{2}$ to $4\frac{1}{2}\%$ copper, 1·8 to 2·3% nickel and 1·2 to 1·7% magnesium. This alloy has the characteristic of retaining a good strength at high temperatures and for this reason is used

* British Standards Institution Specification.

for engine pistons. It is also used largely in the form of sheet and strip, and after proper heat treatment may be brought to a minimum tensile strength of about 350 newtons per square millimetre.

An important and interesting wrought alloy is known as *Duralumin*. This is composed of copper $3\frac{1}{2}$ to $4\frac{1}{2}\%$, manganese $0\cdot4$ to $0\cdot7\%$, magnesium $0\cdot4$ to $0\cdot7\%$, aluminium the remainder. It is used widely in the wrought condition for forgings, stampings, bars, sheets, tubes and rivets. Duralumin has the property of *age-hardening*, and at room temperature its hardness increases rapidly for the first day, and then slowly up to a maximum value after 4 or 5 days. When it is aged it is too hard for work to be done on it, and it must be annealed. This is effected by heating to about 375 °C and cooling in air, water or oil. After annealing, the metal must be worked within the time that its age-hardening will have rendered it too hard for further work, or re-annealing will become necessary. When in the heat-treated and aged condition duralumin may have a tensile strength up to 400 newtons per square millimetre.

Copper

Copper is easily distinguished from all other metals on account of its red colour. The fracture of cast copper is granular, but when forged or rolled it is slightly fibrous. The chief ore of copper is the pyrites, which contains on the average 32% of copper. Extraction may be by the dry or wet process, the former being the most common, carried out in a reverberatory or blast furnace preceded by various stages of ore refinement and followed by various metal refining processes. In the wet process the ore is treated with acids and the metal afterwards precipitated. Electrolytic copper is a pure form obtained by electrolysis from impure lumps of smelted copper.

The metal is malleable and ductile, and because of its high electrical conductivity it is used extensively for wire and cable, and all parts of electrical apparatus which must conduct the current. Copper also is a good conductor of heat and is highly resistant to corrosion by liquids. For this reason it is used for boiler fireboxes, water heating apparatus, water pipes and vessels in brewery and chemical plants. For its high heat conduction it is used for soldering iron bits.

Copper may be cast, forged, rolled and drawn into wire. Rolling and drawing harden it, but it may be softened again by heating to 320 °C. The mechanical properties of copper depend upon its condition. Castings may have a tensile strength of 150 to 170 newtons per square millimetre, which may be increased to 215 to 230 by working (hammering or rolling). The strength of hard-drawn copper wire may be as high as 380 to 460 newtons per square millimetre.

Aluminium bronze. Copper alloys with aluminium to give aluminium bronze and the chief alloys are those containing 6% and 10% of aluminium respectively. These alloys have good strength and working properties and the 6% aluminium alloy has a fine gold colour, being used for imitation jewellery and decorative purposes.

The 10% aluminium alloy, which often contains 5% nickel and 5% iron, is interesting because it can be hardened like high carbon steel. In its softest condition the alloy has a maximum strength of about 380 newtons per square millimetre, but if it is quenched from 900 ° C it is hardened and its strength rises to about twice this amount. After hardening it may be tempered in the same way as steel. Alloys of copper with tin and zinc are important and will be discussed on pages 65 to 68.

Lead

Lead is the heaviest of the common metals, having a relative density of about 11·3 as compared with 7·8 for steel. It has a bluish-grey colour and a dull, metallic lustre, but this is lost on exposure to the air, the surface becoming a dull grey. Lead is very soft and may easily be cut with a knife. It is plastic and malleable, being easily forced cold into mould shapes and rolled into thin sheets.

The chief ore of lead is the sulphide, called 'galena', and smelting may be carried out in reverberatory or vertical furnaces. As galena nearly always contains some silver, the extraction of lead is generally accompanied by the reclamation of silver as a by-product. One of the principal properties of lead is its freedom from any effect due to the action of water and acids, and because of this it is used for water-pipes, roof covering, the sheathing of electric cables and for containers in chemical plants. It has a low melting point (330 °C), and when sheets have to be joined, they may be butted together and the metal fused (burned) with a blowpipe. Two large industrial users of lead are the electrical and paint industries. It is used for the plates of lead-acid electrical accumulators, and its oxides are used largely as the base for lead paint. When added in small quantities to steel and brass, lead improves their machining properties, and the reader may have heard of 'Ledloy' steel and leaded brass in this connection.

Lead alloys with tin to form *solders*, and with other metals to make *bearing white metals*. These alloys will be discussed later.

Tin

Tin is obtained from tinstone, an oxide, which after a preliminary roasting in a reverberatory furnace to give a refined tin oxide is reduced to crude tin in a similar furnace. The crude tin is further refined to remove various impurities.

In appearance the metal has an almost silvery whiteness with a slightly yellowish tinge and its structure is crystalline. Tin is harder than lead and at ordinary temperatures may be beaten and rolled into tinfoil. It is ductile but not very strong. Tin melts at 232 °C but at 200 °C it is so brittle that it may be hammered into powder. A characteristic of this metal is the crinkling sound made when a bar of it is bent. This sound is called 'tin cry' and is easily heard if a thin bar is held close to the ear and bent. It is caused by the crystalline deformation taking place, and is a useful method of judging the quality of solder, as solder rich in tin gives a louder 'cry' than when not much tin is present.

The greatest use we have for tin in its ordinary state is for coating thin steel sheets (tinplate). It is also used for tinning copper wire before the latter is made into cables.

Alloys of tin

Tin is used to a great extent for alloying with other metals and as a result, many useful alloys are obtained.

Tin–lead alloys. The tin–lead alloys constitute the soft solders, and those of chief interest to the reader are shown in the table below. In addition to tin and lead, a little antimony is recommended in some cases. A full list of the solders is given in BS 219.

The proportion of tin in a solder may be estimated by appearance, by judging the 'tin cry' or by watching the time the solder remains in a pasty stage during solidification. Lead-rich solders have the dull bluish colour of that metal whilst those rich in tin show a white surface with a slight yellow tinge. Alloys with more than 25% of lead will mark paper. The long

Table 7. Soft solders

Purpose for which Solder is used	Tin %	Lead %	Antimony %	Melting Range °C	Remarks
Plumber's wiped joints .	30	69	1	180–250	Prolonged pasty stage when melting or solidifying.
Tinsmith's general work and hand soldering .	45	52·5	2·5 ⎫	180–210	Solidifies fairly rapidly.
Tinsmith's and Coppersmith's fine work. Hand soldering	50	47·5	2·5 ⎭		
Steel tube joints and work requiring low melting point . . .	65	35	—	180	Solidifies quickly. No pasty stage.

pasty stage associated with the solidifying of the lead-rich solders enables the plumber to make his 'wiped' joint.

Bearing metals. It has been found by experience that to give an efficient bearing combination the following conditions are necessary: (1) That the shaft and bearing be dissimilar in their natures with the bearing softer than the shaft. (2) That the most efficient bearing metal is one consisting of small pieces of a comparatively hard metal embedded in the softer body of another metal. In general, a metal consisting of one uniform constituent does not serve well as a bearing. The 'white' metals all contain tin alloyed with various other metals, of which antimony is always present. A micrograph of an alloy containing tin with 10% antimony and 3% copper is

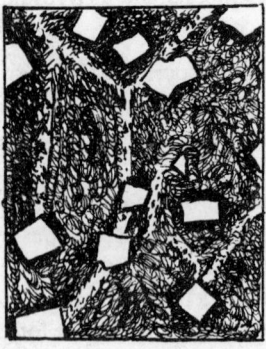

Fig. 35 Microstructure of tin with 10% antimony and 3% copper. Showing hard cubes × 80

shown at Fig. 35, and the hard cube-shaped copper-tin constituent set in the soft body of the general metal can be clearly seen. It is thought that in practice the soft constituent wears away slightly and the shaft is carried on these 'high spots,' the lower level of the remainder of the bearing helping to maintain the film of oil necessary for efficient lubrication. The white metals all have a low melting point which facilitates the formation of the bearing by casting the metal from a ladle. This can be done with the shaft in position if necessary.

The reader will notice from the table below that some of the metals are composed mainly of tin whilst in others lead is the predominant constituent. These are referred to as *tin base* and *lead base* metals respectively and those with a tin base are often called *Babbitt metals*.

Table 8. White bearing metals

Tin	Antimony	Copper	Lead	Uses
93	$3\frac{1}{2}$	$3\frac{1}{2}$	—	Big end bearings, motor and aero engines.
86	$10\frac{1}{2}$	$3\frac{1}{2}$	—	Main bearings, motor and aero engines.
80	11	3	6	Bearings for heavy loads and high speeds.
60	10	$1\frac{1}{2}$	$28\frac{1}{2}$	Bearings for engines and electrical machines, railways and tramways.
40	10	$1\frac{1}{2}$	$48\frac{1}{2}$	Heavy pressure and medium speeds.
20	15	$1\frac{1}{2}$	$63\frac{1}{2}$	Medium pressure and speed.
5	15	—	80	Long bearings with medium load.

Copper—tin alloys—the bronzes

When alloyed with tin, copper forms a set of alloys called bronzes, which are important metals in engineering practice. *Gunmetal* contains 88% copper, 10% tin and 2% zinc. The zinc is added to cleanse the metal and increase its fluidity. The metal is used chiefly for castings which must be strong and resistant to corrosion by water and the atmosphere. Gunmetal is not suitable for cold working but may be forged at about 600 °C (just below red heat). Another common bronze is made of 5% tin, 5% zinc, 5% lead and 85% copper.

Phosphor-bronze. When bronze contains phosphorus it is called phosphor-bronze. The main function of the phosphorus is to act as a cleanser to the metal so that good sound castings can be produced. The composition of this metal varies according to whether it is to be forged and wrought or whether made into castings. A common type of wrought phosphor-bronze is copper 93·7, tin 6, phosphorus 0·3. This may be obtained as rod, sheet and wire. When it is severely worked cold as in the drawing of wire or the rolling of strip, it becomes very hard and springy and is used for springs. Its tensile strength is then in the region of 500 newtons per square millimetre as compared with about 300 newtons in the soft condition.

A variety of phosphor-bronze suitable for casting contains 11% tin and 0·3% phosphorus alloyed with copper. This is used for bearings which must carry heavy loads, worm wheels, gears, nuts for machine lead screws and many other purposes. It may be obtained in the form of cast bar which, for making bearing bushes, may be bought as a hollow tube. This saves the waste of boring out a large part of the metal when turning bushes. Castings of all types may also be made in this metal.

The bronzes, although fairly hard, can be machined fairly easily with good tools, and the chips come off in short pieces. The fracture of the metal has an earthy appearance and is yellow, with a greyish or reddish tinge depending upon the amount of tin it contains, high tin promoting the greyish appearance. When quenched from a little below a red heat it is softened somewhat and made more ductile.

Zinc

The chief ores of zinc are blende (zinc sulphide) and calamine (zinc carbonate). In the extraction of the metal the ore is first roasted in a reverberatory furnace to convert the sulphide to oxide, and in the case of calamine, to drive off carbonic acid and water. The ore is then mixed with some form of carbon and put into long retorts in a special type of furnace. The heat distils off the zinc in the form of its greeny-white vapour and this is condensed to molten zinc.

Zinc is a bluish-white metal which if nearly pure shows large, bright smooth crystals at its fracture. If it is contaminated with iron, dull spots may be seen on the crystal faces, and if much iron is present the fracture becomes granular. The metal may be obtained in rolled sheets, and in cast ingots, the metal in these generally being fairly brittle. The relative density of zinc is about 7·1 and its melting-point 420 °C. The malleability and ductility of zinc are improved by heating it to 100 to 150 °C, but at just over 200 °C it becomes so brittle that it may be powdered.

In its pure state the chief use of zinc is for covering steel sheets to form galvanised iron, the covering being done by dipping the sheets into the molten metal after previous fluxing. Galvanised wire, nails, etc., are also made by this process. When rolled into sheets, zinc is used for roof covering and for providing a damp-proof non-corrosive lining to containers, etc. Zinc casts well and forms the base for various die-casting alloys.

Alloys of copper and zinc—brass

All important use of zinc is for alloying with copper to give the various classes of brass. These alloys are of importance in view of the great variety of mechanical properties that may be obtained, the wide range of production processes to which they lend themselves and their resistance to atmospheric effects and corrosion.

As in the case of aluminium alloys, the brasses may be divided into the cast and the wrought varieties. Suitable types of brass lend themselves to the following processes: casting, hot forging, cold forging, cold rolling into sheets (which may be varied in hardness up to spring constituency), drawing into wire and being extruded through dies to give special shaped bars. By adding small quantities of other elements (aluminium, iron, manganese and tin) the strength of brass may be greatly increased from its normal strength of 300 to 400 newtons per square millimetre, and a range of 'high tensile' brasses is available having ultimate strengths as high as 600 newtons per square millimetre.

The melting-point of brass varies according to its composition, but most of the brasses in the common range liquefy between temperatures of 850 °C and 950 °C. Hard brass may be softened by heating to about 750 °C.

To improve the machining quality of brass 1 or 2% of lead is often added, as in its ordinary condition the metal is soft and ductile and tends

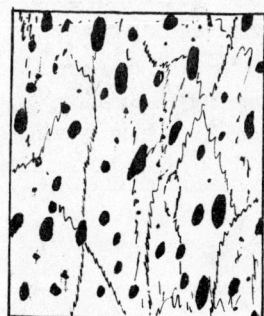

Fig. 36 Structure of a leaded brass showing black specks of lead × 80

to drag under the tool. The lead appears as small specks in the micro-structure and slightly destroying the continuity and causing the turnings to break up into small pieces. Leaded brass is 'hot short'. Small amounts of tin are sometimes added to brass to increase its hardness and add to its resistance against the corrosive action of sea water.

Some of the chief brasses used, together with their applications, are given in Table 9.

The fracture of brass usually has an earthy appearance and its colour is yellow to reddish yellow depending on the copper content.

Table 9. Properties and uses of brasses

Copper	Zinc	Properties and Uses
85	15	Gilding metal. Cheap jewellery.
75	25	Brazing brass. Used where parts must be brazed or silver soldered.
70	30	Great ductility and strength. Drawing into tubes and cartridge cases. Drawing into wire. Cold rolling into strip (up to spring hardness). Cold pressing.
66	34	English standard brass. Casts well and may be rolled and hammered. Little lead added to improve machining.
60	40	Muntz metal. A yellow brass used for general range of articles, e.g. water fittings, household articles, etc.
$58\frac{1}{2}$ Lead	40 $1\frac{1}{2}$	Plastic when red hot. Forging and hot pressing (BS 944).
$57\frac{1}{2}$ Lead	40 $2\frac{1}{2}$	Free cutting. Does not bend well. High speed turning and screwing (BS 249).
50	50	Low melting point. Does not work well. Brazing spelter.
$58\frac{1}{2}$ Aluminium 3 Manganese $1\frac{1}{2}$ Iron 1	36	High tensile brass. Ultimate strength 600 newtons per square millimetre.

Zinc casting alloys

Alloys of zinc are used extensively for making pressure die castings, a method of production that enables large numbers of components to be produced quickly when once the dies have been made. Two common alloys for this purpose are as follows:

	Zinc	Copper	Aluminium
(1) . . .	93·3	2·7	4
(2) . . .	96	—	4

The alloys give best results when the castings have relatively thin sections, as large bodies of metal tend to be porous.

Alloy steels

The steels we have so far dealt with—the plain steels—have been those containing carbon, together with small amounts of manganese and silicon. When certain special properties are required, other elements are added and the steel then becomes known as an alloy steel. As the number of alloy steels is great and their applications very wide, we shall only discuss those which, at this stage, are likely to be of interest to the workshop engineer.

High-speed steel

We have seen that carbon tool steel, after it has been hardened, progressively softens again as it is heated up to the lower critical temperature. If this steel is used for cutting tools where the cutting temperature is likely to be high, softening will occur and the tool will be useless. It has been found that by adding tungsten to steel it is given the property of holding its hardness at a red heat, hence the name 'high speed'. The hardening technique for high-speed steel differs from that of carbon tool steel. The temperature before cooling must be round about 1300 °C—a white-hot, almost melting heat. From this temperature the tool may be oil quenched or cooled off in a blast of cold air. Most high-speed steels have their hardness and general properties improved by re-heating to about 400 to 600 °C after quenching. This is called *secondary hardening*. Owing to the poor heat-conducting capacity of this steel, and the time necessary for the structural transformations to take place, the initial heating must be done carefully. The steel should be brought gradually to about 850° in one furnace, and then transferred quickly to another furnace which is at the quenching heat. The tool is ready for quenching when small bubbles can be seen on its white-hot edge.

Because of its tendency to oxidise and scale when at the hardening temperature, methods which exclude air are preferable when heat-treating high-speed steel. Two methods commonly used are, (1) using a salt bath furnace, where the steel is heated by immersion in molten cyanide, and (2) using a *controlled atmosphere* furnace in which oxygen is excluded from the furnace interior by introducing another gas (e.g. incompletely burned coal gas). This latter principle is also used in bright annealing furnaces where a heat-treatment is necessary for bright bars and strip, but which must not blacken or scale the bright surface of the steel.

Table 10. High-speed steels

Carbon %	Chro- mium %	Tung- sten %	Vana- dium %	Cobalt %	Hardening Tempera- ture °C	Method of Cooling	Remarks
0·6	4	14	—	—	1250	Oil	
0·6	4	18	1	—	1280	Oil or Air	Secondary harden at
0·7	4·5	17	1·5	4	1280	Air	500 to 600 °C

When properly hardened this steel will cut at high speeds when the tool nose is at a dull red heat. High-speed steel may be annealed by heating at 850 °C for about 4 hours and then slowly cooling in the furnace. During the whole of this time it must be protected from the air or it will oxidise and scale badly.

To cope with the difficulties of machining the modern heat-resistant steels and alloys, and other materials coming into use (e.g. titanium alloys) certain properties of high-speed steel have been improved by adding more cobalt and vanadium. The 10% cobalt alloy has 0·82% carbon, 20% tungsten, 5% chromium and 10% cobalt. This has a higher hot-hardness at the expense of some loss of toughness and is useful for heavy continuous-turning. The high vanadium cobalt alloy has 1·5% carbon, 12·5 tungsten, 4·5 chromium, 5 vanadium and 5 cobalt, and possesses maximum wear resistance and hot hardness. Tools and cotters made from this possess good resistance against the highly abrasive action of the metals mentioned above.

'Substitute' high-speed steel. Within limits the tungsten in high-speed steel may be replaced by molybdenum without seriously affecting the properties of the steel. During the last war, when tungsten supplies were seriously curtailed, it became necessary for us to use such molybdenum steels which became known as 'substitute' steels. Two of the chief of these were 'substitute 66' containing 6% each of molybdenum and tungsten and 'substitute 94' which had 9% of molybdenum and 4% tungsten.

Alloy tool steels

For many tools the most important property required is to be able to quench them for hardening without fear of cracking. It is useless, after much expense and labour have been put into the construction of a tool, for it to crack in hardening, and carbon tool steel is very prone to this. A Good *non-shrinking steel* for such purposes has the following composition: carb on 1%, manganese 1%, tungsten 0·5%, chromium 0·75%. It is oil quenched from about 800 °C and tempered up to 250 °C to suit requirements. This steel is useful for making dies and thin tools which should not distort or crack when hardened.

Hot working die steels are steels suitable for punches and dies which have to shape other metal when it is in the red-hot condition. Such treatment would soften ordinary carbon steel. A steel suitable for this purpose contains carbon 0·27%, chromium 1%, vanadium 0·15%, manganese 0·2%. It is hardened by water quenching from 850 °C and tempered for about 2 hours at 230 °C. Special steels for drop forging dies are given in BS 224.

An alloy steel for chisels contains carbon 0·4% and nickel 3%. It must be oil quenched from 900 °C and no tempering is necessary. This material gives an edge superior to that on carbon steel chisels.

Sintered carbide cutting alloys

We have seen that the use of tungsten as an alloying element gives steel the property of retaining its hardness at high temperatures, and this is due to the formation of carbides of iron and carbon in the structure of the hardened steel. This property of the carbides led to investigations which, in 1926, resulted in the successful preparation of *tungsten carbide*, an extremely hard material able to cut metal at much higher speeds than high-speed steel. The extreme brittleness and high melting point (2500 °C) of this material prohibit ordinary methods for its preparation, and it is made by mixing tungsten metal powder and lampblack (carbon) and then heating the mixture to about 1600 °C in an atmosphere of hydrogen. This results in the chemical reaction necessary to produce the carbide which, after the process, is crushed and ground into a powder. Tool tips and other shapes for use as cutting edges are made from this powder by mixing it to a paste with powdered cobalt metal (3% to 20% cobalt) and water, and then compressing the paste to the shape required. This is then rendered hard and homogeneous by a sintering process consisting of heating to about 1500 °C for about an hour in a non-oxidising atmosphere such as hydrogen. The development of tungsten carbide was followed a few years later by titanium carbide which also possessed similar hardness but which, either alone, or mixed with tungsten carbide, has been found more suitable for cutting certain metals.

Because of their high cost, low strength and difficulties of preparation, carbide cutting tools are always made by brazing a tip or block of carbide to a shank of medium carbon steel. Owing to their extreme hardness the carbides cannot be ground with ordinary grinding wheels and special wheels are necessary. In addition, the cutting edge must be lapped to a very high finish so that the grinding process for a tool generally consists of three to five grindings on progressively finer wheels with a final burnishing on a special wheel impregnated with diamond dust.

Tungsten carbide is most suitable for machining the non-ductile materials such as cast-iron, brass, bronze, stone, glass, etc. For cutting steel and ductile metals mixtures of tungsten and titanium carbides are most suitable. Cutting speeds up to three or four times those possible with high-speed steel may be used with up to ten times the life of the tool between re-grinds. Hard spots, scale, etc., which would ruin an ordinary tool, have no effect on this material, and its application has revolutionised cutting practice, not only in the machine shop but in other industries (e.g. stone cutting) where previously no material existed which would stand up to the conditions imposed.

High tensile steels

For many purposes in engineering, high strength and toughness are necessary, and there are many alloy steels having these properties. Some of these steels when properly heat-treated will give strengths as high as 1850 newtons per square millimetre. These high strengths are usually obtained by suitable heating and cooling, followed by tempering.

The elements most commonly put into such steels are nickel and chromium and for this reason they are called *nickel-chrome steels*. The $1\frac{1}{2}\%$ steel of this class contains carbon 0·4%, nickel 1·5%, chromium 1·25%. It is oil quenched from 840 °C and tempered up to 600 °C according to requirements. When tempered at about 250 °C it has a strength of about 1500 newtons per square millimetre. A high tensile *air-hardening steel* has carbon 0·34%, nickel 4·25%, chromium 1·25%. Hardening is effected by air cooling from 830 °C and tempering is carried out at 200 to 220 °C. This steel is used for gears and other intricate parts which must have high strength, and where cracking or distortion caused by quenching must be reduced to a minimum.

Use of the 'rare' metals

Recent nuclear power developments have created a demand for materials to withstand the stringent conditions imposed, and as a result some metals, previously considered 'rare', may come into common use. The student should know of them and follow up the limited details given below.

Titanium is a light metal, used in a range of alloys with aluminium, vanadium, manganese, iron, chromium and tin. Some of its alloys reach a tensile strength of $925/1075$ newton/mm^2, are highly resistant to corrosion and retain a good strength at high temperatures. These alloys are extremely difficult to machine as the metal galls up and work-hardens under the tool.

Uranium is used as a nuclear fuel and is radioactive. It can be machined readily, but severe safety measures are needed against radiation. Copious coolant is necessary when machining to minimise swarf ignition and keep down the dust which is particularly hazardous.

Thorium, also radioactive, has machining properties similar to uranium.

Vanadium is also of nuclear interest, and the pure metal has only recently become available. Its oxide is toxic and precautions must be taken to avoid swarf ignition when machining. Although it picks up rather badly, it can be cut at low speeds with high speed steel tools.

Ceramic materials

Ceramics is the general term applied to a range of materials, basically sintered oxides, of extreme hardness that have been developed as cutting agents. The most common of these is aluminium oxide and it is used in two forms: (a) in its pure state and (b) mixed with small amounts of other compounds (e.g. chromic oxide) to impart some special desired property. In the preparation of the tool tips alumina powder is ball-milled and mixed to a paste with water. The blanks are then pressed to shape afterwards undergoing a long high temperature sintering process followed by slow cooling. The resulting tip is harder and more wear resistant than tungsten carbide but is much weaker in tension. For this reason tools and machines must be extremely rigid and free from chatter to prevent the ceramic tip from disintegrating.

Because of its nature no satisfactory method has been found for brazing the tip on to its shank so that some form of mechanical clamping is necessary. One of these is shown at Fig. 121(c). Ceramic tips are extremely wear resistant and will cut at speeds in the range of 150 to 450 metres per minute. They can only be ground up with diamond wheels but because of their low cost they are often discarded and replaced when the available cutting points have been used up (the tool shown has eight cutting points).

The preparation of metals

Generally, when the metal is delivered for use in the workshop, it is in the form of bar, strip, sheet, wire or some other shape most convenient for the starting-point in the job of forming it. A few notes on how it reaches these forms will interest the reader and extend his knowledge of the properties of material when in its various forms.

Hot rolling

All steel in the form of bar, strip or sheet, having a non-polished reddish-blue surface, has been hot rolled. The process starts with the steel ingot as soon as it has solidified sufficiently for the ingot mould to be stripped from it. This ingot will be about $1\frac{1}{2}$ m long, slightly tapering, and approximately square in cross-section, sizes varying from about $\frac{1}{3}$ m to $\frac{1}{2}$ m square. When taken from the ingot mould or soaking pit, it will be almost at a white heat. The first operation to the ingot is carried out at the blooming mill, where it is reduced to blooms. The action of rolling is shown at Fig. 37(a), and a mill consists of at least two rolls mounted on their necks in bearings and driven through the couplings as shown in Fig. 37(b). When there are only top and bottom rolls the mill is a 'two-high' mill.

If three rolls are mounted so that rolling may be done between the top, or bottom roll, and the centre one, it is called a 'three-high' mill, whilst the combination of a two-high mill with each roll backed up by another roll for strengthening purposes is called a 'four-high' mill (Fig. 38).

The size of the mill is expressed as the centre distance of the rolls, blooming mills ranging from about 0·7 m to 1·2 m. Where mills are used for rolling plate, the length of the rolls also is specified. Two-high mills are made reversing so that the metal being reduced may be fed backwards and forwards, being supported and fed into the rolls by rollers driven by the mill mechanism (Fig. 40).

The main rolls of a mill, which may vary from approximately 180 mm to

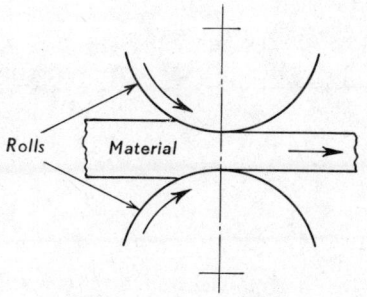

(a) Diagram of rolling

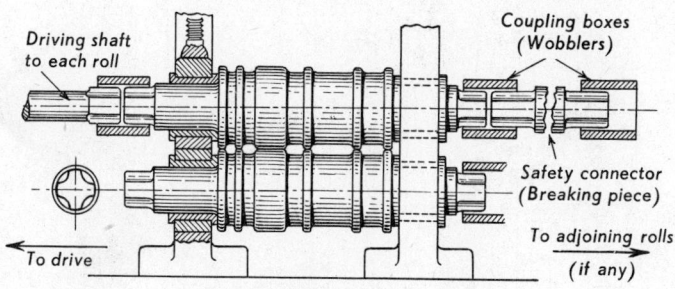

(b) Two-high rolls and stand

Fig. 37 Hot-rolling

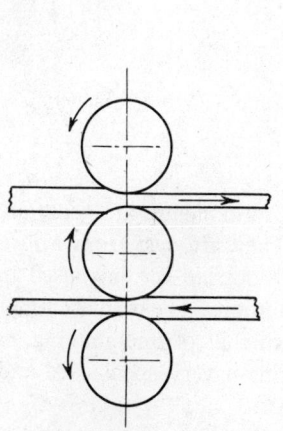

(a) Three-high rolls

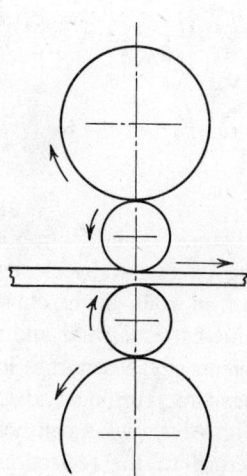

(b) Four-high rolls

Fig. 38

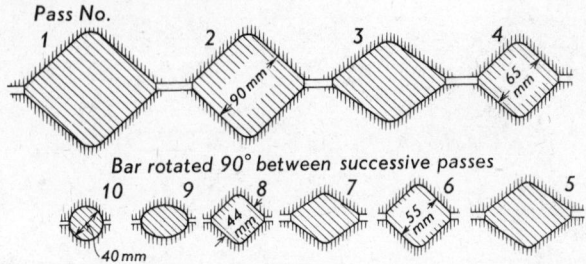

(a) Stages in rolling 40 mm bar from a 100 mm × 100 mm billet in 10 passes

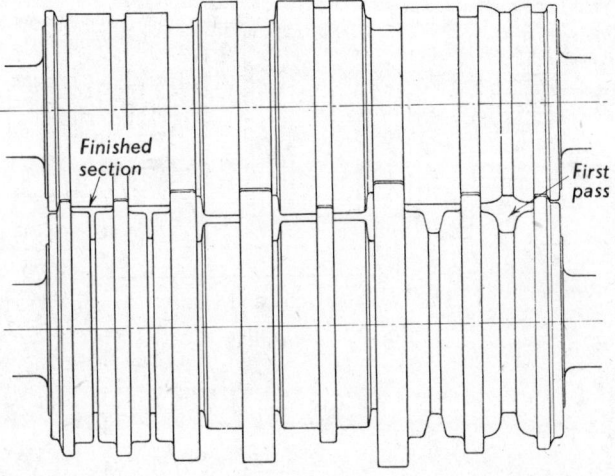

(b) Pair of rolls for producing a 100 mm × 75 mm T section

Fig. 39

1·25 m diameter, and from 300 mm to 5 m in length, are generally driven by steam engine or by electric motor. They are cast from iron or steel, the cast-iron rolls being chilled on their outside for increased hardness against the extreme wear and service conditions. The manufacture of these rolls is generally a separate industry from that of steel making, and their design, casting, turning and grinding forms a very specialised and skilled trade. Fig. 41 shows a roll being turned.

To return to the reduction of the ingot; this will be carried out in a specified number of passes through the rolls, depending upon the cross-section of the blooms being produced.

English Steel Corporation Ltd

Fig. 40 625 mm. two-high reversing mill (outgoing side) showing bloom cogging on the right and billet finishing on the left

Thomas Firth and John Brown Ltd

Fig. 41 Turning a roll (finished mass 20 tonnes)

The blooms are cut up in lengths convenient for the subsequent reducing process into billets* and bars, being re-rolled either in a mill somewhat similar to the blooming mill where the bar being reduced is passed backwards and forwards through progressively varied and reduced roll openings, or it may be reduced in a continuous mill. This consists of a series of 10 or 12 stands of rolls spaced at distances varying from 4 m to 2 m apart and made up of two-high rolls about 350 mm centres with the rolls 380 mm to 460 mm long. The speeds of the rolls are adjusted so that the metal may travel faster as its length is increased by the reduction, and arrangements are made to turn it through 90° occasionally to secure correct reduction in all directions. Fig. 39(a) shows the successive stages in the reduction of a billet to a round bar, whilst at (b) are shown the rolls for a T-section.

Plates and strip. When the final product of the rolling mill is to be sheets, plates or strip, the output of the blooming mill is usually composed of slabs instead of blooms. These are of a flatter form and lend themselves better to re-rolling into flat shapes. The procedure for rolling plates and strip follows the same general lines as for bars, the rolls generally being parallel instead of having forms turned in them.

Cold rolling

For many purposes, particularly where steel strip has to be blanked and formed in presses it is made by a process of cold rolling. The objects of cold rolling are: (1) to secure improved surface finish, (2) to obtain smaller and more uniform thickness than is possible in hot rolling, (3) to give the metal improved physical properties by combining the cold work of rolling with suitable heat treatment. Cold rolled strip can be produced to very fine limits of accuracy in thickness dimension (to within 0·025 mm) and uniformity. By combining a shearing operation for the edges of the strip, widths may also be controlled to within a few hundredths of a millimetre.

The raw material for cold rolling is hot rolled strip from the hot rolling mill, and in its general principles the method is similar to hot rolling except that roll pressures are much greater because the resistance of the cold metal to reduction is much greater than when it is hot. This fact encourages the use of four-high mills for cold rolling, the two outer rolls being of large diameter and serving as backing-up supports for the smaller working rolls. The strip is usually wound in a coil which is supported on a spindle at one side of the rolls. After passing through the rolls the strip is

---

* A bloom is 150 mm square or over, whilst a billet has dimensions between 30 mm and 150 mm square.

wound up on to another power-driven roll whose speed is so controlled that the strip is kept taut as it emerges from the reducing rolls. Cold rolling may also be done in tandem (two mills in line), or continuous.

Before the hot rolled strip can be put through the cold mill all the scale must be removed from its surface by pickling in dilute acid. If this were not done the scale (oxide) would be rolled into the metal during the process. Another problem with which the cold roller has to deal is that of annealing the strip. We have seen that when metal is cold worked it becomes work hardened, and unless it has to be supplied in the work-hardened, springy condition, as copper, brass and bronze often have to be, then it must be annealed after the final pass, and also, if several reductions are necessary, it may, after one or two passes, be too hard for further reduction and must be annealed to enable further rolling to be done. Since the polished surface of the strip must not be tarnished or scaled during the annealing process, it is necessary to exclude all oxygen from the metal whilst it is heated, and for this reason the operation is called 'close' annealing. If the strip is annealed in coils, these are packed in iron boxes along with cast-iron borings, and put into the furnace in this way. The boxes and borings protect the strip from the ingress of air, and any oxygen which is present combines with the carbon in the cast iron, thus preventing oxidation of the surface of the strip. More recent developments in the annealing of bright strip employ annealing furnaces inside which the atmosphere is kept free from oxygen by maintaining a slight pressure of another gas such as coal gas or hydrogen.

Cold drawing

All the wire that is made is produced by cold drawing through dies. This method of production is also used for making bright drawn bars which are extensively used for producing parts on capstan, turret and automatic lathes. As in the case of cold rolling, the raw material for drawing is black rolled bar from the hot rolling mill. When required for drawing into wire this may be in coils of metal, the coils being $\frac{1}{2}$ m to $\frac{3}{4}$ m diameter and the metal 6 mm to 10 mm diameter. For the production of larger sized bars the black bar will be a small amount larger (1 mm to 3 mm) than the finished bars into which it has to be drawn.

The principle of drawing is shown at Fig. 42. The metal to be drawn is reduced at the end (tagged) to a diameter small enough to pass through the die and of sufficient length for it to be gripped by that portion of the draw-bench which pulls it through. The hole in the die is generally conical, and

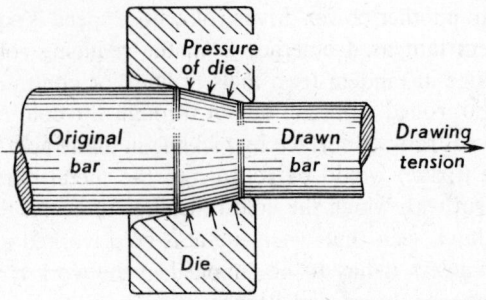

Fig. 42 Principle of drawing

the reduction in diameter is effected by the pressure of the sides of the cone on the rod. This pressure is set up by the drawing pull on the wire, and at the instant of reduction plastic flow takes place in the metal. In view of the enormous pressure between the die and the rod being drawn it is necessary to lubricate the area of contact. In bar drawing this is effected by smearing the bar with a solid lubricant in the form of a grease. This is not a convenient procedure in wire drawing as the wire is coiled in long lengths on drums, and is drawn through the die by being coiled on one drum and uncoiled from the other. Wire drawing is either effected by the 'wet' or the 'dry' process. In the dry process the wire, after pickling, may be washed in lime-water, which on drying off leaves a thin layer of lime on the metal which serves as a lubricant. Alternatively, the wire on its way to the die may be passed through a box filled with dried soap. In the wet process the wire is passed through a solution of copper sulphate followed by a soapy solution. The copper sulphate deposits a thin film of copper which, when combined with the soapy solution, facilitates the drawing of the metal.

Bars are drawn on a drawbench which is a long bed with an endless chain running along the surface and returning underneath. The die is supported against a bracket at one end and a small wagon which carries the gripping jaws for the bar runs along the bed, being provided with a hook ('dog') which can be engaged in the chain. When the tagged end of the bar has been threaded through the die and gripped, the hook is lowered ('dogging in') so that the chain catches it and thus draws the wagon to the other end of the bench (Fig. 43).

The dies for drawing may be made of hardened tool steel, tungsten carbide, diamond or chilled cast iron (for large bars). The primary object is to obtain a die which is intensely hard, and resistant to the extreme wear which takes place.

Worm drive reduction unit   Change gear-box   Motor   Wagon (spare wagon also shown)

Die pedestal

Flexible coupling   Endless chain

The Wellman Engineering Corporation Ltd

Fig. 43 Drawbench

Cold drawn steel has a smooth, bright surface, and for this reason is called 'bright drawn'. As in the case of cold rolling, the drawing work hardenes the metal, and for wire, which has to undergo many reductions to bring it down to size, it is necessary to anneal from time to time during the process. Larger bars are generally only reduced sufficiently to correct out of roundness from hot rolling, and bring up the surface to a good finish. This confers a certain amount of work hardening and for some purposes, e.g. shafting, annealing is not carried out and the hardness is left in the metal. The hardening strains, however, are often left in also, and sometimes when the outer surface of such metal is removed, e.g. by cutting a keyway, the strains will be relieved, causing the bar to warp. Bright bars and wire can be drawn to very close limits as regards size and roundness.

The production of castings

After bar and sheet, the next most common form in which material is received for working is in the form of castings. The casting of metal into the intricate shapes that are possible is an immense advantage in engineering construction, and the work of the foundry and pattern shop is of great importance. Castings may be made in steel, cast-iron, brass, bronze, aluminium, as well as in numerous other alloys. A description of sand casting, which is the most common method employed. will now be given.

A casting is produced by pouring molten metal into a mould made to the shape of the part required. In die casting this mould is made of metal, whilst in sand casting it is made in sand, its form being obtained by means of a wooden pattern having a shape similar to the part required. The first step, therefore, in making a casting, is that of making the pattern, and the patternmaker may be counted as one of the most highly skilled craftsmen in the engineering industry. It is he who first has to interpret a drawing, however complicated, into the solid shape it represents. His experience decides upon the method to be used in the production of the casting, a factor upon which quality and success often depend. We will give a simple and obvious example. When metal is cast it contains gases which rise up through it, leaving at the highest point. The top surface of a casting therefore is liable to be rather spongy with blowholes left by entrapped gases unable to escape before the metal solidified. Now, if a round bar is cast on its side, the chances are that on one part of its circumference it may be a little spongy, whilst if it is cast vertically on its end, the sponginess will be at one end. If we extend this simple case to a complicated casting, we can see that the method of casting has some control over which portions of the casting will be sound and which may be spongy, and this method is under the control of the patternmaker by the manner in which he disposes the pattern. In our round bar, if the curved surface was important we should wish to have any likely blowholes in the end, and the pattern would be made for vertical casting. On the other hand, importance of the flat ends would make side casting preferable. Very often, castings can be so made that the spongy metal is in the portion of the casting which has to be machined, and it is removed by machining.

The wood mainly used for patterns is a well-seasoned soft wood such as yellow pine, and the pattern is made to conform with the outside shape of the casting. To allow for the contraction which takes place when the casting cools, the pattern must be a little larger than the finished casting. The approximate amounts that different metals contract are as follows:

Metal	Approximate Contraction per metre
Cast iron	8 to 9 mm ($\frac{1}{120}$)
Brass and steel	16 to 18 mm ($\frac{1}{60}$)
Aluminium	13 mm ($\frac{1}{77}$)

This contraction is allowed for by the patternmaker employing a special rule called a *contraction rule* (Fig. 44(a)) with which to set off all the dimensions on the pattern. Since for cast-iron a length of 1 m contracts 9 mm and cools to 991 mm, a length of 1009 mm will cool approximately to 1 m. On a contraction rule for cast-iron, therefore, a length of 1009 mm is marked as 1 m and is then divided up exactly similar to an ordinary rule, so that when a patternmaker measures off a length with this rule, he automatically makes the length longer in the ratio $\dfrac{1 \cdot 009}{1}$ and so allows for contraction. For brass and other metals, the rule length is compensated in the same way according to their contraction.

The finished pattern will be smoothed up as much as possible to facilitate its withdrawal from the sand when moulding, and to impart a smooth surface to the casting. This effect is further increased by the shellac varnish with which the pattern is covered. When the casting is of a simple nature the appearance of the pattern is similar to that of the casting, but when the casting is more complicated with holes and cavities it becomes necessary to resort to coring in the mould. This will be best explained when we are discussing the actual moulding process.

Let us commence by following the moulding of a bracket of the form shown in Fig. 45(a). The patternmaker will be supplied with a drawing of this as it should be when finished, and on the drawing will be indicated those faces of the backet which have to be machined. He will make the pattern, leaving contraction allowance, and will leave 3 mm to 5 mm extra on the faces marked for machining. At the foundry the pattern will be used to prepare the mould in moulding boxes, which are steel frames without top or bottom, provided with lugs and holes for lining up when several are put together and with handles for lifting (Fig. 44(b)). The pattern will be placed face downwards on a board and a moulding box-section laid over it. After a little smooth facing sand mixed with coal dust has been sprinkled on the pattern, the box will be filled with moulding sand and this will be well rammed down with a wooden rammer shaped like a dumb-bell. When the sand has been rammed sufficiently to have made a good impression from the pattern, the top of the box is trimmed off level and it will be like Fig. 45(b). The moulding box is now turned over, another section placed on top of it and rammed with sand as before. A hole is now formed through the sand in the top box for pouring the metal (called the 'pouring gate') and the two boxes are parted again. The pattern is now carefully withdrawn from the bottom box, its impression is touched up with smoothing tools if necessary and the sides dusted with fine plumbago to assist the facing sand in giving a good surface to the casting. Channels (called 'runners') are formed to lead the metal from the pouring gate to the mould and the top box is replaced. The mould is now ready for pouring and is shown at Fig. 45(c). The top box is called the 'cope' and the bottom one the 'drag'.

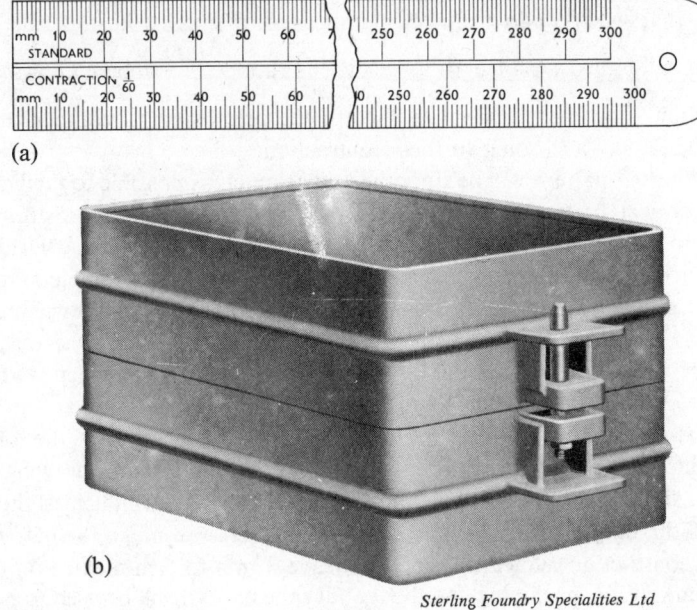

Sterling Foundry Specialities Ltd

(a) A $\frac{1}{60}$ contraction rule
(b) A pair of moulding boxes

Fig. 44

When casting takes place metal is poured in until it rises to the top of the pouring gate and this, together with the runners, is cut off the casting during the process of cleaning it up. To facilitate the withdrawal of the pattern from the cope, its sides are slightly tapered in the direction of moulding, the amount of taper usually being about 1 in 50 of the length. The moulder also loosens the pattern for withdrawal by 'rapping' it. To do this he screws a steel rod into its top face, and taps the rod in all directions so that the mould impression is slightly enlarged. When the bottom box has been turned over, and before the sand is put in the top box, a layer of dry parting sand is added. This preserves the joint, and prevents the sand in the two boxes from congealing into a solid mass which would make it impossible to separate the boxes for the purpose of withdrawing the pattern.

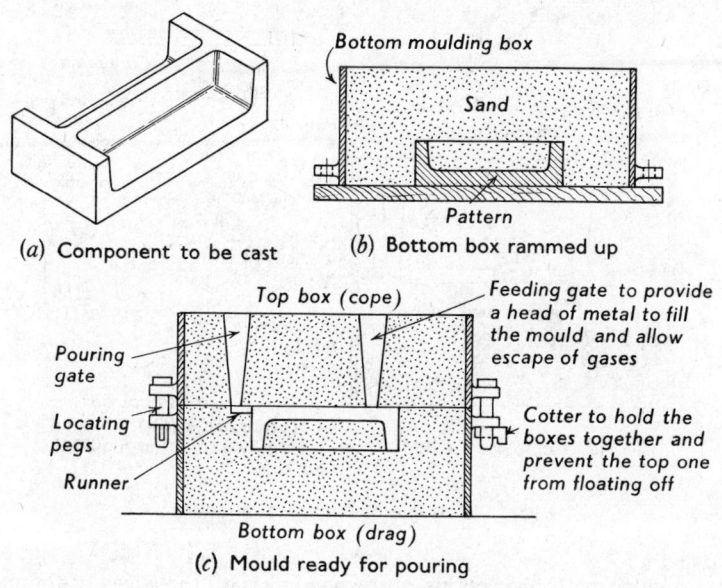

(a) Component to be cast

(b) Bottom box rammed up

Bottom moulding box

Sand

Pattern

Top box (cope)

Feeding gate to provide a head of metal to fill the mould and allow escape of gases

Pouring gate

Locating pegs

Cotter to hold the boxes together and prevent the top one from floating off

Runner

Bottom box (drag)

(c) Mould ready for pouring

Fig. 45 Moulding a bracket

Cores and core boxes

It will be appreciated that there is a limit to the moulding which could be done by the simple method we have just discussed. Take, for instance, the casting of a bush, say, 100 mm outside diameter with a 50 mm hole and 150 mm long. If this is moulded on its side the pattern cannot be withdrawn if sand is to be left to cast a hole in the finished bush, whilst if cast vertically the pillar of sand left to cast the hole would be too weak to withstand the inrush of metal when pouring. When castings of this nature have to be made, a *core*, made of hard, dried sand and moulded in a *core-box*, is inserted in the mould. We will illustrate this by discussing an example which includes coring, and follow the moulding of the component shown in Fig. 46(a).

The inside of this and the hole in the bottom would be formed by coring and the pattern would be as shown at (b). It will be noticed that the pattern is not only solid where the inside faces of the component should be, but is

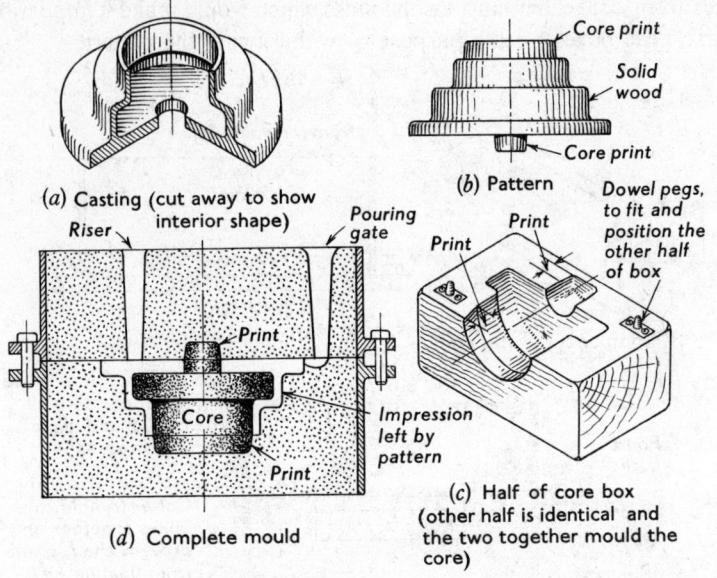

(a) Casting (cut away to show interior shape)

(b) Pattern

Core print

Solid wood

Core print

Riser

Pouring gate

Dowel pegs, to fit and position the other half of box

Print

Print

Print

Print

Core

Impression left by pattern

Print

(c) Half of core box (other half is identical and the two together mould the core)

(d) Complete mould

Fig. 46 Example of coring in moulding

also provided with projections corresponding with the edges of the metal. These projections are called *core-prints* and are for forming recesses in the mould to locate the core. The core for this is made in a separate box called the core-box, which is made by the patternmaker, and is auxiliary to the pattern. The core-box is often made in two halves to facilitate withdrawal of the moulded core and the one for our example is shown at Fig. 46(c). At the foundry a core would be moulded in the core-box from special core sand and then this core would be dried in an oven until it was hard. The moulding of the pattern in the cope and drag would leave an impression having a shape similar to the outside of the casting together with the impressions of the core-prints. The core itself is made with extremities which just fit these prints when the cope and drag are assembled and when the pattern has been withdrawn the boxes are put together with the core fitted in. This completes the preparation of the mould which will be as shown at (d). The core-prints serve an important function in keeping the core in its proper position relative to the outside of the mould. If it were not located properly, or moved during pouring, the sides of the casting would not be of uniform thickness, and the hole not in its proper position.

Split patterns. To facilitate the moulding of some components, the pattern, instead of being made in one piece, is split into two or more parts. These are dowelled so that they may be fitted together accurately to the finished shape. Such an example is given by the flanged pipe shown at Fig. 47(a) and its moulding would be carried out as follows:

1. Ram up one half in bottom box (Fig. 47(c).
2. Turn over, fit the other half and ram top box.
3. Remove pattern and insert core.

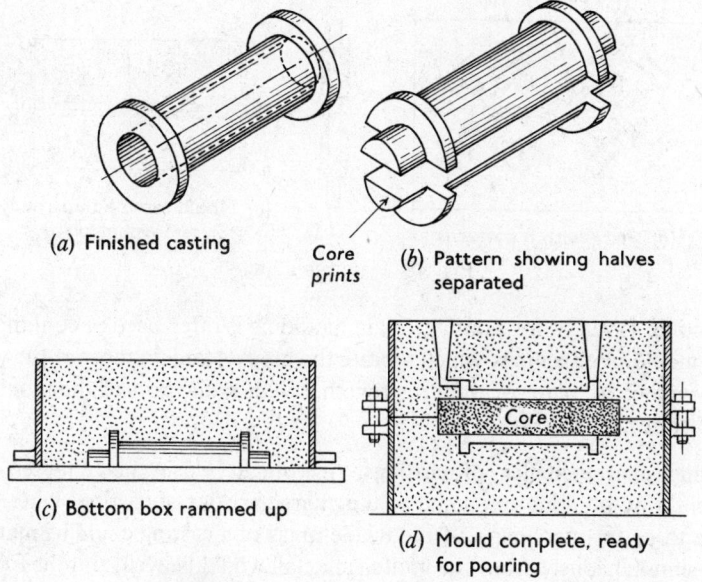

(a) Finished casting

Core prints

(b) Pattern showing halves separated

(c) Bottom box rammed up

(d) Mould complete, ready for pouring

Fig. 47 Moulding a flanged pipe

The completed mould ready for pouring is shown at Fig. 47(d).

Plate moulding. When large quantities of fairly simple castings are required they are produced several at once by a process of plate moulding. The form of the part is divided into two halves and the patterns (often in metal) are attached to each side of a board. One side is moulded in the drag, and after turning over, the cope side is moulded. The boxes are then parted and the board and patterns removed. When the boxes are re-assembled without the board, the moulded impressions match up to give the shape of the part and

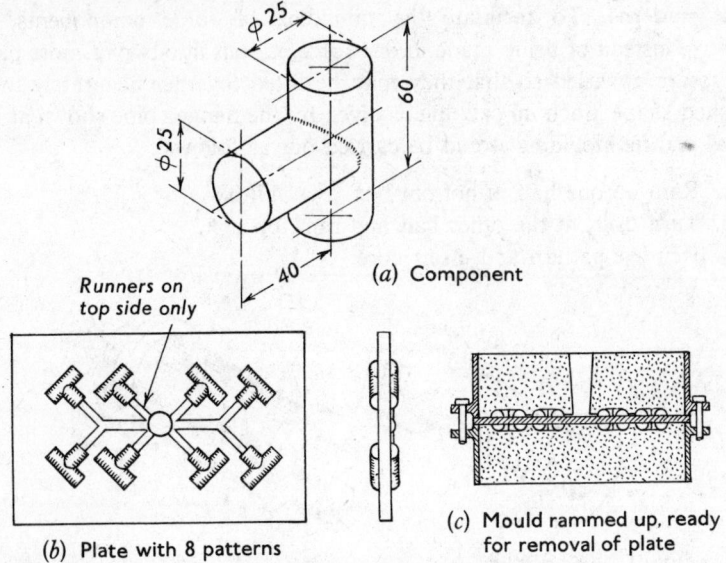

(a) Component

Runners on
top side only

(b) Plate with 8 patterns

(c) Mould rammed up, ready
for removal of plate

Fig. 48 Plate moulding

the mould is ready for pouring. Plate moulding is often used in conjunction with moulding machines which vibrate the moulds and do most of the work of ramming. A plate of patterns together with a diagram of the moulding up is shown at Fig. 48.

Cooling characteristics of castings. We might close this chapter conveniently by pointing out some of the characteristics of castings when they come to us for machining. If the whole mass of a casting could be made to cool simultaneously from the molten state all would be well, and the following remarks unnecessary. Unfortunately, however, castings are nearly always composed of masses of metal having various shapes, volumes and thicknesses, and when these cool and contract, very complex stresses are often set up. Large sections of metal will cool much slower than small, thin sections, and cooling will commence at the outside of the casting where the metal is in contact with the cool sand, and will travel inwards.

Let us examine the effect of this cooling on the two parts of a casting shown at Fig. 49(a) and (c). At (a) the sides of the L will commence to cool first and cooling will proceed inwards. As cast iron cools the crystals tend to set themselves in the direction of cooling. This is shown shaded at (a) and results in a plane of weakness at the corner. For this reason, and to

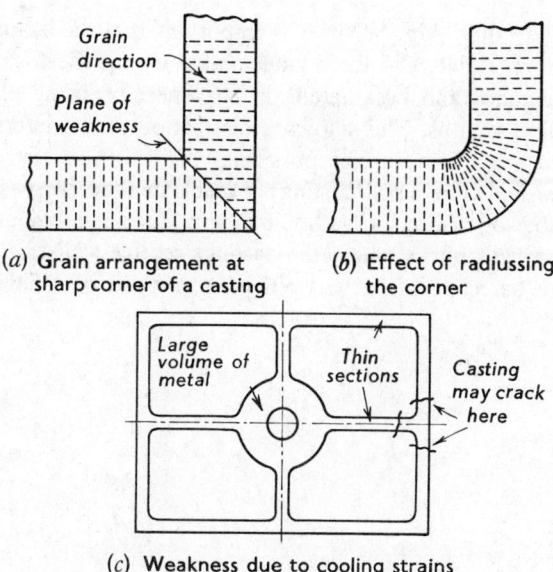

(a) Grain arrangement at
sharp corner of a casting

(b) Effect of radiussing
the corner

(c) Weakness due to cooling strains

Fig. 49 Cooling effects on castings

help in moulding, the corners of castings are well rounded off by fillets as shown at (b). At Fig. 49(c) a large boss of metal is joined by arms to a much thinner section. When cooling from the molten state, the thin sections will have solidified and be well cooled down before the large volume of metal at the boss has cooled very much. Ultimately this will cool down and contract, but as the arms have now cooled and are rigid, the force exerted by the contraction of the large boss cannot be relieved by any movement on the part of the two thin sections. This means that a permanent internal stress is left in the casting between the large boss and the outer frame, and even if the casting appeared to be all right, a sudden blow or shock might cause it to fracture at one of the sections shown. Machining the hard and rigid skin from the casting on its outer face would relieve some of the stress and probably allow a certain amount of spring to take place.

These two simple cases will serve to show how internal stresses and weaknesses may exist in a casting, and although the patternmaker endeavours to minimise these effects by studying the method of casting, they cannot be prevented altogether. Castings with internal stresses will often warp when these stresses have been relieved by machining off the hard and rigid outer skin. When it is important that the shape of a casting shall remain stable, e.g. for a gauge, large worm-wheel, etc., it should be rough machined and 'weathered' or 'seasoned' by being placed in the open for a few months. This allows a dissipation of the internal stresses to take place and minimises the possibility of distortion after the casting has been finished. For small castings a quick method of seasoning is to place a number of castings in a tumbling barrel for about half an hour. The vibration and light blows which the castings receive whilst being tumbled round in the barrel forms a very effective method of relieving internal stresses.

4 Workshop processes requiring heat—forging, riveting, soldering and brazing

The hot forging of metals is an important subsidiary process to workshop production and possesses many advantages. We have seen how the hot working of metal is beneficial to its granular structure and how, if it is properly carried out, it may confer a strengthening effect owing to the directional property of the grains. In addition to these aspects, the use of forgings may save a great deal of time and expense, because instead of cutting the metal from the solid to give the shape required, it may often be forged to a shape very near to the finished one. This economises in machining time, and avoids an undue proportion of the metal being cut into swarf.

We may roughly divide forging into the processes of hand forging, machine forging and drop forging, the two latter processes being of types adaptable for large quantity production of similar articles.

Hand forging

In hand forging the shaping of the metal is carried out under hand control, and accuracy depends upon the skill of the smith. For large work, some form of power-operated hammer or press must be used. We will deal with the smaller class of work first, and it will be advisable at the outset to consider the essential forging equipment, together with details of certain fundamental forming operations.

The forge. For heating the metal some form of forge or furnace is necessary. For small work the blacksmith's forge is commonly used and consists of a hearth for holding combustible coke, a tuyère for leading forced air into the fire, a blower for supplying air under a slight pressure to the tuyère and a chimney for carrying away smoke and gases. Usually a 'bosh' (tank) is also provided for holding water for quenching purposes. Because of the

Alldays & Onions Ltd

Fig. 50 Smith's hearth

high temperature under which the tuyère operates, this is often surrounded by cooling water contained in a tank fixed at the rear of the forge.

The most substantial and satisfactory forges are brick built, incorporating a flue and chimney, but many in use are of the portable type and if these are well constructed they give satisfactory results. Fig. 50 shows the front and rear views of a cast-iron double smith's hearth where all the details mentioned above may be seen, together with the electric motor for driving the blast fan, and the blast regulating valves. To meet the requirements of the Clean Air Bill it is now customary to incorporate a grit arrester in the chimney of the forge. This is not shown in Fig. 50. The best fuel for forge fires is coke 'breeze', which is gas coke crushed into small pieces about 10 to 12 mm diameter. Small soft coal may be used but it should not contain sulphur, and when burnt should adhere into a mass and not fall to pieces. The temperature of the fire is governed by the amount of blast supplied and may be varied up to a white heat.

Forging tools

The *anvil* (Fig. 51) is for supporting the work whilst it is being struck with the hammer, as well as providing means for other forging operations. The body of the anvil is of mild steel and, to give a hard top face, a piece of high carbon steel about 20 mm to 25 mm thick is welded on. The beak is soft like the anvil body and its shape makes it useful for bending metal to rounded formations. The ledge between the beak and the anvil face is also soft and may be used for resting metal when cutting through with a chisel;

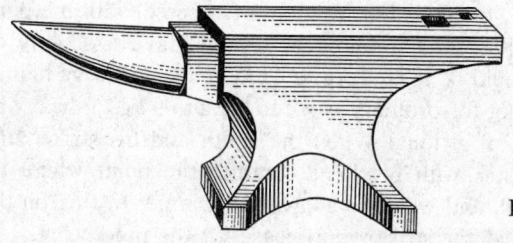

Fig. 51 Anvil

John Hall (Tools), Ltd

the soft underneath metal does not damage the chisel edge. In the top of the anvil is a square hole to take the shank of various tools to be described later. Anvils vary up to about 150 kilogrammes and should stand with the top face about 600 mm from the floor. This height may be attained by resting the anvil on a cast-iron or wood base.

Two kinds of hammer are used in hand forging:

1. the *hand hammer* used by the smith himself; and
2. the *sledge hammer* used by the *striker*, the smith's assistant.

Hammers particular to the smith are shown at Fig. 52, whilst engineer's hammers, which he also uses, are on Fig. 169. Hammer-heads should be

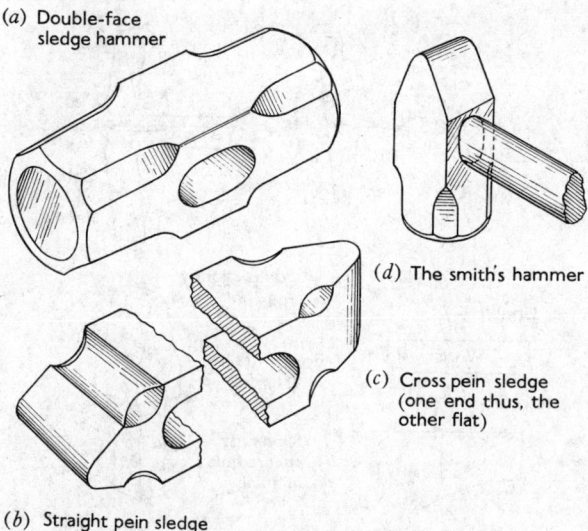

(*a*) Double-face sledge hammer

(*d*) The smith's hammer

(*c*) Cross pein sledge (one end thus, the other flat)

(*b*) Straight pein sledge (one end as shown, other end flat)

Fig. 52 Blacksmith's hammers

of cast steel, the ends hardened and tempered, with the portion round the eye soft. Smith's hand-hammers usually have a slightly convex striking face and should be from 1 kg to $1\frac{1}{2}$ kg in mass. Sledge-hammers vary from $4\frac{1}{2}$ kg to $5\frac{1}{2}$ kg for ordinary work to 7 kg to 9 kg for heavy blows, the shaft being about 1 m long. When the smith and his striker are at work, the smith indicates with his hand-hammer the point where he requires the sledge to fall, and when he allows his hammer to ring on the anvil it is an indication that the striker is to cease for the time being.

Hand tools

For roughly forming the metal, direct blows from the hammer alone may be used, but for cutting off, forming and other finishing operations, various hand tools are necessary. The most common of these are as follows:

Chisels. These are used for cutting metals and for nicking prior to breaking. They may be *hot*, or *cold*, depending on whether the metal to be cut is hot or cold, and the main difference between the two is in the edge. The cold chisel has its edge hardened and tempered with an angle of about 60°, whilst the angle of the hot chisel is 30° and hardening is not necessary,

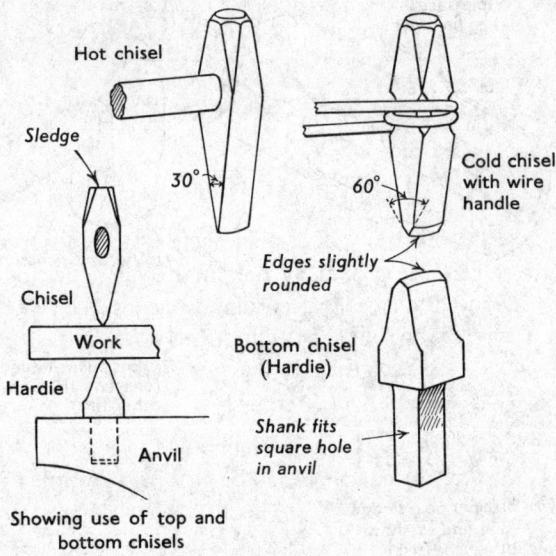

Fig. 53 Blacksmith's chisels

since in any case the hot metal would re-soften it. The edge of a chisel should not be quite straight but slightly rounded as shown (Fig. 53). Chisels are generally used in pairs, a pair comprising a *top tool* and a *bottom tool* (often called the *hardie*). The hardie has a square shank and fits in the square hardie hole in the anvil face, whilst the top chisel, which is held by the smith and hit by the striker, may be fitted with either a wooden handle or a metal wire handle as shown in Fig. 53.

Fullers. Fullers are used for necking down a piece of work, the reduction often serving as the starting-point for a reduction (Fig. 54). They are made

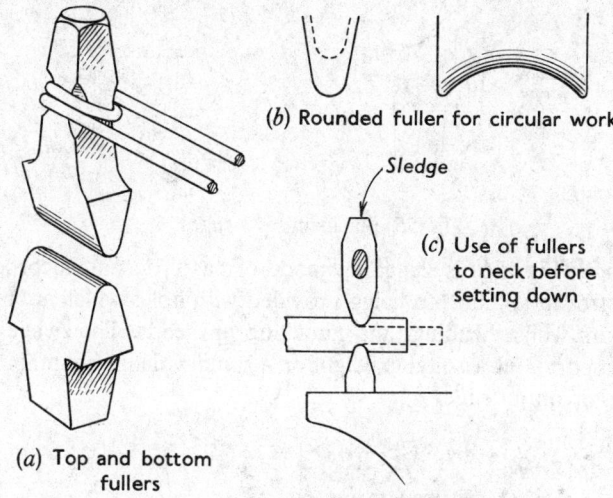

(b) Rounded fuller for circular work

Sledge

(c) Use of fullers to neck before setting down

(a) Top and bottom fullers

Fig. 54 Blacksmith's fullers

in top and bottom tools as in the case of chisels, the bottom tool fitting in the hardie hole and the top held by the smith and struck by the striker. Fullers are made in various sizes according to needs, the size denoting the width of the fuller edge. Thus a 12 mm fuller would have a semi-circular edge 12 mm wide. For shouldering round work fullers may have their edges hollowed out as shown at (b).

Swages (Fig. 55) are used for work which has to be reduced and finished to round or hexagonal form and are made with half grooves of dimensions to suit the work being reduced. Swages may be in separate top and bottom halves, or the two halves may be connected by a strip of spring steel as shown, thus enabling them to be used by the smith single-handed. The

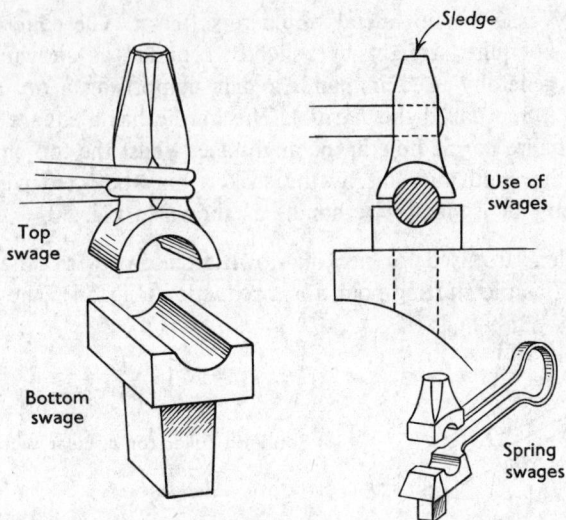

Fig. 55 Blacksmith's swages

swage block (Fig. 56) is generally made of cast-iron and incorporates a range of sizes in addition to being provided with holes which are useful for holding bars whilst bending, and knocking up heads. The swage block is usually supported at a suitable height on a stand which is adaptable to hold it flatwise, or on its edge.

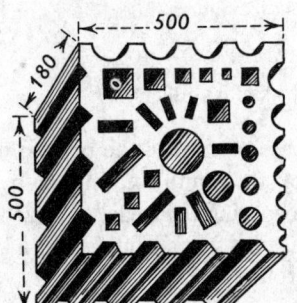

John Hall (Tools), Ltd

Fig. 56 Swage block

Flatters (Fig. 57) are used for finishing flat surfaces and are made with a perfectly flat face about 75 mm square (or round). The *set-hammer* is a similar but smaller tool used for finishing in corners and confined spaces. As the work is supported directly on the anvil for flatting, only the top, handled tool is necessary.

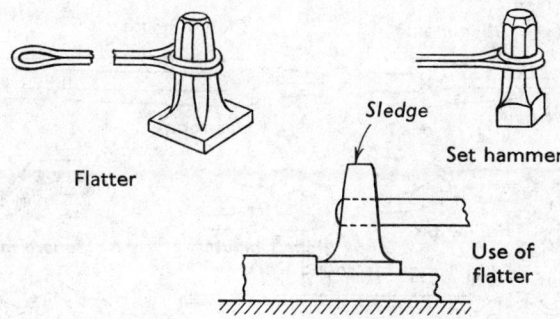

Fig. 57 Flatter and set hammer

Punches and drifts. When metal is at forging heat holes may easily be put in it by punching, and opened out by driving through a larger tapered punch called a drift. Examples of punches and drifts are shown at Fig. 58. When punching a hole it should be carried out from both sides as shown to avoid driving the punch on to the hard anvil face.

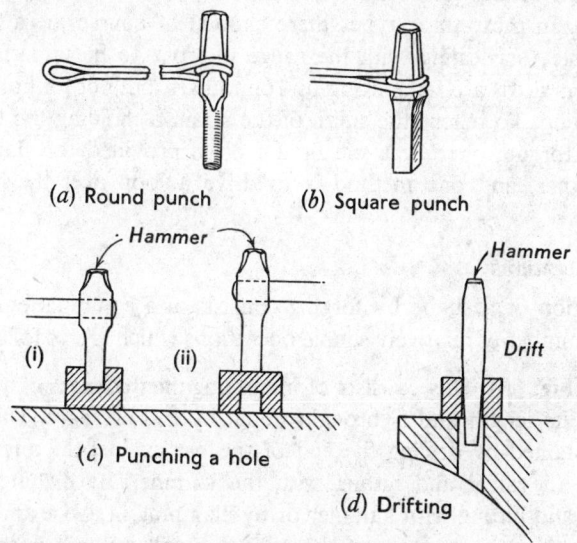

Fig. 58 Punches and drifting

Tongs. The smith requires a good selection of tongs with which to hold his work during the various operations. The chief types of tongs for holding work are shown at Fig. 59(a), (b) and (c).

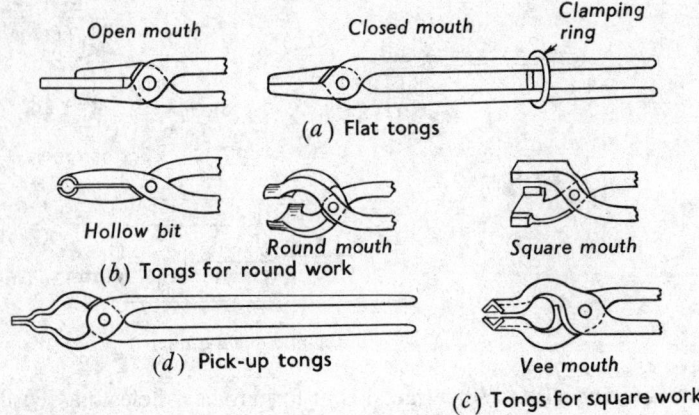

Fig. 59 Blacksmith's tongs

Those at (a) are the flat tongs and should hold along the entire length of the jaws as shown. If they grip at the front or back only the grip will not be secure. Hollow tongs for round work are shown at (b) and tongs for square bars at (c). In these three types there should, of course, be a variety of sizes suitable for dealing with the range of work to be undertaken. The tongs shown at (d) are for picking up round bars, but not for holding work during forging. To relieve the smith of the strain of holding the tongs during a long forging operation, means are often provided for clamping the tongs together, and one method is to drive a loop over the handles as shown.

Forging operations

The formation of a shape by forging consists of a combination of two or more of a number of relatively simple operations which are as follows:

Upsetting (Fig. 60). This consists of increasing the thickness of a bar at the expense of its length and is brought about by end pressure. The pressure may be obtained by driving the end of the bar against the anvil, by supporting on the anvil and hitting with the hammer, by placing in swage block hole and hitting with hammer or by clamping in a vise and hammering. The position and nature of the upsetting will depend on the heating, and upon the type of blow delivered. In general, the increase in lateral swelling will be greatest at the parts where the metal is hottest (more plastic), so that for upsetting the end of a bar only that end should be heated. If a short bar is heated uniformly over its whole length, heavy blows will cause a fairly uniform degree of swelling, but light blows will have a more local effect at the ends only.

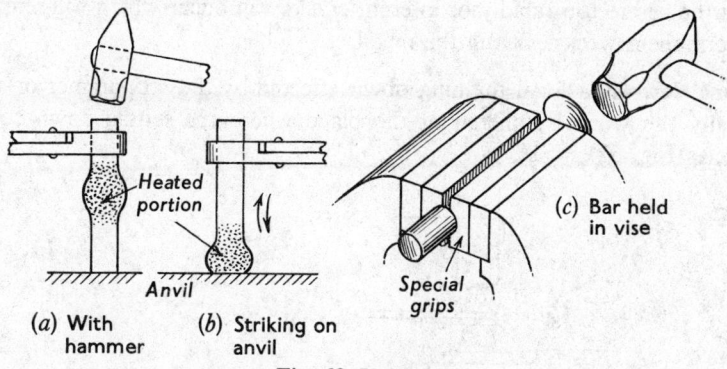

(a) With hammer (b) Striking on anvil (c) Bar held in vise

Fig. 60 Upsetting

Drawing down is the process of increasing the length of a bar at the expense of its width or thickness or both (Fig. 61). A good method of drawing down square or rectangular bars is to use the edge or beak of the anvil as shown, turning the bar through 90° if the thickness in both directions is to be reduced. When assisted by a striker, a pair of fullers may be used as shown. When the preliminary drawing has been done in this way the work may be finished off with the flatter.

When round bars have to be drawn down a considerable amount they should first be brought down to a square, drawn by the above method and then squared up again. The square is then taken to an octagon by taking off

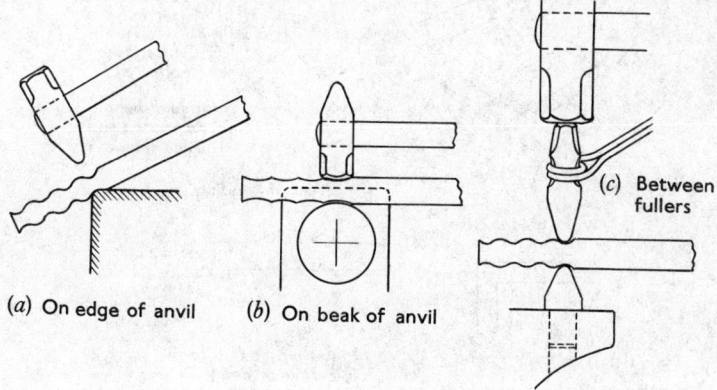

(a) On edge of anvil (b) On beak of anvil (c) Between fullers

Fig. 61 Drawing down

the corners and finally rounded and finished between swages. When reducing the thickness of metal care should be taken not to drive it outwards from the centre too rapidly or internal cracks will occur which will remain as permanent weaknesses in the metal.

Setting down is a local thinning down effected by the set-hammer or set. Usually the work is fullered at the place where the setting down commences (Fig. 62).

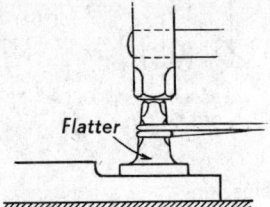

Fig. 62 Setting down

Bending. Bending is an important operation in forging and is one very frequently used. Bends may be either sharp-cornered angle bends or they may be composed of a more gradual curve. Angle bends may be made by hammering the metal over the edge of the anvil, over a block of metal held in the hardie hole or in a vise, or over a vise jaw itself, whilst the metal is being gripped (Fig. 63(a)). When metal is bent the layers of metal on the inside are shortened and those on the outside are stretched. This causes a

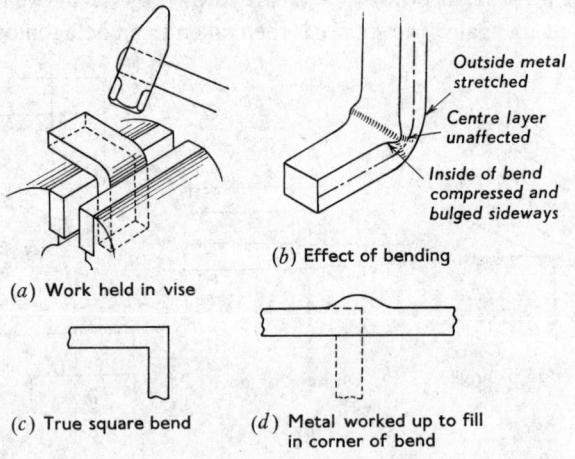

Fig. 63 Bending (sharp bends)

bulging of the sides at the inside, and a radius on the outside of the bend as shown at (b). If a true square bend is required as shown at (c), additional metal must be worked to the place where the bend occurs as shown at (d). When this is bent the additional metal will go to make up the corner. If this is not done it may be possible to work up the square corner by blows from a light hammer, working metal into the bend from nearby portions of the work. Whenever possible, a sharp corner on the inside of a bend should be avoided as it constitutes a weakness which may lead to a fracture of the corner. When double bends have to be made a simple bending fixture will often save time and enable more uniform results to be obtained. Such a

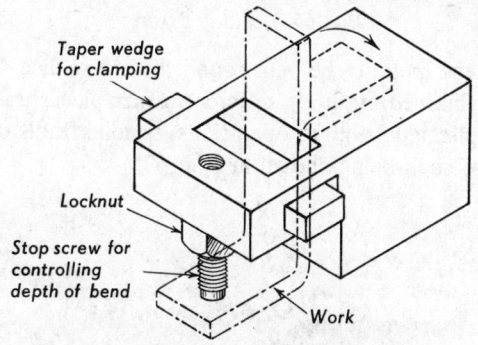

Fig. 64 Simple bending fixture

fixture is shown at Fig. 64 and its operation is self-explanatory from the diagram.

Gradual bends may be made by using the beak of the anvil as a former as shown at Fig. 65(a), or the metal may be bent round a bar of the correct radius held in a vise as at (b). When a quantity of bending has to be done, the use of some type of bending fixture will save time and produce better and uniform bends. At Fig. 66 the handle carrying the roller is swung

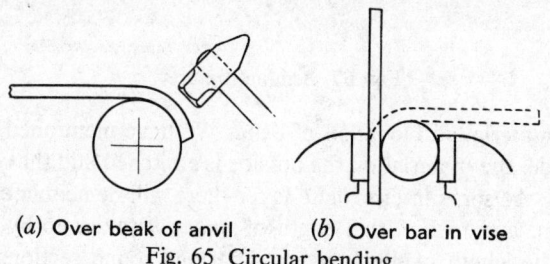

(*a*) Over beak of anvil (*b*) Over bar in vise
Fig. 65 Circular bending

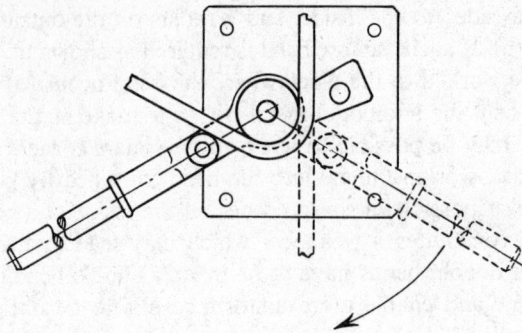

Fig. 66 Bending fixture

round, causing the metal to be bent round the centre disc. Varying sizes of bends may be obtained by fitting different centre pieces and rollers to suit. For more complicated bending operations, bending tools of various types are used, one of such being shown at Fig. 67.

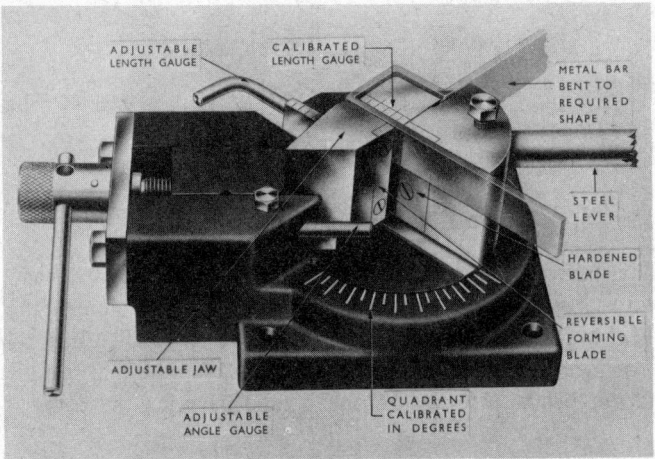

Chas. Taylor Ltd. (Kennedy Products)

Fig. 67 Bending fixture

Length of material and location of bend. We have mentioned that when a bend is made, the material on the outside is stretched and that on the inside compressed. At some intermediate layer there will be neither extension nor compression, and to find the length of material taken up by the bend we must find the length of this layer. For flat and round sections this neutral layer is at the centre,* so that if we find the bent length of the *centre line* for such bars it will give us the length of material in the bend, and by adding the lengths of any adjoining straight portions we may obtain the total length of material required. Thus in Fig. 68(a) the total length of material is

$$52 \text{ mm} + 40 \text{ mm} + (\text{curved length AB}) = 52 + 40 + (\tfrac{1}{4} \text{ of 15 mm rad-circle})$$
$$= 52 + 40 + (\tfrac{1}{4} \times 2\pi \times 15) = 52 + 40 + 24 = 116 \text{ mm approx.}$$

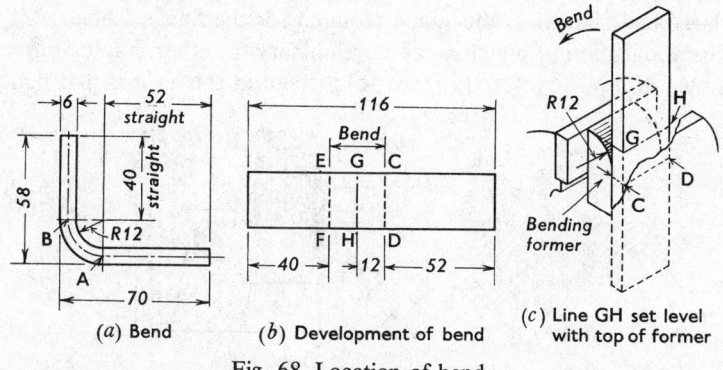

(a) Bend (b) Development of bend (c) Line GH set level with top of former

Fig. 68 Location of bend

The other point which deserves some attention is that of making the bend to come in the correct place, with 52 mm of straight at one end, and 40 mm at the other. This may be achieved by bending the bar over a shaped former held in the vise, and locating its position relative to the radius on the former. Thus in Fig. 68(b) the bend must start at line CD and finish at EF. If line CD is set level with the commencement of the bend as shown at Fig. 68(c), a line at GH will be level with the top of the bending former if GH is distant from CD by the inside radius. Hence mark a line where the bend must commence, and strike another line at a distance from it equal to the inside radius of the bend. Set this second line level with the top edge of the former and then bend the bar over the former.

Punching and drifting. As well as being used for its obvious purpose of producing holes, punching may often be employed as a shaping process. When performed without a die, punching should be carried out in two stages. In the first operation the metal is held flat on the anvil and the hole punched approximately half through. The work is then turned over and the hole completed by supporting the bar as shown at Fig. 58(c) (page 97). If

---

* For other sections see the treatment in the author's *Elementary Workshop Calculations*.

the hole is pierced right through from one side a bulge is thrown upon the underside as shown at Fig. 69(a). If a die is used having a hole with just clearance for the punch, the hole may be pierced in one operation and the arrangement is shown at Fig. 69(b). The taper in the die hole is to allow the punching to fall through and a top plate may be added as shown, to align the punch with the hole in the die. After a hole has been punched it may be opened out to any other size or shape by driving through it a tapered drift, the large end shaped to the shape required for the finished hole.

The application of punching as a preliminary for other shaping processes is shown at Fig. 69(c) and (d). At (c) the slot in the lever is first punched

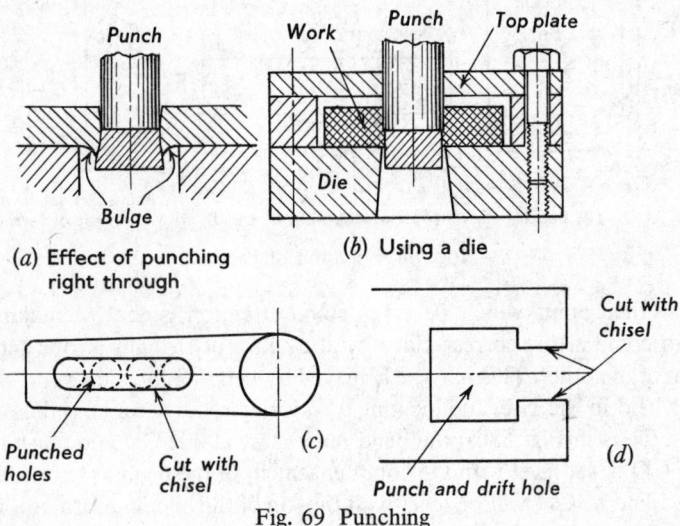

Fig. 69 Punching

in a number of places as shown, after which it is cut out with a hot chisel. It may be finished by drifting or it may be filed when cold. For the component shown at (d) the hole may be punched and then drifted out square, followed by cutting out the opening to form the jaws.

Hand welding

Wrought iron and mild steel may be welded by pressing together two surfaces of the metal after they have been raised to the correct welding heat. The welding heat for iron is at a temperature of about 1350 °C, when the metal is white hot, and in a condition bordering on to the pasty stage. For mild steel the temperature should be a little lower than this, the best point being when its colour is yellow, and before merging to white. It is important that the temperature is correct; if it is too low no amount of pressure or hammering will cause the weld to take place, whilst a temperature too high will ruin the metal by burning it. The second essential for a good weld is that the surfaces to be welded are perfectly clean. When iron or steel at a high temperature is exposed to the air it oxidises and becomes covered with a film of scale (oxide), and if this, or dirt and ashes from the fire, is allowed to remain on the surfaces to be welded, the result will be a failure. To counteract this a flux must be used which melts at a high tempeature and dissolves the scale and ash to form a liquid slag, at the same time acting as a protective covering against any further oxidising action by the air. For wrought iron, sand is a suitable flux, whilst calcined borax serves the same purpose for mild steel. When the two parts of the joint have been fluxed and welding commences, the operation must be so controlled that the action commences at the centre and the joining of the metals proceeds outwards. By this means the slag and impurities are expelled outwards from the two surfaces and a clean, efficient weld is produced. If welding starts from the outside and spreads inwards the slag will be enclosed within the metal. In the case of wrought iron, the natural slag in the metal is influential in helping to flux a weld and form a liquid slag with the impurities.

Types of joint. The three principal types of weld are: (1) the butt weld, (2) the scarf weld, and (3) the vee or splice. These are shown at Fig. 70(a), (b) and (c). The *butt weld* is difficult to make by hammering because of the difficulty of bringing force to bear in the direction of the contacting faces, but it may be employed when facilities are available for pressing the faces together. This is made possible by employing a welding machine in which the bars are gripped, and their ends forced together by hand or power operation. Whilst they are so held, the sides of the weld may be smoothed up by hammering.

The *scarf weld* is the most straightforward one to carry out, and when preparing the ends of the pieces for welding they should not be flat, but should be rounded as shown at Fig. 70(d). This ensures that the union of the metals commences at the centre and travels outwards, expelling the slag to the outside of the bar. If the ends were flat, or hollow, there would be the risk of slag being left in as explained above. When the ends have been prepared they must be raised to the welding heat in the fire and then sprinkled with a little flux to take up the scale and ash. They must then be pressed together and hammered, and the first few blows of the hammer should expel the slag, after which attention may be directed to the completion of the weld and its smoothing up.

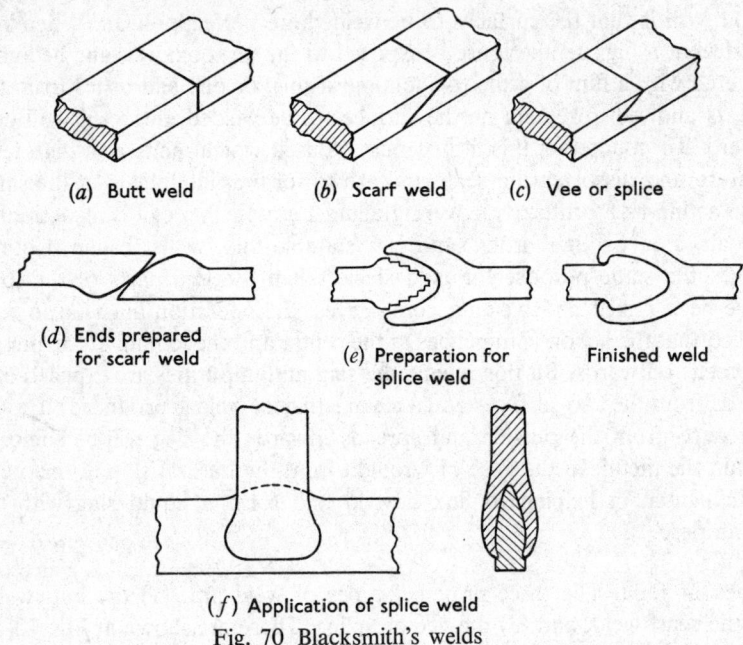

(a) Butt weld   (b) Scarf weld   (c) Vee or splice

(d) Ends prepared for scarf weld

(e) Preparation for splice weld   Finished weld

(f) Application of splice weld
Fig. 70 Blacksmith's welds

The *vee weld* results in a very strong job and should be used where the thickness of the metal permits of preparing the vee. The scarfed end of the bar should be roughened as shown in Fig. 70(e) and when hammered, the ends of the vee'd portion should be hammered over the shoulders of the upset end of the scarfed bar. This type of weld should be employed, if possible, when a bar must be welded to another bar as shown at (f).

Examples of forging

We have discussed in detail the forging operations because most jobs are produced by a combination of these processes. We will now consider the production of one or two typically shaped parts, and, as far as possible, convey the information by means of sketches. It must always be borne in mind, however, that forging is a skilled trade and cannot be learned from books. We can only indicate the lines on which the work should be done and leave its execution to patient practice on the part of the student.

Example 1. Upsetting a head on a bar (Fig. 71).

1. Heat one end of bar for a length of sufficient to make the head.
2. Jump up heated end on anvil.
3. Flatten head by hammering against the end of a bush through which the shank will pass.
4. Swage to size.
5. Finish flatten.

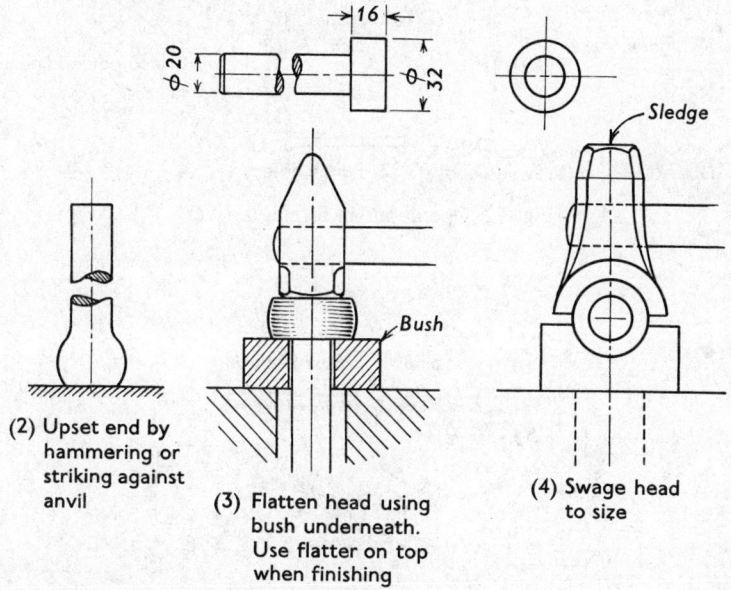

(2) Upset end by hammering or striking against anvil

(3) Flatten head using bush underneath. Use flatter on top when finishing

(4) Swage head to size

Fig. 71 Upsetting a head on a bar

Example 2. Small lever with boss (Fig. 72).

Use material with section large enough to make the boss.

1. Fuller on flat sides.
2. Draw out lever leaving sufficient for final flatting.
3. Roughly flat faces of lever.
4. With hot chisel take off corners of boss; cut taper sides of lever and rough-shape end.

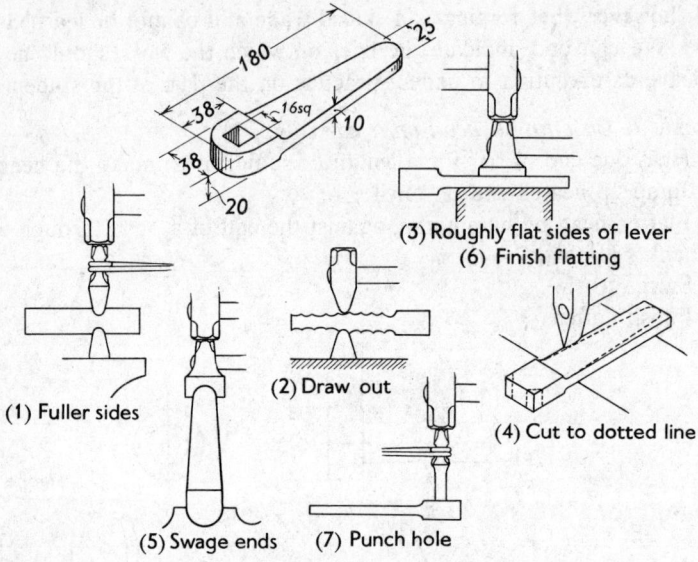

Fig. 72 Operations in forging a lever

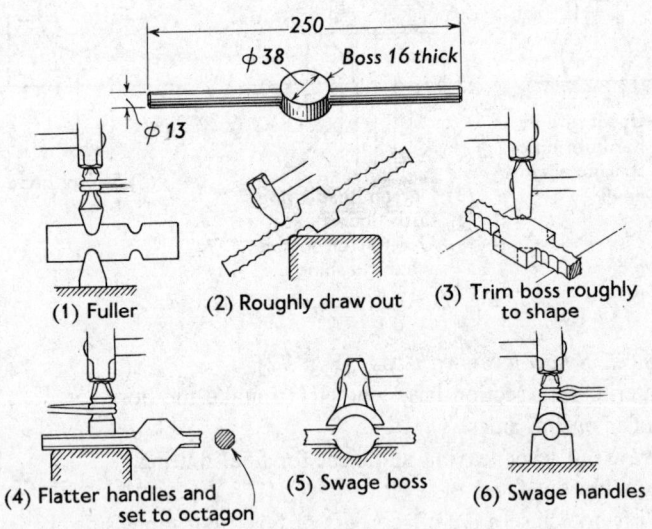

Fig. 73 Forging a small set of die stocks

5. Swage semi-circular ends of boss and handle.

6. Finish flatting sides and edges of lever.

7. Punch hole in boss.

8. Drift hole out square.

9. Smooth up all over.

Example 3. Small die stocks (Fig. 73).

Use material with section large enough to make the boss.

1. Fuller on edges to just above thickness of handles.

2. Draw down handles approximately to square.

3. Trim boss roughly with hot chisel.

4. Flatten corners of handles, making them roughly octagonal in cross-section.

5. Swage boss.

6. Swage handles.

7. Flatten boss faces and smooth up.

After forging, the handles would be turned smooth in a lathe, and the boss bored to suit the die.

Power hammers

When forgings are large, and considerable volumes of metal have to be moved about, the work becomes too big for a sledge-hammer and has to be

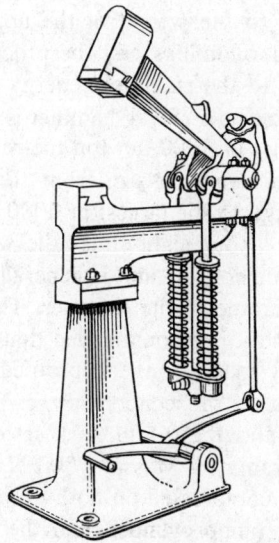

Fig. 74 The Oliver hammer

done by some form of power hammer. For very many years, considerable thought has been given to the introduction of mechanical hammers, so as to dispense with the need of a blacksmith's striker, and one of the first mechanical hammers was the 'Oliver', a foot-operated contraption which gave an effect similar to a sledge wielded by a striker. The Oliver hammer is now almost extinct, but there may be still a few working on special nut and bolt forging jobs in the Black Country, and Fig. 74 is a sketch of one of them. Following the principle of this hammer, various power-operated types have been developed and these are useful for lighter and more specialised classes of work.

The Pneumatic Hammer

To meet the more general need for a heavier power operated hammer the upright type is now the most popular. This class of tool may employ a crank and spring, or steam pressure or pneumatic pressure as the means of delivering the blow. The last two types are the most common, and in their general layout they are similar, consisting of an anvil block for supporting the work, a tup (hammer) which strikes the work and which is connected to the piston by means of a piston-rod, the frame, and the mechanism for the control of the fluid pressure on the piston. The tup and anvil are faced with carbon steel pallets which may be changed for any special shaped tools that may be required. The size of these hammers used to be designated according to the weight of the upper moving parts, but this system has various shortcomings and the practice is growing to specify the hammer according to the maximum energy of the blow delivered. Thus, what used to be classed as a 250 kg hammer is now designated as one with a maximum blow energy of 7000 newton metres (700 kgfm).

Hammers vary in size from the smallest, rated at 100 N m (moving parts 50 kg), to the largest at 6000 N m (moving parts 2000 kg). The anvil should be about eight times the weight of the moving parts and on the larger hammers this unit is generally set upon a foundation separate from the main frame of the hammer. The foundation for these hammers is important and it is recommended that the anvil block should be bedded onto a specially made synthetic proofed cotton fabric mat, supported through timber, on to the concrete base. For the hammer illustrated, the timber would be about 150 mm thick set on $1\frac{1}{2}$ m of concrete.

A diagram of a Massey 7000 N m pneumatic hammer is shown at Fig. 75. Air is compressed on both upward and downward strokes of the piston in the air pump cylinder, and the method of feeding this air to the ram

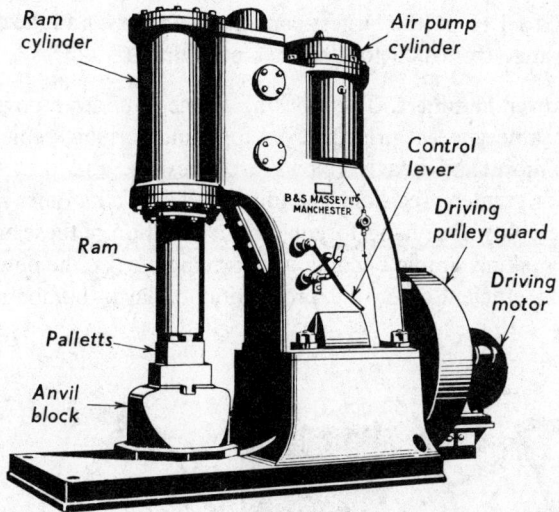

Ram cylinder

Air pump cylinder

Control lever

Driving pulley guard

Driving motor

Ram

Palletts

Anvil block

B&S MASSEY L MANCHESTER

B. & S. Massey Ltd

Fig. 75. Massey 7000 N m (700 kgfm) pneumatic hammer ('clear space' type).

cylinder is controlled by a long valve between the cylinders, the valve being moved by the control lever.

When the lever is in the 'neutral' position, the spaces above and below the pump piston are open to the atmosphere, so that the pump reciprocates idly and the ram piston rests at the bottom of its stroke with the top pallet resting on the anvil. Depressing the lever lifts the valve from its neutral position, closes the openings to atmosphere, and connects the spaces above and below the pump piston with corresponding spaces above and below the hammer piston until eventually, with the valve in its highest position for 'full work', the hammer piston lifts as the pump piston descends, and vice versa (full stroke automatic blows). If the control lever is raised to its full extent from the neutral position the valve is lowered, the upper part of the hammer cylinder is opened to atmosphere and the lower part connected to the air reservoir. This lifts the ram to the top of its stroke and holds it there (hold up). Raising the lever just a little above the neutral position reverses this condition and holds the hammer down (hold down). Single blows may be obtained by moving the lever from the 'hold up' to the 'hold down' positions, the force of the blow depending upon the speed and extent of the movement.

On the Massey 'Clear Space' hammers the strokes vary from 350 mm on the smallest, to 1 m on the largest capacity hammers, and the corresponding speeds range from 200 to 80 blows per minute.

Tools for power hammers. The working of metal under a power hammer follows the same general principles as for hand forging, remembering of course that more and heavier blows are possible, and that the blow is delivered with greater precision. The tools used with a hammer are different from those employed in hand forging and a selection of these are shown at Fig. 76. By making simple bending and forming blocks, the power hammer may be used efficiently for the production of large numbers of similar forgings.

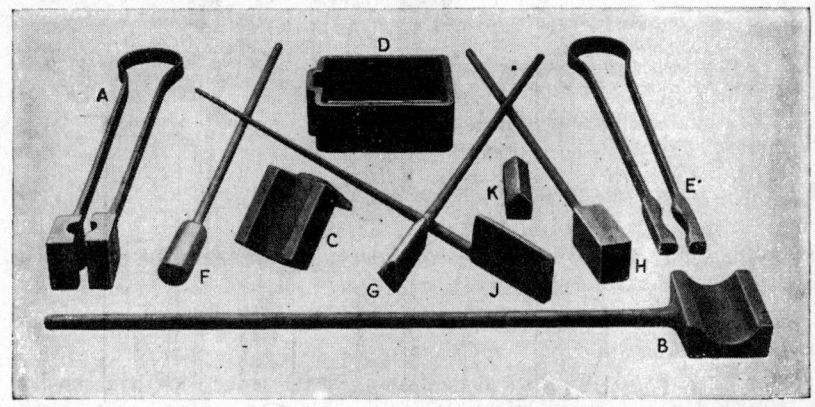

B. & S. Massey, Ltd

Fig. 76 Smithy tools (power hammer)

A, Spring swages; *B* and *C*, Top and bottom swages; *D*, Ring for dropping over anvil pallet for holding swages; *E*, Spring necking tools; *F*, Round necking tool; *G*, Vee tool; *H*, Flatting tool; *J*, Hot cutter; *K*, Cold cutter.

Drop and press forging

Hand forging is a useful and indispensable process, but it is not suitable for the production of large numbers of identical forgings such as are necessary in mass methods of manufacture. For one reason, it would be too slow and costly to use hand methods of production, and also the forgings would not be uniform enough in size to enable the subsequent machining operations to be performed by mass production methods. When large quantities of a certain shaped component have to be made, and the mechanical properties required are such that a casting would not be suitable, drop forging and press forging are very common methods of making the part. These forgings can be produced from nearly all qualities of steel, from some aluminium alloys and from certain brasses and bronzes. The hot working that is given to the metal makes for soundness and high quality in its structure, so that drop forged parts are used in all cases where strength and reliability are the prime considerations.

Drop forgings are made by squeezing the metal at forging heat into

B. & S. Massey, Ltd

Fig. 77 4 tonne bridge type friction drop stamp. (Blow energy 110 kN m)

shaped impressions cut in heavy steel blocks called dies. Generally half the impression is cut in one die and the remainder in the other, so that when the faces of the dies meet with metal squeezed into the cavities, a complete stamping has been formed. The method used for squeezing the metal is to allow one die to fall from a height of 3 to 6 m on to the other, with the metal in between, and it is from this that the name 'drop' forging is derived. The dies for drop forging are made from large blocks of steel, the quality used being a medium carbon steel or one of the alloy die steels (BS 224), and the size of the drop stamp is expressed as the mass of the top die block (the tup) and the energy of the blow. In practice, stamp sizes vary from 50 kg to 16 tonnes. A drop stamp consists of the bottom die, held by set screws on to the base, the top die carried on the tup, together with the mechanism for raising the tup and allowing it to fall. For raising the tup two methods are in general use: the English pattern, which makes use of a belt passing over a pulley, and the American type, in which a board, attached to the tup, is raised when it is held between two

B. & S. Massey, Ltd

Fig. 78 Headgear for raising the tup of a friction drop stamp

rollers rotating in opposite directions. A general diagram of a belt lift drop stamp is shown at Fig. 77, and a head gear for raising the tup at Fig. 78. The upper, lifting shaft, carried in bronze-lined, ring-oiled bearings, carries two cast iron water-cooled friction drums which are keyed to it. Encircling these drums are asbestos-lined friction bands and between them is a lifter arm and belt pulley, both freely supported on the shaft. This upper shaft is gear driven from the lower shaft which carries the flywheel, driven by vee belts from the motor. When the control cord (Fig. 77) is pulled a mechanism, hydraulically operated, causes the friction bands to grip their drums and rotate. In doing so they carry round the lifter arm to which is attached the lifting belt of the tup. This belt winds itself on to the centre pulley and so raises the tup. If, when the tup is fully raised, the cord is released entirely, all the friction is taken off and the tup falls freely, but by manipulating the cord some friction may be retained and a variation in the speed of drop occurs. The buffer supported on rubber cushions, seen in the centre of the gear, acts as a stop for the lifter arm at each end of its stroke.

A pair of dies for forging a component will have approximately half the shape of the component formed in each one, and the process of cutting these impressions is called *die-sinking*. Die-sinking is a very skilled trade and requires not only the skill and patience to do accurate work but also a thorough knowledge of all classes of bench and machine tool technique. In planning and making the dies for a particular forging the following conditions have to be looked after:

1. *The size of the forging and the material of which it is to be made.*
   This will determine what size of stamp must be used to produce it.

2. *The direction in which it may be forged.*

For some articles, e.g. a flat disc, this is obvious, but for others a choice is possible (e.g. the component Fig. 79(a) could be drop forged with the dies meeting on line XX, or on YY).

The considerations which chiefly decide this point are (i) the relative efficiency with which the metal will flow to the alternative die impressions, and (ii) where taper may or may not be tolerated on the finished component. This is shown at Fig. 79(b) and (c), where the taper that would be on the component sides is drawn rather exaggerated for illustration, and it can be seen that if, for example, the boss of the lever were required with flat faces, then it would have to be stamped on YY.

3. *The pre-forming of the raw material for the forging.*

Sometimes the bar or blank may be introduced between the dies without

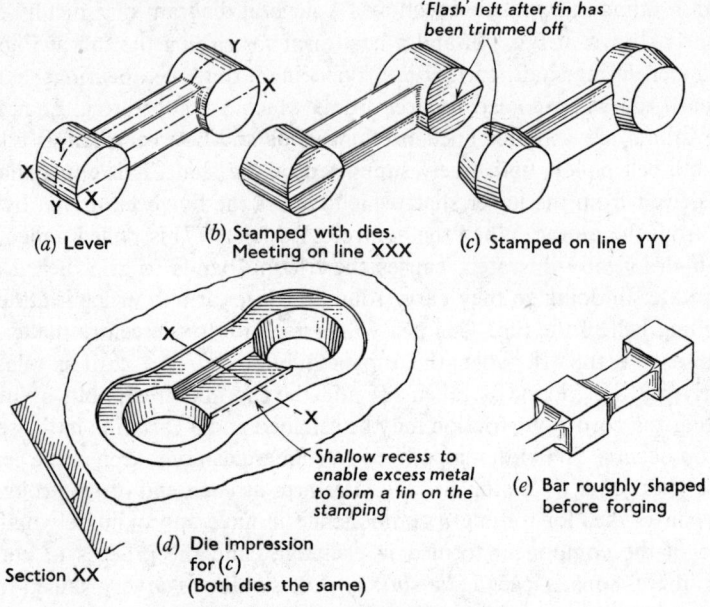

'Flash' left after fin has
been trimmed off

Y
X

Y
X
X
Y
X

(a) Lever

(b) Stamped with dies.
Meeting on line XXX

(c) Stamped on line YYY

X

X

Shallow recess to
enable excess metal
to form a fin on the
stamping

(e) Bar roughly shaped
before forging

Section XX

(d) Die impression
for (c)
(Both dies the same)

Fig. 79 Drop stamping a lever

any previous shaping, but in many cases some preliminary forming is
necessary if the die impressions are to be everywhere filled up. This would
be necessary for Fig. 79(a), and it can easily be imagined what would be
the result if a parallel bar were stamped between the dies made for this
component. Pre-forming may be carried out in impressions on the actual
dies or it may be done separately under a power hammer.

4. *The number of stampings to be produced and the life of the dies.*
The life of a pair of dies is a variable factor depending upon the form of
the impression, the importance of the forging, the material of the die and
the material being forged. If large numbers of stampings have to be made,
long die life is aimed at and expense need not be spared on the material
and finish of the dies. For small orders cost should be kept low and life is
not so important. As a rough idea it may be taken that a pair of medium
carbon steel dies stamping an average mild steel component will give a life
of 20 000 to 30 000 stampings.

When the foregoing points have been settled satisfactorily, a special
drawing of the stamping is usually made, being somewhat similar to the
finished drawing of the component but having allowances made for machin-

ing and contraction, as well as taper shown on it. From this drawing the die-sinker will cut the impressions in the die blocks, using hand or machine-tool methods as may be necessary. When finished, the impressions must be polished, as the forging will receive a good or bad outer finish according to the finish put on the dies. On the face of the die blocks, and round the pro-file of the impressions, are now cut shallow grooves or depressions. These are to allow a fin to be formed on the stamping round the line where the dies meet. The reason for this is that it is impossible to judge the exact volume of metal to fill the die impressions when the dies meet, and a little extra is allowed which is squeezed out into a fin or flash when the impres-sions have filled up. This flash is afterwards trimmed off on a trimming (clipping) press by pressing the stamping whilst still hot through a sharp die having an opening the same profile as the finished stamping.

In the process of forging the metal to be forged, which may be in the form of bars, blanks or pre-shaped pieces, is heated in oil, gas or coal-fired furnaces. (In the case of forging direct from bar, the end is heated for a suf-ficient length to make the forging.) The stamper controls the process, having one assistant to manipulate the controls of the drop stamp, and others if necessary to help in handling the work. The dies, having previ-ously been set with a dummy stamping so that the two half impressions are

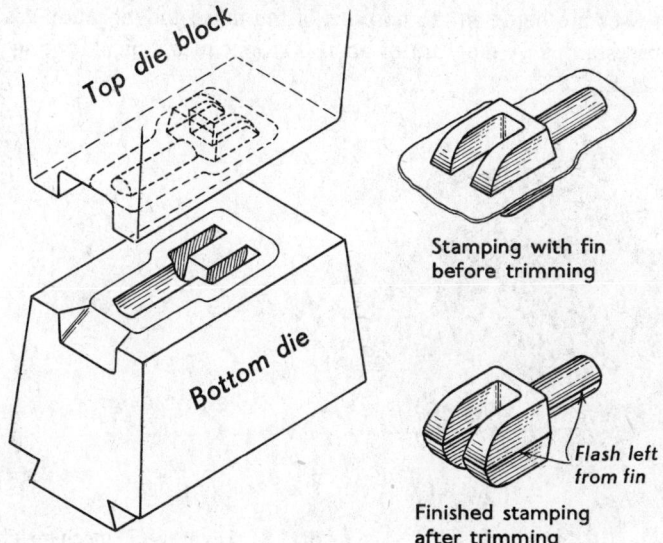

Top die block

Bottom die

Stamping with fin
before trimming

Flash left
from fin

Finished stamping
after trimming

Fig. 80 Drop forging a fork end

an exact match, are now ready for their work, and according to the experience of the stamper the top die is raised and allowed to fall until the forging is completed.

Fig. 80 shows a component, together with the dies for stamping it, and a sketch of the forging before and after finish trimming.

Press forging

Although the drop hammer is still the most widely used method of producing forgings, mechanical forging presses have been developed to meet demands from the automobile industry for long runs of forgings to close dimensional tolerances. The main difference between drop and press forging is that whilst on drop forging any number of blows may be given until the die cavities are filled, on press forging only one blow may be made between a pair of die impressions. As a result of this, press forgings are usually made in two or three stages with only the last stage consisting of the die impression in the final form. Often, also, a measure of preforming is necessary as indicated at Fig. 79(e). The design of the dies and their impressions follows the same principles as we have discussed for drop forging. The capital cost of a press is roughly three times that of a drop stamp of similar capacity, but production rates are higher and less skilled operators can be employed. An additional factor is that working conditions with presses are better in the absence of the noise and vibration associated with drop stamps. A diagram of an 1800 tonne mechanical forging press is shown in Fig. 81.

B. & S. Massey, Ltd

Fig. 81 High speed mechanical forging
press (1800 tonne)

The Joining of Metals

Riveting

When plates have to be fastened together to form a permanent joint, riveting is a satisfactory method of making the joint and is extensively used on boilerwork, shipbuilding and structural construction. Rivets are classified according to the shape of their head, and Fig. 82 shows the usual

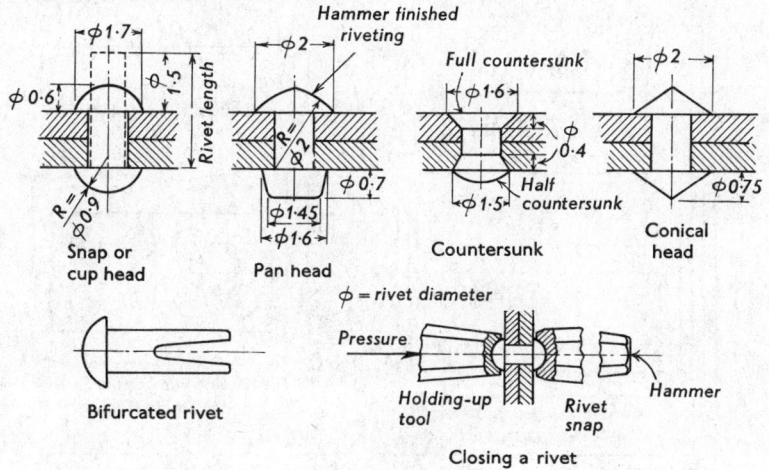

Fig. 82 Rivets

types together with the general size proportions. The round (snap) head is the most commonly used, but if the projecting head is an inconvenience, the countersunk type enables flush head conditions to be attained but does not give such an efficient joint. Steel rivets put in and riveted up hot give the best results, but for light work, copper, brass or aluminium rivets may be used cold. For very light work, hollow and bifurcated rivets may be obtained, but these cannot be expected to give a joint as satisfactory as a solid rivet.

A good guide for the rivet size to use is to make $d = 1\frac{1}{2}t$, where $d =$ rivet diameter, and $t =$ thickness of plates being joined. For thin plates this will give a rivet too small and must be used with discretion. For riveting up the end of a snap head rivet a length equal to $1\frac{1}{2}d$ should be allowed, whilst d should be allowed on the length of a countersunk rivet.

Riveted joints to plates may be made either by lapping over the edges of the plates and fastening with one or two rows of rivets (lap joint), or the edges of the plates may be butted together and then completed by holding them together with one or two cover straps and riveting (butt joint). Examples of these joints are shown at Fig. 83. When preparing the plates for the joint they should, if possible, be clamped together with the top plate marked out for the holes. The holes may then be drilled in all the plates at

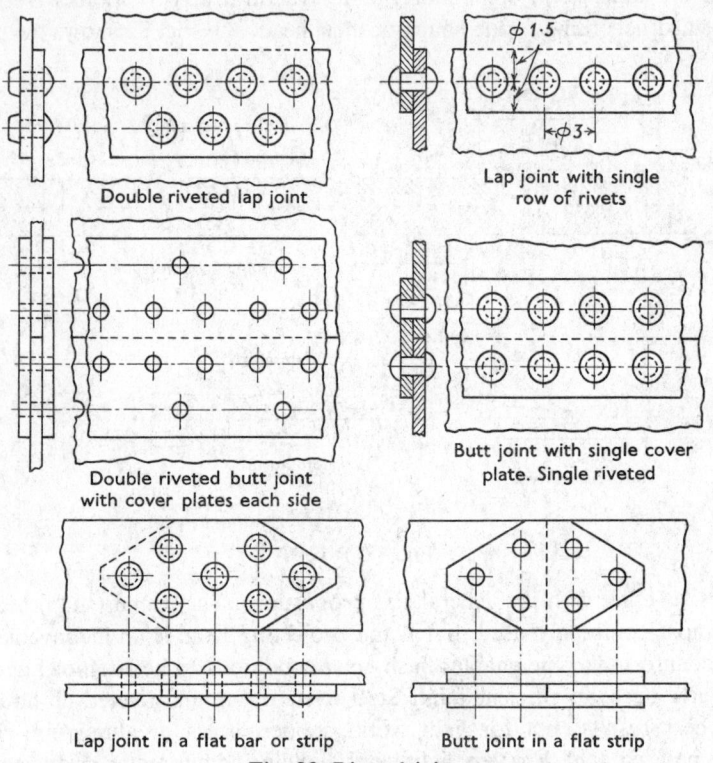

Double riveted lap joint

Lap joint with single row of rivets

Double riveted butt joint with cover plates each side

Butt joint with single cover plate. Single riveted

Lap joint in a flat bar or strip

Butt joint in a flat strip

Fig. 83 Riveted joints

once and there will be no doubt about all the holes being in alignment for insertion of the rivets. If one plate already has holes, it may be clamped in position and the holes marked through for drilling in the other plate, or the holes used as a guide for the drill itself. The holes should be slightly larger than the rivet, an allowance of about $\frac{1}{16}d$ being about suitable (e.g. for a 12 mm rivet; allowance $= \frac{1}{16} \times 12 = \frac{3}{4}$ mm clearance). In some classes of work rivet holes are put in the plate by punching with a punch and die. When this is carried out on the cold plate the edges of the holes are severely cold worked and the work hardening may be severe enough to start small cracks round the edges of the holes. This defect may be removed to a large extent by punching the holes a little smaller than their finished size and opening them out to size with a 3 or 4 flute drill. This removes the hardened metal and any small cracks that may have started in it.

For countersunk rivets the holes in the plates must be chamfered on their outer edges until they are the same diameter as the rivet head, the angle of the chamfer being the same as that under the rivet head (see Fig. 82). This chamfering is called countersinking, and is usually done with a special fluted cutter.

When the plates are ready for riveting they should be clamped together and located with the respective rivet-holes in alignment. If hot riveting is being carried out the rivet should be at a forging heat, and the operation should be completed before it cools too much. For snap head rivets a punch, or tool (snap), having a half-spherical cavity similar to the rivet-head must support the head, whilst the plain end of the rivet is riveted over by a similar punch held to the rivet end and struck with a hammer (Fig. 82). If the work is small enough to be handled it may be rested on the supporting snap whilst the riveting is completed, otherwise an assistant (a 'holder up') must hold the supporting snap whilst the head of the rivet is closed. The aim in riveting should be to swell the body of the rivet by the hammering until it completely fills the hole and to complete the process whilst the rivet is as hot as possible. This avoids any risk of cold working (and brittleness) of the rivet, and as the rivet cools it contracts, drawing the plates tighter together. Care should be exercised to ensure that the rivet end is spread evenly in all directions, and not bent over one way. This is often helped by giving the rivet end a few preliminary blows with the ball end of the hammer. Countersunk rivets may be held up with a flat-ended punch, and when only a few have to be dealt with the ball end of the hammer may be used for the initial spreading, followed by finishing blows with the flat end.

For cold riveting the closing of the rivet follows the same lines as when the rivet is hot, except that the metal is not as plastic, and more difficulty will be met in swelling it to fill the hole, also if the plates are not tightly together after riveting there will be no cooling contraction to pull them to.

When a considerable number of rivets have to be put in it is advisable not to start at one place and fill the holes in order from that place but to rivet the extremities first. For example, rivet holes in a line should have the end ones filled first, then the centre, after which the order does not matter much. If the formation is square, fill opposite corners first, then the other corners, followed by the remainder. By working this way the embarrassing effects of the plates creeping are eliminated. In the same way, if the rivet holes have to be drilled, the operation is facilitated by drilling and riveting two extreme holes first, afterwards drilling the remainder with the plates fastened together.

For structural work, boilermaking, etc., where large numbers of rivets have to be dealt with, there are various types of machines for closing rivets. The pneumatic riveter—probably known to readers by the noise it makes—closes the end of the rivet by a quick succession of blows. Other forms of machine such as the hydraulic riveter squeeze the rivet by hydraulic pressure.

Soldering

Soft soldering

Soldering is a quick and useful method of making joints in light articles made from steel, copper and brass, and for wire joints such as occur in electrical work. It should not be used where much strength is required, or in cases where the joint will be subjected to vibration or heat, as solder is comparatively weak and has a low melting point. When joints must stand these treatments they should be riveted, welded or brazed.

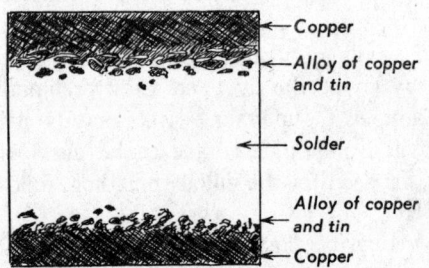

Fig. 84 To illustrate a soldered joint × 300 approx.

When conditions are made suitable for molten solder to wet the surface of the metals being soldered, a thin layer of an alloy of the tin in the solder is formed with the metals, and this union, with the solder in between, causes a joint between the two metals. The conditions are shown in Fig. 84, which illustrates the joint between the two copper surfaces, the layer of metal where the solder joins the copper being an alloy of copper and tin. Now in order that this union shall take place, intimate union between the metal and the solder must be made possible by rendering them scrupulously clean. If molten solder is placed on the surface of an uncleaned metal it will not wet the surface, but will remain in a globular state. This is because the dirt and oxide film prevent the intimate union necessary for the solder to spread over and wet the surface to form the union required. For successful soldering then, the principal requirement is cleanliness, but however careful we may be in cleaning the surfaces to be joined, we should still be unable to make a joint because immediately a polished metallic surface is warmed a thin film of oxide spreads over it and this would prevent the solder from running.

Fluxes. To assist in maintaining the necessary cleanliness, a flux must be used to protect the cleaned surfaces from fumes and atmospheric action whilst the process is taking place. Fluxes are of two kinds: (1) those which not only protect the surface, but play an active chemical part in cleaning it, and (2) those which merely protect a previously cleaned surface. The first class of fluxes are the most efficient, the chief being zinc chloride ('killed spirits'), ammonium chloride, and zinc ammonium chloride, a combination of the two. In the 'protective' group of fluxes (under (2) above) there are tallow, resin, vaseline, olive oil, etc., and in addition to these there are various patent fluxes of which 'Fluxite' is a well-known example.

Killed spirits gives the best all-round results. It is prepared by adding zinc to hydrochloric acid until the acid will dissolve no more and some undissolved metal is left in the acid, the action taking some hours to complete. The main disadvantage to this flux is its corrosive after-effects, and joints made with it should be well washed with, or dipped in, a weak solution of an alkali such as ammonia to neutralise the acid.

For light, continuous work such as the soldering of electric connections, the problem of fluxing is greatly facilitated by using one of the strip solders with the flux incorporated. One such solder is resin cored, being in the form of a small thin tube having a core of resin which melts and fluxes the work as the solder is consumed.

Making the joint. A soldered joint may be made by heating and fluxing the parts to be joined and adding the solder, by dipping the previously fluxed parts into a bath of molten solder or by the use of a soldering iron. For the general run of work the use of a soldering iron is the most common, and irons are made of copper. This metal is used (1) because it is a very good conductor of heat and rapidly transmits heat from itself to the metal of the joint, and (2) it readily alloys with tin and this facilitates the operation of coating the end of the iron with a layer of solder, known as 'tinning' the iron. A good way of heating an iron is by placing it inside an iron or steel box, open at the ends and bottom, and heated inside by the flames

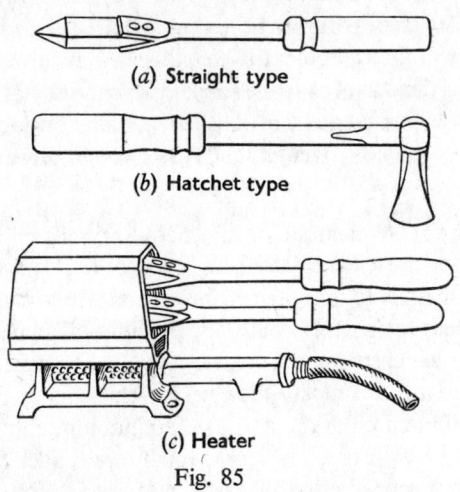

(a) Straight type

(b) Hatchet type

(c) Heater
Fig. 85

from a rectangular gas heater (Fig. 85). Electric and gas heated irons may also be obtained and are preferred by some people. Sketches of straight and hatchet-shaped irons are shown at Fig. 85, the hatchet type being useful where working with the straight iron is likely to be awkward.

The first operation is to 'tin' the iron and this is done by heating it to a good temperature but less than a red heat, quickly cleaning its end with a file or emery cloth, dipping it in flux, and rubbing solder on it. If the temperature and other conditions are right, the iron will take on a thin film of solder, and if it is held against the stick of solder, a certain amount in the form of a globule will adhere to it (suitable solders have been discussed on page 64). The work must now be cleaned, fluxed and tinned. With a file or emery cloth thoroughly clean the surfaces to be joined and cover them with a film of flux. Take the iron in one hand and a brush or spill containing flux in the other. Slowly stroke the surface with the tinned iron and if necessary with more flux until a thin film of solder spreads over the surface. When both surfaces have been tinned, place them together after adding a little flux, and slowly move the iron over them endeavouring to melt and fuse together the tinning on the surfaces, at the same time adding a little solder from the iron if this is necessary. The object should be to use the least amount of solder possible and to make the heat of the iron do the job after previous tinning rather than to feed a large amount of extra solder.

A good joint is characterised by a small amount of solder and perfect adhesion, rather than by large unsightly masses of solder, and the reader should practise making joints and pulling them apart, until he is satisfied with what he is doing. From tests made on joints it has been established that the best joints are those in which the thickness of solder varies between 0·08 and 0·15 mm. Experimental work has also shown that the thinner the layer of solder, the higher is the soldering temperature necessary for maximum strength. Thus for a solder made up of 56% tin to 44% lead it was found that for maximum strength the relationship between joint thickness and soldering temperature was approximately as follows:

Joint thickness	Soldering temperature
0·03 mm	400 °C
0·10 mm	270 °C
0·18 mm	230 °C
0·25 mm	220 °C

The ideal then, in soldering, should be to start with a pair of perfectly clean and fluxed members, so that the solder will flow in under capillary action without much aid from the iron. The following additional hints are given in the hope that they may be useful.

(a) Always use an iron as large as can be handled and err in the direction of having it too hot rather than not hot enough (it should not, of course, be red hot).

(b) A better joint can be made if the work is warm rather than cold.

(c) Iron tinning is facilitated by having some blobs of solder in a tin lid with a little spirits, and touching both the spirits and the solder at the same time.

(d) Quenching the hot joint in spirits, or painting on spirits whilst hot, will often effect remarkably thorough cleaning.

Brazing

At its best, solder is only a soft metal with a strength of about 45 newtons per square millimetre, so that where much strength is required an alternative form of joint must be used. This alternative is provided in brazing, where the joining metal employed is brass, a harder, stronger and more rigid metal than solder. As in the case of soft soldering, an alloy of the brazing metal and the metal of the joint is formed at the surface of the joint metal.

The brass used for making the joint in brazing is generally called 'spelter' and its composition depends upon the metal being brazed because it is essential that the spelter shall have a lower melting point than the material being joined. The following table gives particulars of brazing spelters.

Table 11. Brazing spelters

Metal being joined	Melting Point	Brazing Spelter.		Melting Point
		Copper	Zinc	
Steel	1530 °C	65	35	915 °C
Copper	1080 °C	60	40	900 °C

Spelter may be obtained in the form of sticks, or in a granular state when it may be mixed with the flux before being applied to the joint. If it is used in this form the granules should not be too fine or they may fuse and oxidise away instead of melting into the joint. When buying spelter it is advisable to specify the material for which it is to be used, and, for the same reason, if brazing a delicate job with unknown spelter a test should be made to ascertain that the spelter will melt and run well before the work is too near its point of fusion.

Borax is practically the only flux used for brazing. It dissolves oxides which form on the surface of the work and when a red heat is attained the borax vitrifies and forms a film over the metal which prevents any further atmospheric action. When granular spelter is used the powdered borax may be mixed with the spelter, but if the spelter is in stick form the flux is best applied as a paste.

The heat for brazing is obtained by a blowpipe fed with towns gas, and air at a slight pressure. The air supply may be obtained by tapping off from the blower used for the blacksmith's hearth, from a compressor and pres-

Fig. 86 Blowpipe

sure tank or, for a small installation, from foot bellows. The mixing of the air and gas is arranged for in the blowpipe, which may have separate air and gas regulating valves, or the gas only may have a valve. A diagram of a brazing blowpipe is shown at Fig. 86. When the heating of the work is taking place it is advisable to make the most of the heat given out from the blowpipe flame, and for this reason a small sheet-metal hearth containing firebrick or coke should be used. If the work is placed upon such a substance heat is reflected back on its underside and this conserves the heat during the operation. A brazing hearth, with motor-driven rotary blower, is shown at Fig. 87.

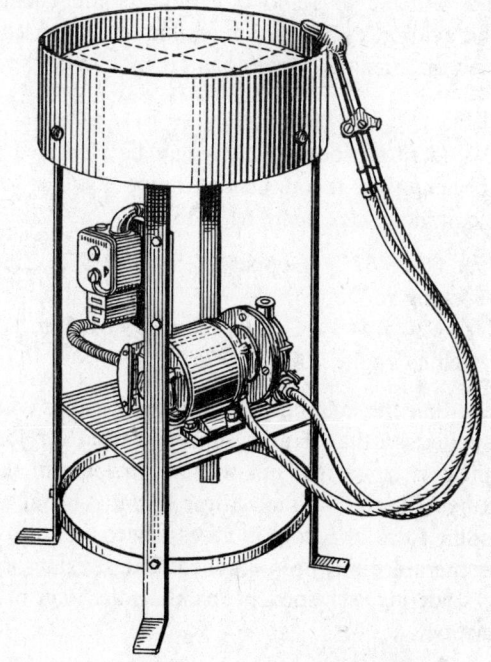

Alldays & Onions, Ltd

Fig. 87 Brazing hearth

The rule of cleanliness applies with equal force to brazing as to soldering, and before commencing any brazing, the work should be clean and polished at all points where brazing is to occur. Where one part fits over another the fit should be neither too tight nor too loose, but just sufficient to allow a film of molten spelter to find its way between the surfaces. In cases where one tube has to be brazed on to another, if conditions will allow, an additional security is given by drilling, and driving one or more steel pegs through the two tubes before brazing. Having prepared the surfaces for brazing, they should be given a coat of borax paste, placed on the hearth, and raised to a red heat as quickly as possible. Care must be exercised when brass or copper parts are being brazed in order that they shall not be fused. When the correct temperature is reached a mixture of borax and spelter is applied to the joint and the spelter will melt and permeate the surfaces, forming a hard brazed junction. The work should be allowed to cool off normally, as quenching may lead to distortion of the joint or cracking of the spelter. When the borax is cold it becomes hard and glassy and difficult to remove without some chemical assistance. Pickling in weak acid or a hot solution of alum may be used for this, or salt may be sprinkled on after the spelter has set but whilst the joint is still hot.

Silver soldering

This is a hard soldering process which may be classified in between soft soldering and brazing. The recommended range of silver solders is given in BS 1845, two common ones being as follows:

Type 4 . Silver 60 to 62%. Copper 27·5 to 29·5%. Zinc 9 to 11%.
Melting range 690° to 735 °C.
Type 5 . Silver 42 to 44%. Copper 36 to 38%. Zinc 18·5 to 20·5%.
Melting range 700° to 775 °C.

It will be seen that the melting point of this solder is considerably lower than that of spelter, so that it may safely be used for joining brass and copper. The process is carried out with a blowpipe in the same way as brazing, borax being used as a flux. Silver solder is usually supplied in thin strip so that some form of holder is necessary to feed it to the joint when the required temperature has been reached. The satisfactory execution of a joint by silver soldering is helped by pickling the joint in dilute sulphuric acid before starting.

5 Power, safety and care in the workshop

Before discussing the various workshop processes and operations we will note a few items which are of general interest to all who are connected with the shop.

Industrial safety

Every day a large number of accidents occur in the factories of this country. These sometimes result in death, sometimes in permanent disablement and in many cases, fortunately, in nothing worse than a few days' or weeks' absence from work. Even if an accident does not render the victim unfit for work it makes him liable to infection, or any other of the ills which may be contracted as the result of injury and shock. On an average three people are killed and 750 injured *every day* in industrial accidents. The Government takes a very lively interest in these accidents, and, through the Factory Inspectors, exerts every means in its power to keep them as low as possible. Statistics gathered of the accidents show, in general, that of every three accidents which occur, two are caused by the personal element of the victim, and one by means beyond his control. To put it briefly, we may say that two out of three are the victim's own fault, and the third was his employer's fault for not making safe working conditions. We will discuss, first, accidents which are the worker's own fault and which might have been prevented by his own precautions.

Machines. Power-driven machines are always a fruitful source of accidents, however well the moving parts may be protected. An operator who is careful at first may gradually, through familiarity, take risks he ought not to take, and ultimately he will meet with a misfortune which he will call 'bad luck'. Confidence is a necessary part of our characters and without it we should be of little use in a machine shop. Over-confidence, however, is to be guarded against at all costs.

The first thing the operator of a machine should find out is the quickest way to stop the machine, and he should practise this until it can be performed instantly, and without thinking. If we are involved in an awkward situation on a power-driven machine, a fraction of a second may mean the difference between a humorous and a serious result. The rotating bolt on the driving plate of a lathe is a danger-spot, and if filing *must* be done on work between centres it is safer to file left-handed. (When the writer was learning his turning he was not permitted to have a file.) Rotating work, whether between centres or held in the chuck, is generally rather rough after a cut has been taken over it, the roughness often being sufficient to pick up rag, or waste held to it. If the waste alone is taken round no harm is done, but if the hand holding the waste cannot be disentangled in time the result is likely to be serious.

On many machine operations the work must be clamped to the table of the machine. If the clamping has been carelessly done the work may move when the force of the cut comes on it. In drilling, particularly, we may imagine that a hole can be drilled with the work only lightly supported, or even held from rotating by the left hand, but sooner or later we shall regret it. Also the waste metal (swarf) which is removed by the cutting tool on machines should not be handled. Turnings particularly are sharp and ragged, being capable of inflicting deep and painful cuts. A good plan is to keep handy a metal hook or rake with which to handle these.

Driving belts. Some machines are still driven by belts from overhead shafting. When these belts have to be moved from one pulley to another for the purpose of changing the speed of the machine, the belt should not be touched by hand when it is moving, but should be manipulated by a bar of metal or a spanner. In addition to the risk of the hand being entrapped between the belt and the pulley, there is the danger of a projecting part of the steel belt lacing catching the hand and tearing the flesh. The guarding of dangerous belts is the concern of the management of the works and is generally well looked after according to the advice of the Factory Inspector.

Loose and flapping clothing may at any time become entangled in a belt or moving part which is not adequately covered, and for this reason overalls without loose ends are to be recommended. If a pair of combination overalls is worn and the ends of the sleeves are buttoned up fairly tightly, or secured with elastic bands, the risk of accidents through loose clothing is greatly minimised.

Non-mechanical accidents

The presence of machines or moving belts is not the only cause of serious accidents. Badly fitting spanners may slip with disastrous results to the knuckles, and it should always be ascertained that a spanner fits well before putting the full load on to it. A loose file haft may result in the spike at the end of the file being driven into the hand if the haft slips off on the draw-stroke of the file. Metal or tools struck with a hammer without sufficient regard to their support against the blow may cause the striker to wish he had been more observant.

The reader should cultivate the 'safety-first' habit and then, if misfortune does come his way, he can truly say it was bad luck, but if he takes simple and elementary precautions it is unlikely that anything very serious will happen.

Accidents of all types are rendered more probable by overtime and tiredness, the speeding up of production beyond the economic rate, bad lighting, cold working conditions, etc., and especial care should be exercised under any such conditions.

Mechanical accidents

Let us now examine some of the causes of accidents which are a result of dangerous working conditions, e.g. unguarded machinery, etc., and against which it is the responsibility of the management to provide safeguards. The Factory Acts administered by the Home Office through the Factory Inspectors contain elaborate and far-reaching provisions for the prevention of accidents, and in the limited space at our disposal we can only indicate the bare outline of these accidents. The reader's experience may be concerned with rotating shafts, belts and pulleys, power presses, milling machines and lifting tackle. All these, as well as many other classes of machinery, are subjected to careful scrutiny by the Factory Inspector. That portion of the Act relating to the fencing of machinery states that 'every part of the transmission machinery, and every dangerous part of any other machinery, must be securely fenced unless it is in such a position, or of such construction, as to be as safe to every person, as it would be if securely fenced', i.e. if a shaft is high up in the air it is 'in such a position as to be safe'. This part of the Act alone puts considerable burden and expense on the Management, in fulfilling its requirements.

However smooth a shaft may be there is always the risk of loose clothing being caught and wound up, and if there are rotating projections such as bolt or key heads the dangers are very much greater. If such shafts are in such a position as to be touched by clothing they must be guarded. Belting which may break and fall on persons underneath, open belt drives which may catch the clothing of passers-by, and any moving parts which are deemed to be dangerous, must be provided with some type of guard. The reader, if he uses his observation in the workshop, will see many such guards and he should take an intelligent interest in them, so that when his turn comes to manage a workshop, he will have the safety of his subordinates at heart.

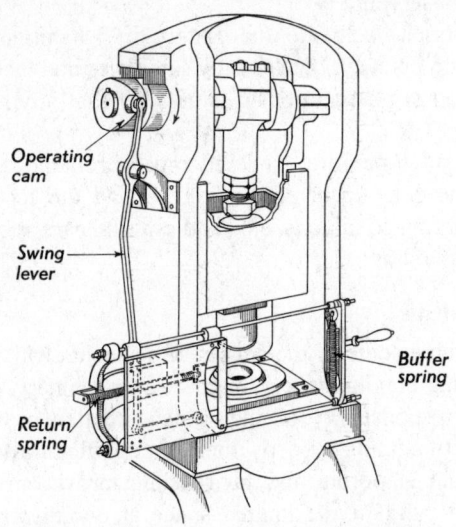

By permission of H.M.S.O.

Fig. 88 Gate-type guard for power press

Power presses and milling machines generally receive the attention of the Factory Inspector because they are considered as especially dangerous machines. When work must be put under a press by hand, it is not possible to fit a permanent guard, so that various safety devices have been devised which close as the press ram descends. These guards are so designed that in closing they would sweep the hands away if they were in danger. One example of such a guard is shown at Fig. 88, the functioning of which will be obvious. The vertical bar at the right-hand end of the gate is a rubber-cased spring to soften the blow should the gate strike the operator's hand.

The danger-point on a milling machine is the cutter, and Fig. 89 shows an example of a guard for keeping the hands away from this.

Fig. 89 Guard for milling cutters

Brayshaw Furnaces and Tools, Ltd

In connection with cranes and lifting tackle the Act lays down that the equipment must be marked with its safe working load, that it must be examined within specified periods, that it must be tested and certified before purchase, and that the chains must be periodically annealed (see notes on work and strain hardening).

Besides the actual provision of guards, safety is promoted in various other ways, of which the following are examples:

1. Foot actuated presses are operated by a clutch which trips out when the foot is removed from the pedal. If an accident were imminent, the operator would instinctively remove his foot from the pedal and so stop the machine.

2. Very large presses operated by two attendants can only be set in motion by pressing two switch-buttons, one for each man. To reach his button each may must move well away from the danger-point.

3. The provision of interlocked devices whereby the machine cannot be started until the guard is in position, or until the machine is safe, helps to reduce accidents. An example of this occurs in connection with laundry machinery, where a horizontal revolving washing cylinder might be started when the operator has the cover off, and his arms inside the cylinder. To obviate this risk the cylinder cannot be run until the cover is in position.

A similar device is incorporated in electrical switchboxes in which the front of the box cannot be opened until the switch is in the off position.

Electrical accidents

One main difference between an electric shock and any other form of accident is that whilst some warning is given of ordinary accidents, the electric shock gives no warning—it just knocks you out. Electrical apparatus should always be treated with the utmost respect and never tampered with. If the wires seem frayed or loose, do not tamper with them but report the matter immediately. Apparatus should not be lifted or pulled about by the wires as the connections are not made strong enough for such treatment. Electrical apparatus should not be used whilst the user is standing on damp or stone floors unless it has the special 3-core cable with earthing wire; in fact it is now essential that all apparatus should be fitted with this arrangement.

All electrical apparatus should be fitted with an isolating switch in order that it may be completely switched off from the mains independently of the usual starting switch. If the appliance is plugged in with a loose length of wire, then this switch is located at the plug socket and it should be switched off before inserting the plug, and after withdrawing it. Remember always, that electricity is not a thing to be played with; its effects are sudden, extremely painful and sometimes fatal.

Power in the workshop

The processes of cutting and forming metal generally require some power assistance so that the application of power to workshop processes is important. Energy is required in various forms by machine tools, e.g. reciprocating motion to a planer table, longitudinal feed to a lathe carriage, rotational motion to a drill, etc., but it is nearly always supplied to the machine in the form of rotational energy at the pulley. One of the functions of the machine is to convert the input of rotational energy into the various forms necessary for doing the work required. As supplied to most of our workshops, energy is in the form of electricity, and this is converted to rotational energy suitable for transmitting to the machines by means of an electric motor. Machines may be driven by either of two methods:

1. Each machine may have its own electric motor which drives through

belt, chain, gearing, or by direct coupling. This is usually termed *individual or self-contained* drive.

2. The machines may be arranged in groups, each machine being driven from a *countershaft*, the countershafts being driven from the *mainshaft*.

Very often the mainshaft is set down the centre of the shop and belt drives are taken off to the machine countershafts which are arranged on either side. Sometimes, when a machine is driven by a single pulley, and the stop and start arrangements are contained in the machine itself, the countershaft is dispensed with and the machine driven direct from a pulley on the mainshaft.

The mainshaft is driven from the electric motor by belt or chain, the motor being placed either on the floor (belt drive) or slung above (belt or chain drive). The use of shafting in workshops is rapidly declining, and in modern workshops is extinct, since most of the machine tools now being installed are provided with individual motor drive. The details we have given on line shaft driving, however, will be of historical interest.

H. W. Ward & Co., Ltd

Fig. 90 Individual drive adapted to a capstan lathe

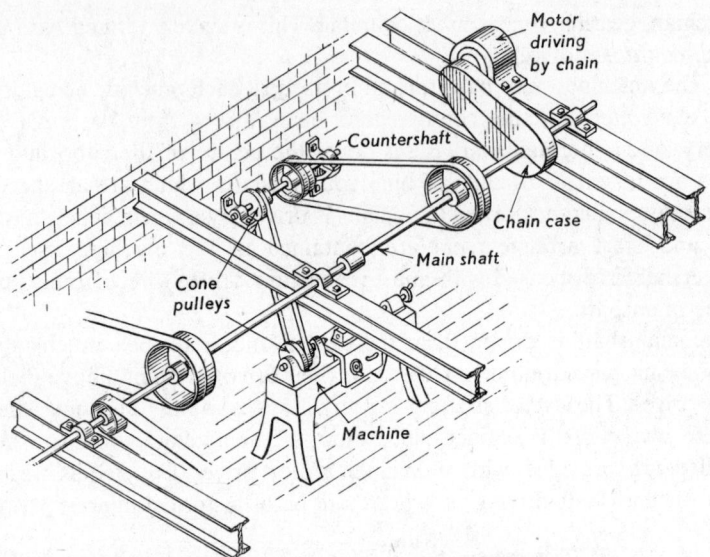

Fig. 91 Driving by mainshaft and countershaft

Diagrams showing individual and countershaft drives are shown at Figs. 90 and 91. For disconnecting the drive and stopping the countershaft and machine, fast and loose pulleys, or a clutch, may be used (Fig. 92(a) and (b)).

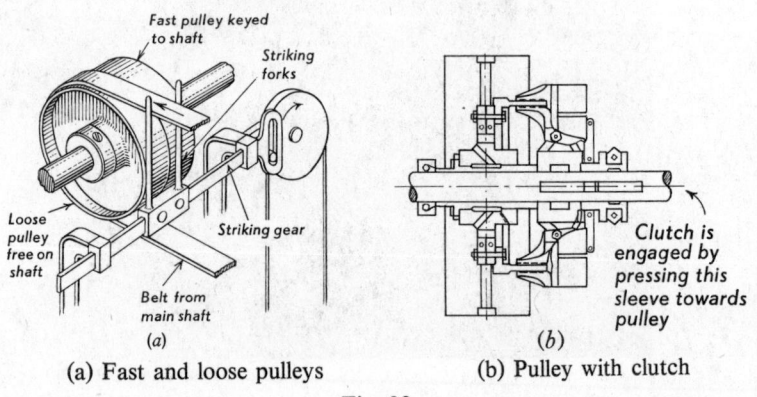

(a) Fast and loose pulleys   (b) Pulley with clutch

Fig. 92

Machine speeds

In order that they may be able to deal with a reasonable range of work sizes, it is necessary for most machines to have a range of speeds varying according to the work, and the type of machine. With a few exceptions electric motors run at a fixed, constant speed, so that speed-changing arrangements must be made, either on the machine itself, or somewhere between the mainshaft and the machine. Machines which are individually driven, and those driven by belt to a single pulley, have the speed-changing arrangements incorporated in the machine itself—generally by means of gears. Those driven from a countershaft often obtain their speed range by making use of a *cone pulley sometimes with the addition of back gear.*

A diagram of a 5-step cone pulley drive applied to a bench drilling machine is shown at Fig. 93. The cone pulley is fixed on the extension of the motor shaft and this drives another cone on the machine spindle. The vee belt is tensioned by the two springs, and adjustment is effected by the

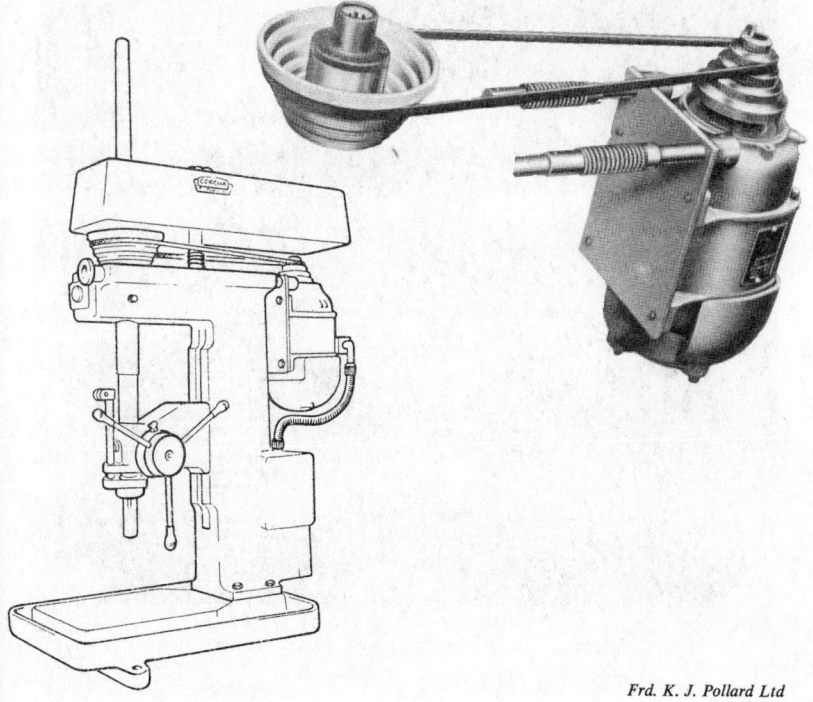

Frd. K. J. Pollard Ltd

Fig. 93 Cone pulley drive on a bench drill

small thumbwheel which is fixed to the end of the RH spring shaft, and which can be seen near the cone pulley on the machine shown alongside.

The all-gear drive

An all-gear lathe head giving eight speeds is shown at Fig. 94. Gear A is solid with the pulley shaft, but this is free of the shaft to the right unless coupled, by sliding B to engage with A through a dog clutch. B, C and D are keyed to and slide on this shaft. E, K, G and J are keyed to the intermediate shaft, but L and M are free to rotate on it as a unit. F and H are free on the main spindle, but either may be locked to it with the sliding clutch unit P. (The ribs on P are really rack teeth engaging with a gear underneath to effect the movement by a handle at the front of the head—

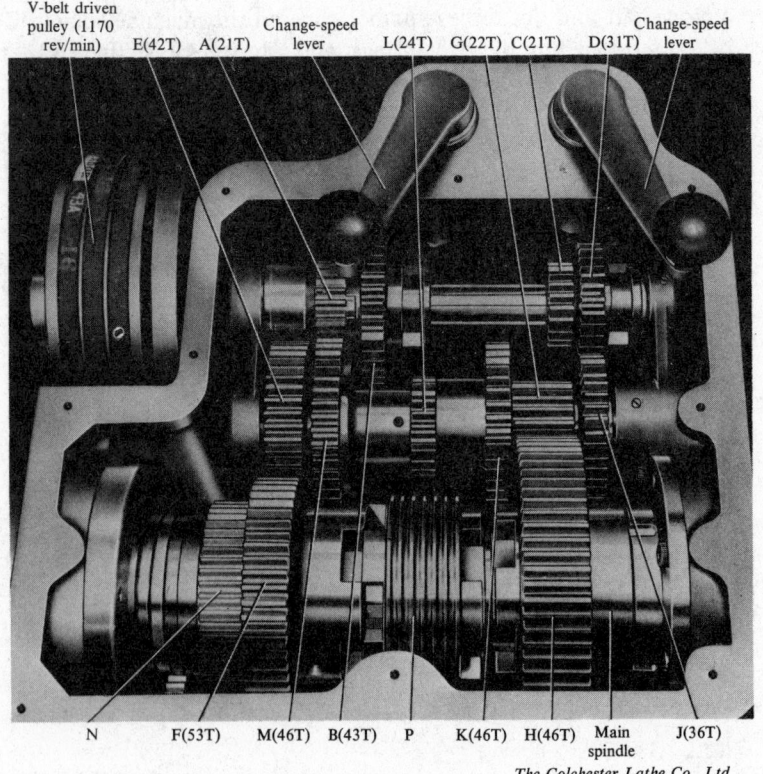

V-belt driven pulley (1170 rev/min)   E(42T)   A(21T)   Change-speed lever   L(24T)   G(22T)   C(21T)   D(31T)   Change-speed lever

N   F(53T)   M(46T)   B(43T)   P   K(46T)   H(46T)   Main spindle   J(36T)

The Colchester Lathe Co., Ltd

Fig. 94 All-gear head

see Fig. 228.) F and H are in constant mesh with E and G respectively. N is a gear for obtaining the reverse drive for the screw-cutting and feed motion. The eight speeds are obtained as follows (lowest upwards):

G driving H: (1) A to M, L to B, C to K; (2) A to M, L to B, D to J;
               (3) A locked to B, C to K; (4) A locked to B, D to J.
E driving F: (1) A to M, L to B, C to K; (2) A to M, L to B, D to J;
               (3) A locked to B, C to K;
               (4) A locked to B, D to J.

As an exercise the reader might calculate the speeds from the pulley speed and gear sizes given on the diagram. On this head starting and reversing are arranged by electrical switching of the motor, and a brake is incorporated in the pulley for quickly bringing the spindle to rest.

The all-gear head, either motorised or shaft driven, has now almost completely superseded the cone drive, which, in a few years, will probably be of historic interest only.

Amongst the advantages of the gear drive are:

1. Easier and more efficient operation of machine, not belt changing, etc.
2. The elimination of cumbersome and unsightly shafts and belts, giving more attractive surroundings and greater safety.
3. Machines are more independent and when motorised may be set down in any position where an electric supply is available. If only one or a few machines are required it is not necessary to drive the whole shafting.
4. Driving pulley runs at a constant speed giving a constant power input. On the cone drive the belt speed (and power input) is lower on the large cone steps

Belt driving

Since belt driving is still used for various purposes, it is worthy of some attention. The drive from one pulley to another is caused by the frictional resistance between the belt and the pulleys. In practice a little belt slip takes place and there is a small amount of belt creep. In Fig. 95, if two pulleys of diameter d and D are connected, and if d makes N rev/min, then if we neglect slip and creep:

    πd millimetres of belt pass over the small pulley per turn
or πdN millimetres per minute.

This length will also pass over the large pulley.

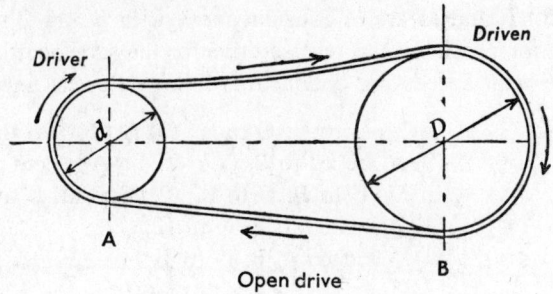

Open drive

Fig. 95 Open-belt driving

Hence, speed of large pulley $= \dfrac{\pi dN}{\text{Circum of } D} = \dfrac{\pi dN}{\pi D} = \dfrac{N.d}{D}$;

i.e. pulley speeds are proportional to their diameters, with small pulleys rotating faster than larger ones in the same drive.

Example 1. The drive to a machine is as follows: A 400 mm pulley on the mainshaft drives a 300 mm countershaft pulley. From the countershaft a 300 mm pulley drives a 320 pulley on the machine. If the mainshaft speed is 250 rev/min, estimate the speed of the machine (a) without belt slip, (b) with 5% slip.

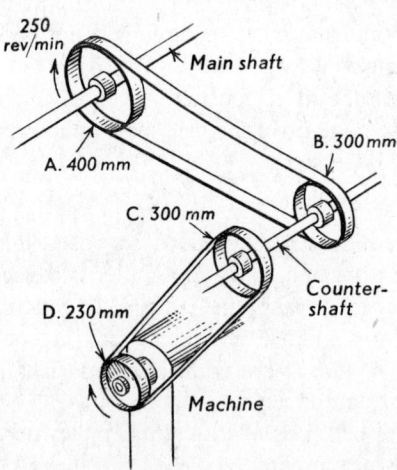

Fig. 96 Drive for Ex. 1

The drive is shown diagrammatically at Fig. 96, where A, B, C and D are the pulleys, A and C being the drivers.

Then with no slip:

$$\text{Speed of B (and C)} = 250 \times \frac{400}{300} = 333 \text{ rev/min.}$$

$$\text{Speed of D} \qquad = 333 \times \frac{300}{230} = \underline{436 \text{ rev/min}}$$

If there is 5% slip:

$$\text{Speed of B and C} = 333 \times \frac{95}{100}$$

$$= 316 \text{ rev/min.}$$

$$\text{Speed of D} \qquad = 316 \times \frac{300}{230} \times \frac{95}{100}$$

$$= \underline{390 \text{ rev/min}}$$

Example 2. A machine with a 230 mm pulley is driven from a 300 mm countershaft pulley. The countershaft pulley which takes the drive from the mainshaft is 380 mm. If the mainshaft speed is 250 rev/min, what size of pulley must be fitted to drive the countershaft if the machine speed is to be 320 rev/min?

For 320 rev/min machine speed, the countershaft speed must be:

$$320 \times \frac{230}{300} = 245 \text{ rev/min}$$

We must now find the size of pulley running at 250 rev/min which will drive a 380 mm pulley at 245 rev/min, and consideration tells us that it will be smaller than 380 mm.

$$\text{Hence mainshaft pulley diameter} = 380 \times \frac{245}{250} = \underline{372 \text{ mm}}$$

Types of belt drive

Being flexible, a belt may be twisted in more than one plane and this renders the belt drive suitable for connecting shafts which are not parallel. Sometimes guiding pulleys are necessary to direct the belt, but these present no disadvantages and may be used to adjust the belt tension. Examples of common drives additional to the open drive are shown at Fig. 97(a), (b) and

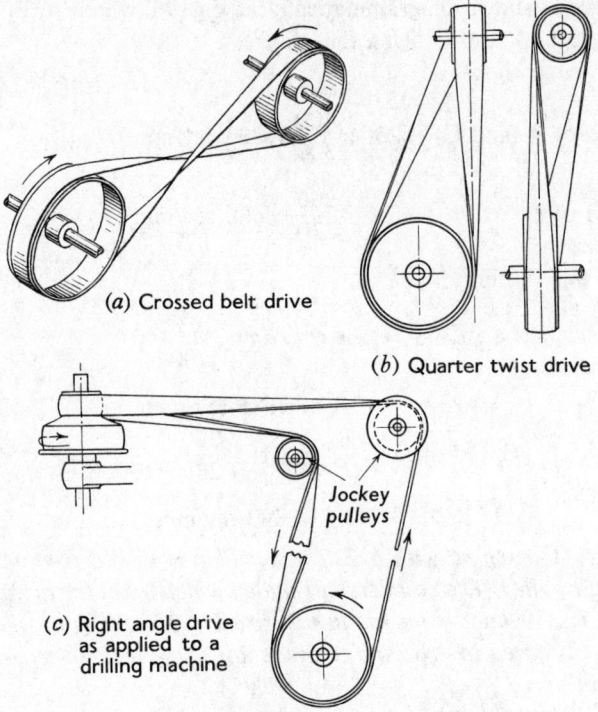

(a) Crossed belt drive

(b) Quarter twist drive

Jockey pulleys

(c) Right angle drive as applied to a drilling machine

Fig. 97 Belt drives

(c). At (a) is the crossed drive used between parallel shafts to reverse the rotation, (b) and (c) are for drives where the shafts are not parallel. When shafts are parallel the pulleys must be lined up by a straightedge placed across their sides. Angular drives (as at (b)) must have the pulleys so disposed that the belt runs *on* to each in a line with the centre plane of the pulley at right angles to the shaft. The belt may run *off* at any angle but will remain on the pulleys if led on straight. When jockey pulleys have to be used as at (c), they should be as large as the smaller pulley of the drive, well balanced and rigidly held, but adjustable in position. Jockey pulleys have no effect on the speed ratio of the drive, which depends only on the sizes of the driving and driven pulleys.

General notes on belting and drives

The most common types of belting are:

(1) *Leather* belts made from the 'butt' of the hide. These are generally of single thickness but two or more may be cemented or sewn together. Single belts average from 5 mm to 7 mm in thickness and double ones from 8 mm to 10 mm. The average breaking strength *per millimetre of width* for single leather belting is: solid leather 157 N (16 kgf); at riveted joint 108 N (11 kgf); at laced joint 64 N (6·5 kgf).

(2) *Cotton and canvas* belts are woven and folded longitudinally to form the plies which are stuck together with cement or rubber solution. These belts are made in various thicknesses up to about 10 ply on the largest widths. When they have to pass through belt shifting forks the wear on the belt edges tends to cause the plies to come apart. The strength of this belting is about the same as leather.

(3) *Rubber* belts are useful out of doors and in damp places. This belting, in common with the cotton and canvas varieties, is not reliable in the presence of oil and grease, as these cause the plies to come apart. The fastenings for woven belts require more care than those for leather, since the canvas is easily torn.

Leather belts should be kept supple, and free from cracks by applying some form of dressing. Tallow is good for this, or, alternatively, one of the patent dressings sold for the purpose. On no account should greasy dressing be applied to woven belting. If belt slip becomes troublesome there are many compositions on the market for increasing the friction between belt and pulley. A cheaper and better known remedy is powdered resin.

The centre distance between the pulleys should not be less than three times the larger pulley diameter, and the ratio of pulley diameter should not exceed six to one. If conditions are such that these ratios must be exceeded, a jockey pulley should be fitted to increase the arc of contact on the smaller pulley (Fig. 98).

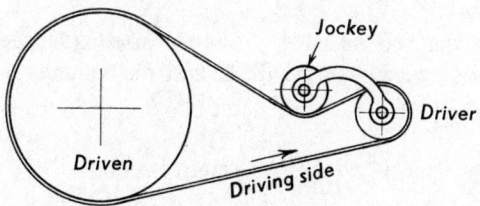

Fig. 98 Jockey pulley to increase arc of contact

Horizontal drives give the best results, and if the belt is open the driving side should be underneath. This brings the slack, sagging side of the belt uppermost and increases the efficiency because the belt is in contact with the pulleys over a longer arc. For this reason a crossed belt drive is superior to an open one, but wear is caused by the belt sides rubbing together.

Power transmitted by a belt

When a belt is stretched over a pair of pulleys and the arrangement is stationary, the tension is the same in all portions of the belt. If one pulley is now rotated so that it drives the other and overcomes resistance, the tensions in the two sides of the belt re-adjust themselves to the requirements of the drive, more tension being in the driving (tight) side, and less in the free (slack) side. It will be appreciated that unless the driving side of the belt is more heavily loaded than its free run there is no difference of tension to drive the pulley, and it is this difference in tensions which is available to turn the driven pulley. The amount of this difference will depend, of course, upon how much resistance there is at the driven pulley; if this pulley is only lightly loaded the tension difference is small, but if it is very difficult to turn, the difference may be so great, and the tight side tension so high, that the belt will either slip, or break.

For average belting it is found that when the belt is driving normally without slip it can safely be loaded to a tension difference of about 9 N per millimetre of width. Knowing this and the belt speed, we may obtain an expression for the greatest power that any given belt can transmit. For example, if the effective driving tension is 9 N per millimetre of width, then the total driving tension $= 9 \times w$ newton [w (mm) = width of belt] and if V metres per second is the belt speed:

$$\text{Work done per second} = 9 \times w \times V \text{ N m.}$$
and $\qquad$ Power $\qquad = 9wV$ W.

We can find the belt speed V from the size and speed of the driving pulley. If its speed is N rev/min, and its diameter is d millimetre, then

$$V = \frac{\pi d N}{1000 \times 60} \text{ metre/second}$$

and since power $= 9wV$ from above

Our expression for power now becomes

$$\text{Power } (P) = \frac{9w\pi dN}{1000 \times 60}$$

$$= \frac{33wdN}{7 \times 10^4} \text{ W } \left(\text{taking } \pi \text{ as } \frac{22}{7} \right)$$

Example 3. Estimate the power that can be transmitted by a leather belt 65 mm wide when driven by a 250 mm pulley at 350 rev/min.
We have

$$\text{Power } (P) \frac{33wdN}{7 \times 10^4} = \frac{33 \times 65 \times 250 \times 350}{7 \times 10^4} = \underline{2680 \text{ W}}$$

Example 4. What width of belt would be required to transmit 3750 watt when driven by a 300 mm pulley rotating at 425 rev/min?

We have that power $(P) = \dfrac{33wdN}{7 \times 10^4}$ and transposing

the formula for w gives:

$$w = \frac{7 \times 10^4 \times P}{33dN}$$

Putting in values for $P = 3750$; $d = 300$; $N = 425$

$$w = \frac{7 \times 10^4 \times 3750}{33 \times 300 \times 425}$$

$$= \underline{62 \cdot 5 \text{ mm}}$$

Pulleys

Mainshaft pulleys are generally made of wood or pressed steel, which gives them suitable strength combined with lightness. They must be fixed rigidly to the shaft, and to facilitate their assembly they are made in two halves which are bolted together. The hole in the pulley is made large enough to accommodate the largest shaft likely to be encountered, and adaptation to smaller shafts is made by using liner bushes made in halves (Fig. 99).

Countershaft pulleys which are narrower and often smaller than those on the mainshaft are made of cast iron (see Fig. 92(a)). The tight pulley is keyed to the shaft and the loose one fitted with a bush which may be renewed when worn. To help the belt run on the centre of the countershaft

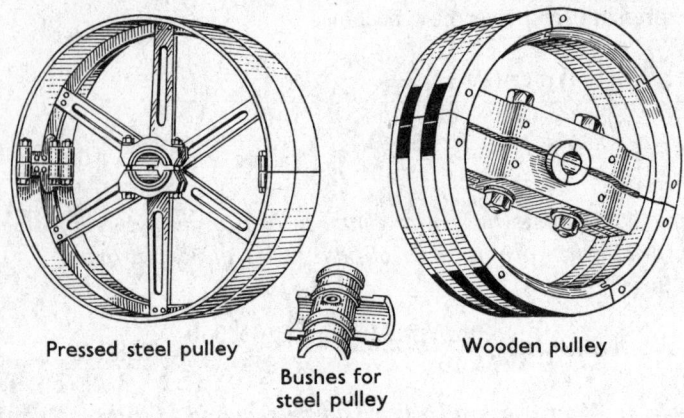

Pressed steel pulley

Bushes for
steel pulley

Wooden pulley

Fig. 99 Mainshaft pulleys

pulleys their top diameter is made slightly larger than at the edges. This is called 'crowning'.

Shafting and spindles—couplings

The shafting used in machine shops is made of mild steel, and up to about 100 mm diameter it is generally supplied and used in the cold drawn or cold rolled bright condition. Above this size it must be turned. Cold drawn shafting is comparatively cheap and the drawing confers a certain amount of work hardening to the skin which is an advantage. This cold working, however, tends to leave internal stresses in the shaft which may be released by the cutting of a keyway or by the removal of the outer skin by wear or machining. The release of such stresses may cause the shaft to warp and result in worn or seized bearings if these are not flexible enough to accommodate themselves to the changed conditions of the shaft. Machine and motor spindles are not likely to suffer this defect since they are turned to size, and this will remove any stresses and correct their after effects before the spindle is put into service.

Shafting is supplied in reasonable lengths, and to make up a long length for a workshop several pieces may be joined together by couplings. When a length of shafting is supported in bearings and coupled up in this way the coupling should preferably be near to a bearing (see Fig. 91) and should support and align the two ends of the shaft rigidly so as to give an effect as near possible equivalent to a solid shaft. The muff coupling at Fig. 100(a) envelops and grips the ends of the shafts, whilst the flanged coupling at (b)

(a)

Buck and Hickman Ltd

(b)

(c)

Renold & Coventry Chain Co., Ltd

(d)

Fig. 100 Couplings

relies upon the bolting up of the flanges to give stiffness. When two spindles each carried on two or more independent bearings are coupled in line, it is necessary to introduce slight flexibility in case a rigid coupling might cause excessive bearing wear when there is a slight mis-alignment of the two shafts. Examples of this practice occur when a motor shaft is directly end coupled to the machine shaft it is driving, and for such purposes couplings having some degree of flexibility are used. Each member of the chain coupling at Fig. 100(c) has a chain wheel cut on a flange at its end and the pair are coupled by a length of double roller chain which just encircles the cut teeth. This type allows for a slight mis-alignment and has the advantage that the two halves of the coupling may easily be disconnected by removing the chain. In the flexible coupling shown at (d) the drive is transmitted between the projecting pins through the laminated rubber disc, which absorbs minor irregularities in the drive as well as allowing for slight mis-alignment.

For the satisfactory operation of all couplings consisting of an end-flanged member fitted to each shaft it is essential that each of these should be a very good driving fit on its shaft before being keyed.

Bearings and support for shafting

Shafting must be carried in bearings spaced at suitable distances and these are supported by brackets or hangers according to the prevailing conditions. The bearings may be in the form of plummer blocks or pedestals or some form of housing to accommodate the bearing bush or ball races, this housing then being clamped to the supporting bracket or incorporated in it. Fig. 101 shows at (a) a plain bearing plummer block and at (b) a ball-bearing pedestal. The suspension of a length of shafting from beams by hangers is shown at (c), and a length being carried on columns by brackets at (d). It is always preferable for the bearing itself to be capable of a small swivelling movement in all directions so that it may accommodate itself to slight deviations in the shaft. With plain bearings this is often achieved by clamping the bearing housing between large spherical-ended plugs in the hanger or bracket. Ball race bearings are made self-aligning by making the outer ball race shaped to a portion of a sphere, thus allowing the inner member carrying the shaft to rotate on its own axis. The use of plain bearings for shafting is giving way to anti-friction bearings of the ball and roller types as the latter have various advantages. The frictional loss in a ball race is only a fraction of that in a plain bearing and its resistance against starting from rest is negligible. When properly designed with felt-

(a) Plummer block (ring oiled)

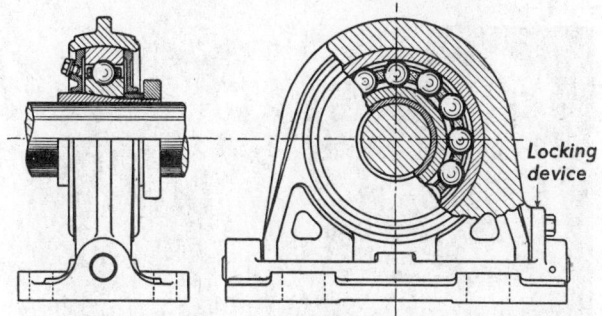

Locking device

(b) Ball-bearing pedestal

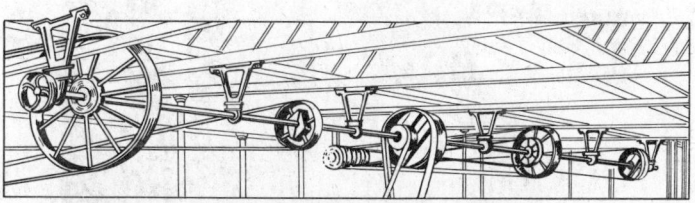

(c) Shafting carried on hangers

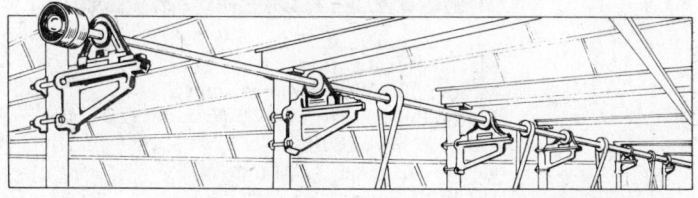

(d) Shafting carried on brackets

Fig. 101

retaining washers the anti-friction unit when packed with grease will operate for long periods without attention. Normally ball races are used for shaft sizes to about 70 mm with roller races for sizes above that.

The vee belt drive

When a machine is driven by its own individual motor it is necessary, in the interests of compactness, to mount the motor on, or inside the machine, as close as possible to the driven pulley. Such an arrangement necessitates pulley centre distances much less than the minimum satisfactory distance mentioned above for a flat belt, and for these conditions such a belt would

J. H. Fenner & Co. Ltd

Machine: Tangye Pump
Motor: 30 kW at 1400 rev/min
Motor Pulley: 190 mm × 5 V
Pump Pulley: 1270 mm × 5 V
Centre Distance: 1280 mm
Width of Pulleys: 95 mm

Fig. 102(a) A wedge vee-belt drive

generally be unsuitable. This has led to the development of the vee belt drive using endless belts of a rubber base with rayon tension cords, and a rubber impregnated woven jacket, working at pulley centre distances in general of not less than the larger pulley diameter, nor more than the sum of the two pulley diameters. Details of the most usual belt sizes are given in Table 12. It will be seen from the Table that two varieties of belt are available. The 'classical' section is the usual industrial type and the wedge belt is a later development having a slightly deeper section and being constructed with neoprene rubber and terylene cording. This gives a rather more compact drive and the increased depth, allied to the construction of the belt, permits higher powers to be transmitted.

The belt runs in 40° vee grooves turned in the pulleys and according to the power and speed several belts may be used for any given drive. Since

Table 12. Details of Vee Belts.

Section designation

(a) Classical Industrial Sections (BS 1440: 1971)	Z	A	B	C	D
A—mm	10	13	17	22	32
B—mm	6	8	11	14	19
Max. kilowatts at belt speed 10 m/sec	1·25	2·68	4·18	8·27	20·2
Max. kilowatts at belt speed 20 m/sec	2·01	4·66	7·0	14·9	33·5
Max. kilowatts at belt speed 30 m/sec	—	5·45	7·85	18·27	39·1

(b) Wedge Belt Sections (BS 3790: 1972)	3V	4V	5V	7V	8V
A—mm	10	13	16	22	25
B—mm	8	10	13	18	23
Max. kilowatts at belt speed 10 m/sec	3·85	6·36	11·17	21·0	21·2
Max. kilowatts at belt speed 20 m/sec	7·15	11·63	20·9	37·9	37·1
Max. kilowatts at belt speed 30 m/sec	9·25	14·9	28·7	54·6	46·9

(J. H. Fenner Ltd)

(Power values above are for largest listed motor pulleys.)

the belts are endless and fixed in length, adjustment is necessary for the motor position to obtain the correct tension and to allow for stretch. When calculating speed ratios for vee belt drives with industrial belt sections it is advisable to assume pulley diameters measured to the centre of the belt, since contact between belt and pulley extends over an appreciable distance. When calculating wedge belt drives, however, the pitch diameter may be taken as the pulley outside diameter since the load carrying cords are almost at the top of the belt section and the belt actually rides slightly proud of the groove. Fig. 102(a) shows a vee belt drive.

The roller chain drive

The reader will be acquainted with the roller chain drive on a bicycle or motor-cycle and will know it for a reliable and efficient method of transmitting power. In its industrial form this drive has various applications in the works and gives a compact method of driving when the shafts are not too far apart (range of about 30 to 80 times the pitch of the chain used). This drive, as for a pair of gears, gives a rigid link between the shafts it connects, and lacks the flexibility of the belt which will slip in the event of an excessive overload. For some purposes this is an advantage since a slipping belt may prevent what would be an expensive smash if the drive were solid. There are occasions, however, when a large amount of power has to be transmitted in a restricted space, where a belt drive would be cumbersome and impracticable, or where a fixed ratio has to be maintained between the driver and driven shafts. For such cases the chain provides a drive which will operate for long periods without attention, provided the arrangement for lubricating the chain is efficient. Normally, an oil-tight casing enclosing the drive incoporates an oil bath. The speed ratio between the shafts of a chain drive depends on the teeth in the chainwheels employed and is determined in the same way as we discuss for gear driving in the next paragraph. A roller chain drive is shown at Fig. 102(b), and the component parts of the outer and inner links at (c). As the chain articulates motion occurs between the plate sides, between the bearing pins and bush bores and between the roller bores and bush diameters. For this reason the protection and lubrication of these chains is important and the best condition is when they are totally enclosed in an oiltight case with bath lubrication. As wear occurs at the joints the chain stretches and means must be provided for this to be taken up. This can be done either by a movement of one of the sprocket shafts or by providing a spring-loaded jockey sprocket which presses on the chain somewhat in the same way as the belt tensioner

Renold and Coventry Chain Co., Ltd

Fig. 102(b) Roller chain drive to special
purpose lathe

Triplex Chain. Ratio 4 : 1. Motor
11·2 kW at 1460 rev/min.)

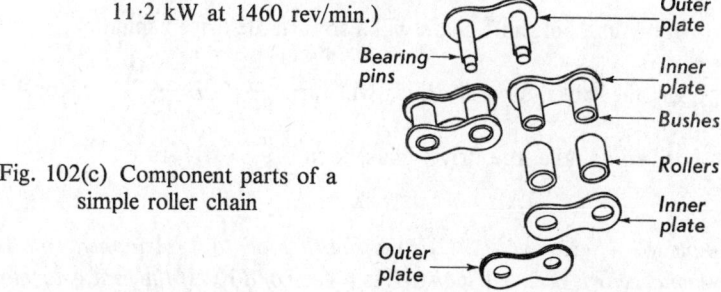

Fig. 102(c) Component parts of a
simple roller chain

shown at Fig. 98. Normally a chain should be replaced when its length has
increased 2% by wear.

Gear driving

A geared connection between two shafts gives a solid drive and a fixed
speed ratio. It is not flexible in the same way as the belt, and whilst this

property of the gear drive is necessary when a fixed unvarying ratio is required such as in screw-cutting, the solid property of the drive is sometimes the cause of trouble. If anything occurs to make one of the shafts in a gear drive to become locked the drive cannot slip, so a smash-up takes place somewhere; in a belt drive the belt would merely slip off.

The speed ratio between the shafts carrying a pair of gears depends upon the numbers of teeth in the gears and, as in the case of a belt drive, the smaller wheel turns the faster.

For example: If a 20-tooth gear were driving a gear of 50 teeth the smaller gear would need to turn $2\frac{1}{2}$ times, or $\frac{50}{20}$, to cause the larger one to make 1 turn.

Thus if t and T represent the number of teeth in the small and large gears respectively, we may say

$$\frac{\text{turns of small gear}}{\text{turns of large gear}} = \frac{T}{t}$$

Example 5. If, in a reduction back gear arrangement, the two small gears have 25 teeth, and the large ones 80 teeth, find the reduction given by the back gear.

When the gear is engaged, 1 turn of the driving spindle will cause the gears on the backshaft to make $\frac{25}{80}$ turn.

1 turn of the backshaft gears will also turn the driven spindle $\frac{25}{80}$; so that for $\frac{25}{80}$ of the backshaft the driver will turn $\frac{25}{80} \times \frac{25}{80} = \frac{25}{256}$; i.e. for 1 turn of the driven spindle, the driver must turn $\frac{256}{25} = 10\frac{6}{25}$ turns.

Example 6. A gear of 22 T drives another of 46 T. Attached solidly to the second gear is a 32 T which drives a gear of 80 T. If the first gear makes 100 rev/min, find the speed of the last one.

The drive is shown at Fig. 103.

The middle shaft turns at $100 \times \frac{22}{46}$ rev/min and the last gear makes

$$100 \times \frac{22}{46} \times \frac{32}{80} = 19\frac{3}{23} \text{ rev/min.}$$

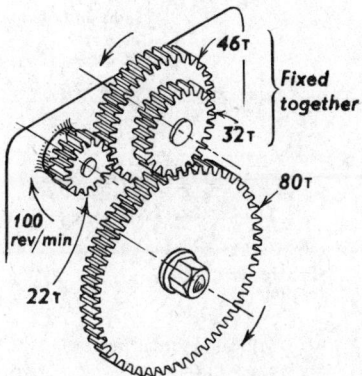

Fig. 103 Gear drive for Ex. 6

Bearings and lubrication. A bearing consists of two members, the moving member, and the stationary one which supports it and carries the load. The surfaces of bearings are generally flat or cylindrical in form and they may consist of metal surfaces (plain bearings) or hardened races supported by balls or rollers. When a bearing supports a cylindrical shaft it is called a *journal* bearing, and the bearing which takes the longitudinal load on a shaft is called a *thrust* bearing. Flat bearings have many forms and the reader may study these for himself by observing the bed and carriage of a lathe, the table of a planer, the ram and slides of a shaper, and so on.

Fig. 104 shows some of the more common examples of bearings. In a journal bearing the bush which supports the shaft may be made as a complete bush (solid bearing) or it may be made in two halves (split bearing). The reason for making a bearing split is for purposes of assembly, e.g. a shaft with a bearing at each end and an enlargement in the centre could not be assembled in two solid bearings. If the reader examines the machines in the shop he will find many examples of bearings and he should make careful note of the relative sizes of various bearings on a machine, the metals of which they are composed, how they are lubricated, and so on.

Materials for bearing. Generally, the only portion of a journal bearing for which we have a choice of materials is the bush. The shaft, almost without exception, is of steel, and may be soft or hardened according to the conditions it has to encounter in service. From a given shaft, then, we have to decide upon the most suitable mating material for the bearing bush, and the materials available are cast iron, bronze and one of the white metals.

For moderate speeds and loads cast iron makes a suitable bearing metal if the lubrication is not neglected. It has the property of taking on a surface

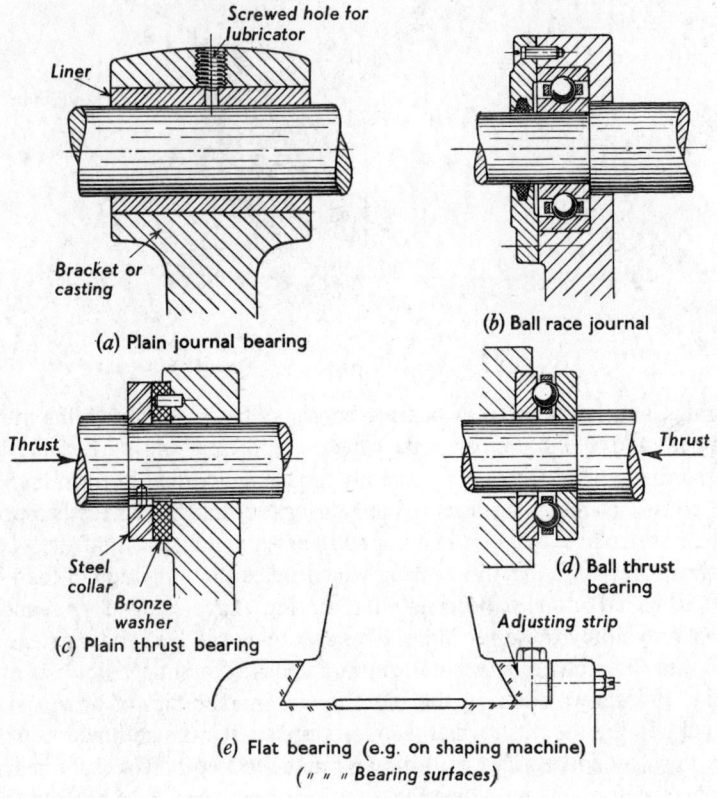

Fig. 104 Types of bearings

which lasts a long time without any appreciable wear. It is cheap and easily obtained.

Bronze is an excellent metal for lining bearings which have to carry high loads at medium speeds, and gives good results if the metal in the shaft is not too soft.

White metal bearings are commonly used for high-speed hard or soft shafts, especially if of fairly large diameter. If the loading is high, care should be exercised to get a suitable white metal as a soft one may be squeezed out of the bearing under the load. The soft nature of the metal enables it to adapt itself more closely to the shape of the shaft, and if the bearing does become overheated the metal melts and runs, leaving the shaft undamaged and giving warning of the happening. White metal bearings are

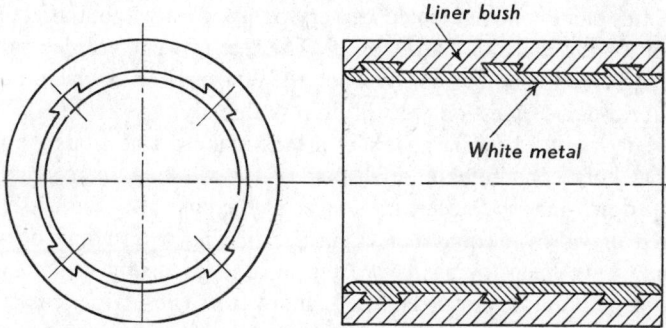

Fig. 105 Dovetail grooves for anchoring white metal to liner bush

commonly made of the split type and the white metal is run in as a thin lining to a shell of cast iron, steel or bronze (preferable). To help anchor the white metal to the liner, grooves or holes are made into which the metal secures itself (Fig. 105). After the lining shells have been cleaned, tinned and fluxed the assembly may be set up with the shaft in position, and the molten metal poured in from a ladle.

The materials forming a sliding bearing such as a planer table or shaper ram are the cast-iron of which the mating components are made. Fortunately, two cast-iron surfaces will run together and make an excellent bearing; if it were not so, the construction of our machine tools would be rendered much more difficult and expensive. In this property cast-iron stands almost alone (i.e. no other metal will run with itself as a bearing), and it is probably due to the free graphite in its structure which helps as a lubricant.

The action in ball and roller bearings is not that of rubbing, but of rolling between the balls or rollers and their races, the inner of which is fixed to the shaft, and the outer to the housing supporting the bearing. These bearings are made from high-grade hardened steel and their manufacture is a highly skilled trade, to be undertaken only by those who specialise in it.

Wear of bearings. During the course of its service a bearing wears and must ultimately be renewed. The rate at which wear takes place, and the amount that can be tolerated before renewal is necessary, depend upon each individual bearing, and no hard-and-fast rule can be laid down. For example, a small amount of wear in the main bearing of a lathe headstock would render the machine unserviceable on account of chattering, and lack of accuracy in the work produced, but the end bearing supporting the feed-shaft of this same machine could be very badly worn without affecting the operation of the machine hardly at all. The rate of wear will depend upon the loading conditions, but chiefly upon the efficiency with which the bearing is lubricated, a question we will discuss directly.

The renewal of a bearing generally means making a new bush, but very often that only constitutes a small part of the work. A large amount of stripping down may be necessary to get at the bush; the shaft itself may be scored or worn, needing a new surface turning or grinding on it; the shaft may have been taking its bearing in a solid casting which must be bored out before a bush can be fitted, and so on. These are aspects of his work which the reader must learn from observation, experience, and from the advice of others.

When the flat bearing surfaces of machine tools become worn they lose their flatness and accuracy. According to the arrangement and service conditions of the bearing the wear may be fairly uniform or it may be local. For example, the table of a planing machine which slides on a bed longer than itself will wear fairly uniformly on its whole length, but the bed will have the greatest wear at the centre position, where, for the majority of the work, the table is located. In the same way the bed of a lathe will wear hollow in the centre, because the carriage is used at that position much more than anywhere else. The wear of such flat bearing surfaces can only be corrected by machining, grinding or scraping the whole of the surface until it is brought back to its flat condition.

Lubrication of bearings. Efficient lubrication plays such a vital part in the reduction of wear that its importance cannot be overstressed. The object of lubricating bearings is to eliminate the solid friction between the metal surfaces and to substitute a fluid whose internal friction is much less. When a bearing is properly lubricated a thin film of oil separates the two metallic surfaces and the upper surface actually floats on this film. When this occurs there is not the metallic wear and damage there would be if the metals were actually in contact, and if the metals do come into contact for any length of time, the bearing very soon seizes up. A good lubricant is one which is able to maintain this film and not be squeezed out by the load on the bearing, whilst at the same time it must not offer undue internal resistance to the motion. This second property depends upon what is called the *viscosity* of the oil; a thick oil has high viscosity and offers greater resistance to motion than a thin one. We might imagine that a thick oil would be less likely to have its film broken and be squeezed out of a bearing than an oil of thinner consistency, but this is not so. The oil film is very thin—probably less than $\frac{1}{50}$ mm—and there are qualities besides its 'body' which make an oil efficient or otherwise in maintaining the film under load.

Action of journal bearing. When a shaft is at rest it will lie in contact with the bush at the side farthest from the load on the shaft, i.e. at the bottom if the shaft load is on top, and as soon as it is started up it will climb to a position which will depend on the fluid friction in the bearing. If friction is high the shaft will climb higher than it would for low friction. Between the two surfaces will be a thin film of oil, and in the regions of closest contact there will be a considerable fluid pressure. At the portions of the bearing opposite to those of greatest pressure there will be very little pressure—in fact there might be a slight vacuum–and it is in this region we must feed the oil to the bearing (Fig. 106).

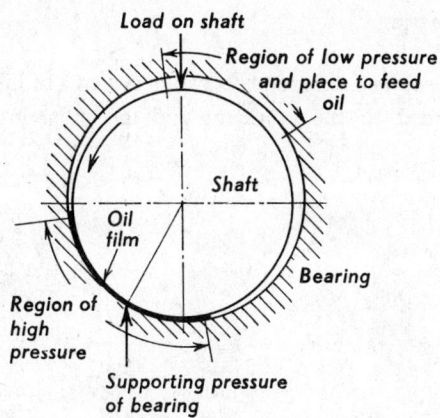

Fig. 106 Action of a journal bearing

Oiliness. The oil film is induced by the motion of the shaft, so that when this is at rest, and is stationary at the bottom of the bush, the film is squeezed out, leaving the metals in contact but having on them an oily wetness. When the shaft is started up, and before the film is formed, it is this wetness which must lubricate the bearing, and the efficiency with which it performs the task depends on the capacity of the oil to leave an oily skin on the metals or its 'oiliness'. It is during the few moments after starting up that the greatest wear is liable to occur in a bearing, so that the 'oiliness' of a lubricant is an important property.

Methods of applying lubricant. In view of the necessity of efficient lubrication, the methods employed for it are worthy of our attention. Broadly classified, these methods may be divided in (a) gravity feed, (b) force feed, and (c) splash methods. Force and splash feed methods are much more efficient than gravity, and should be used on the important bearings of a machine tool.

(*a*) *Gravity feed.* There are numerous methods employing this principle, varying from the simple oil hole to the more elaborate wick and glass sided drip feed lubricators in which the flow of oil may be controlled, and observed through the glass. A selection of these lubricators is shown at Fig. 107, the most efficient being the wick, and drop feed types. Every oil hole should be provided with a lubricator of some description, as otherwise the hole becomes blocked with dirt, and then takes no further part in lubricating the bearing.

(*b*) *Force* (*pressure*) *feed.* There are various systems of lubrication employing a pressure feed to the lubricant and the most important of those

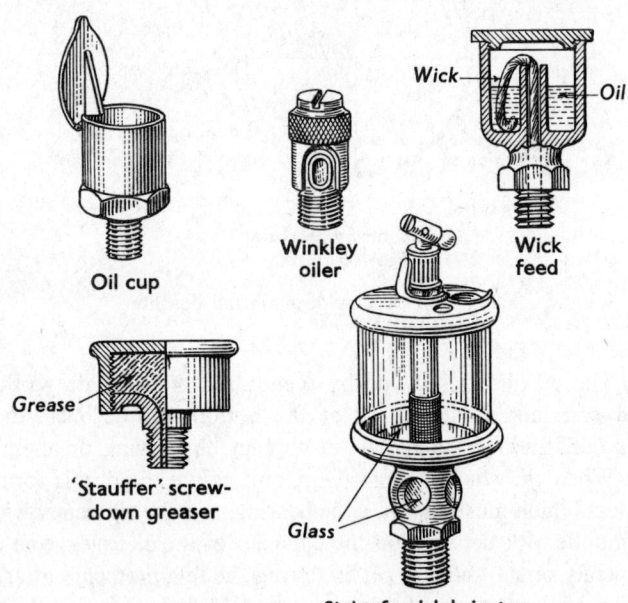

Fig. 107 Lubricators

employed may be divided roughly into the following:

1. Continuous feed of oil under pressure to each bearing concerned. In this method an oil pump driven by the machine delivers oil to the bearings, the oil then draining from the bearings to a sump from which it is drawn by the pump (Fig. 108(a)).

2. Pressure feed by hand pump in which a charge of oil is delivered to each bearing at intervals (once or twice a day), by the machine operator (Fig. 108(b)).

3. Oil or grease gun method. The oil hole leading to each bearing is fitted with a nipple and by pressing the nose of the gun against this the lubricant is forced through to the bearing (Fig. 108(c)).

The first method is the most efficient as (b) and (c) depend on the human element for the regularity with which oil is delivered.

(c) *Splash lubrication.* In this method the shaft, or something attached to it, actually dips into the oil and a stream of lubricant is continually splashing round the parts requiring it. This method is employed for the gears and bearings inside all gear drives, the lower parts of the gears actually dipping in the oil. The oil must be maintained at the proper level, and this is ensured by a filling spout which must be filled up until no more oil can be poured in. A common method of employing splash lubrication is in *ring oiling*. In this method the bearing is cut away in the centre, and a ring passes over the shaft and dips into a reservoir of oil. As the shaft rotates it causes the ring to turn also, and the oil adhering to it is brought up and deposited in the bearing, from which it is led back into the reservoir (Fig. 108(d)).

Grease. Ball and roller bearings, as well as certain other parts of machinery, may be lubricated by grease. Only those plain bearings which move at low speeds and do not carry heavy loads should be lubricated this way, and care should be taken to use grease having a quality beyond question. Thick and unknown greases may set hard in the course of time, and lead to the destruction of those bearings relying upon them for lubrication. It is a popular delusion that a good thick grease is an efficient lubricating agent, but with his knowledge of oil film lubrication, the reader should now realise that it is a delusion. Grease may be applied to bearings by means of a grease gun and nipples, or by a screw-down greaser which is filled, and the top screwed down, forcing the grease into the bearing. Caution should be exercised not to force grease into nipples or lubricators intended for oil, or to use grease of an unsuitable grade; where doubt exists err on the safe side and use oil.

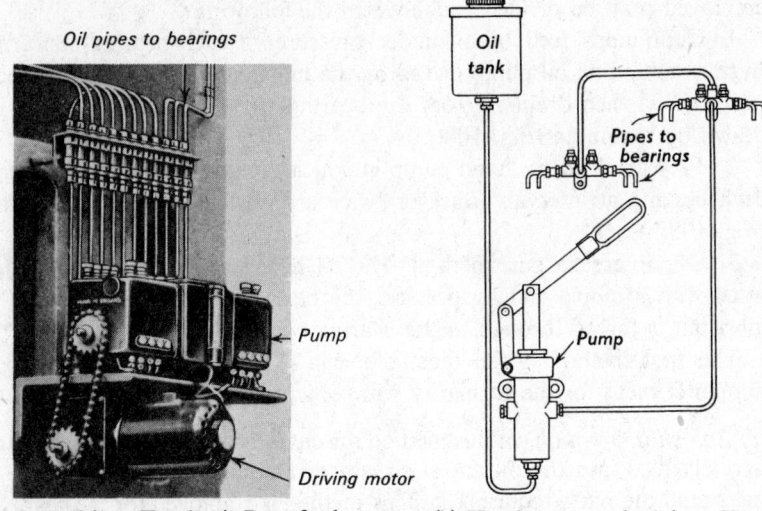

(a) Motor-driven Tecalemit Brantford oil pump with oil flow indicator

(b) Hand pressure feed (one Shot)

Tecalemit Ltd

(c) Grease gun and nipple

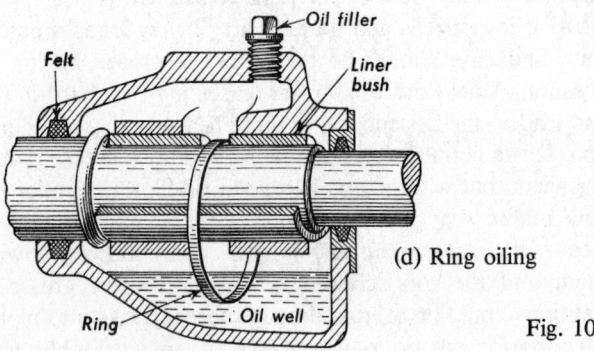

(d) Ring oiling

Fig. 108

Theoretically, the rolling action of ball and roller bearings requires no lubrication, but they should be kept supplied with a thin grease principally to lubricate the cages which space the balls apart.

It will be well to close this section by warning the reader against using cheap unknown qualities of oil. The leading oil refiners have built up a high reputation by the quality of their products, and it is false economy, when such oils are available, to use an inferior grade. Two defects which may develop in an oil after it has been in use for a time under warm conditions are: 1. it may go gummy, and 2. it may turn slightly acidic. If the first fault develops the oil will lose much of its lubricating power and may clog the hole leading to the bearing. Free acid in the oil will corrode the bearings, and the oil may be tested for this as follows: soak a string in the oil and wrap it round a highly polished bar of steel. Leave the string on the bar for several weeks, and if any acid is present, a permanent mark will be left on the bar when the string is removed.

Care and order in the workshop

At the beginning of this chapter we discussed the safety and care of the reader's person when working in the shop. As a contributory factor towards his ultimate emergence as a skilled workshop engineer, the care with which he utilises his tools is of no less importance. We cannot convey by the written word the niceties of control and touch which go to make up the expert user of a pair of calipers, a hammer and chisel, or even a spanner. Such instincts can only come as the result of long years of thoughtful experience, with the early days often being filled with delusions and disappointments. However, let the reader maintain a good heart and a sense of humour amidst such apparent setbacks, and providing he thoughtfully learns the lesson of his early failures he will live to look back on his first efforts with a well-earned degree of pride.

Even although we cannot convey the instinct necessary for such a simple job as tightening a nut, we can offer a few words of advice on matters of a more tangible nature and do so in the hope that they may help to form the right type of habits to commence with. If good habits can be formed early in one's career, they are never left behind and their value cannot be estimated.

Tidiness

Habits of tidiness are no less important in the workshop than in any other aspect of life, and one may form a very reliable judgment regarding the capabilities of a workman by the tidiness and order or otherwise of his methods of working. When we say tidiness we mean more than the obvious habits of tidiness exhibited by keeping one's tools in their proper place, replacing a tool as soon as it is finished with, and so on. Tidiness implies such order, cleanliness and method in one's working that the job seems to proceed smoothly without any obvious effort. Amongst the workshop associates of his younger days the writer still carries the memory of two tool-turners with whom he worked. Both were rather stout and slow of moving, and neither ever appeared to be in a hurry. If one watched them carefully, however, it was obvious that they wasted neither a movement of their body nor a second of time. Their tools were always where they looked for them, they were fastidious in the cleanliness of their machines, and their job never seemed to go wrong—in short, they were two first-class artisans who had trained their minds and hands to lathework.

A first-class mechanic should regard soil, dirt or superfluous articles about his machine, tools, or work with as much concern as if they were on his own person. If he has developed orderly habits it will not be necessary for him to fritter his energy looking for tools, or clearing unwanted articles, he will have his mind clear for attending to the job, and at the end of the day he will be fresher than another of untidy habits who probably has not done nearly as much useful work.

The care of tools and machines

From the very start of his career the reader will need the help of tools and machines in his work, and as he advances these will get more numerous and costly. Some of the tools will belong to the employer, others will be the reader's own property. To the employer, the equipment he provides represents so much capital being used in the business for obtaining output, and earning profits, whilst to the worker, the equipment represents the means of earning his living, and a better living may be earned if the equipment is kept in good condition. We have already discussed cleanliness, but we might add that the accuracy and life of a machine tool is helped exceedingly if it is kept clean. Machines are fitted with accurately made spindles and slides, and it is upon their accuracy that the quality of the work depends. The cutting done on such machines releases grit and other foreign matter from the metal, and unless the machine is cleaned regularly, this grit finds its way into the working parts of the machine and acts like sandpaper.

Spanners, files and other heavy objects should not be dropped on to the machined surfaces of tables and slides. If we must store spanners and tools on a machine bed a piece of board placed under them will look more tidy, and leave the machine undamaged.

Often it is necessary to use a hammer to loosen, tighten or withdraw some part of a machine. If possible a lead or copper hammer should be used to avoid bruising the part being hit. Failing that, a block of hard wood or soft metal interposed between the hammer and the part being hit will save damage. It is a shame to see, in some workshops, parts of machines mutilated by constant hammering; the handles of vises and jigs are particularly unfortunate in this respect; actually these are hammered up far more than they need be to hold the work securely.

Grinding tools

In every machine shop a considerable amount of capital is locked up in cutters, cutting tools, drills and various other cutting equipment, and when we consider that these are made from steel which may cost up to £3 per kg we realise that the annual cost of them must be heavy. Because they do not have the appearance of being worth much, and because they are the general property of everyone, they are not always treated with the respect due to them. As a tool is used its edge gradually becomes dulled and needs resharpening. If this is delayed the tool rapidly deteriorates and might in a short time be ruined. The reader should learn as soon as possible the proper methods of grinding tools, as a tool may be as easily spoiled by faulty grinding as by being left unground. We must leave the technique of the actual holding and shape grinding of the tool for the reader to learn from being shown, but we might advise that a tool should either be ground with a *heavy* stream of water flowing on it or it should be ground gently with no water at all. When a tool is being ground, a large amount of heat is generated, and if plenty of water is flowing it carries the heat away, thus preventing the tool from overheating. If the tool is ground gently without water the heat is not generated as rapidly, and has time to be conducted away to the body of the tool before the edge becomes burned. If the supply of water is sparse and intermittent the outside of the tool becomes heated, and when a splash of water falls on to it a sudden local quenching takes place leading to minute cracks which sooner or later will lead to the failure of the tool. Grinding a tool dry, and plunging it into cold water from time to time, will lead to similar effects, and is equally as bad as using an insufficient supply of water. A general purpose pedestal tool grinder is shown at Fig. 118.

Tool failure in cutting

A frequent cause of tool failure is running the machine too fast. This causes rapid overheating and deterioration of the cutting edge, and if the tool seems to be making hard work of its job the speed should be reduced and plenty of cutting lubricant applied to the tool. Cast iron should not be cut with a lubricant, but can generally be machined if the first cut gets well beneath the hard outer scale. If the tool is allowed to ride over the hard scale its nose will be rubbed away.

Some of the special steels which must be machined have the property of hardening up under the action of the tool. This is accentuated if the tool is allowed to 'loiter' on its job, and if the tool is kept well up to its work on a low speed, it will bite into the material and cut efficiently.

We might close this chapter by urging the reader, in matters of safety and care, not only to develop good habits himself, but also to take initiative in persuading others of the importance of doing the same. If everyone who reads these notes and acts on the advice contained can claim a convert to their principles, he will have achieved a creditable piece of work.

6 Metal cutting

A large portion of the training and skill of a workshop engineer is devoted to the shaping of metal, and to stopping the process at the proper time. In the workshop, most of the shaping involves cutting the metal, and stopping this at the correct place involves a knowledge of measurement. The next two chapters, then, deal with these two very important fundamentals of our knowledge.

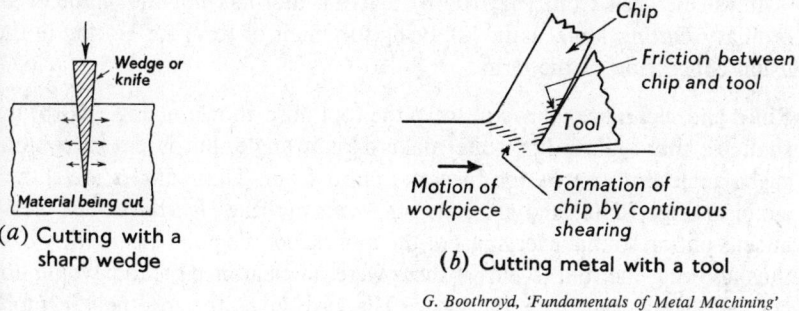

(a) Cutting with a sharp wedge

(b) Cutting metal with a tool

G. Boothroyd, 'Fundamentals of Metal Machining'

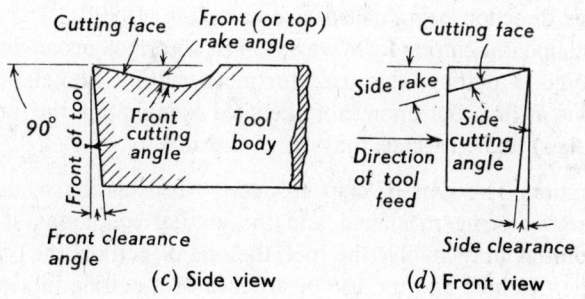

(c) Side view

(d) Front view

Fig. 109 Cutting and cutting tool angles

The cutting of metal. The usual conception of cutting suggests cleaving the substance apart with a thin knife or wedge as represented in Fig. 109(a). When we cut metal the action is different from this, being more in the nature of a shearing than a cutting process. A diagram of a metal chip being removed is shown at Fig. 109(b). When the cut is under way the chip presses heavily on the top face of the tool and a continuous shearing takes place across the zone which marks the severance of the chip from the work-piece. At the same time the chip suffers severe pressure and this, in conjunction with the shearing action, results in the chip being compressed to a shorter and thicker length as it leaves the tool. The occurrence of the heavy pressure on the tool can often be observed by examining the nose of a tool which has been cutting for a long time without having been reground. Under favourable conditions a small cavity may be seen which has been worn in the hard tool by the severe rubbing of the chip.

It will be seen from Fig. 109(b) that the tool has not the shape of an ordinary cutting knife as in (a), being too blunt of form for cutting in the usual conception of the term.

Rake and clearance. If we observe the tool nose shown at Fig. 109(b) we shall see that it does not quite make a right angle, but is cut away to a slight angle both on its top face and on its front. These angles are shown again at Fig. 109(c) and are called the *rake* and the *clearance*. The clearance is put on so that every part of the tool except the point clears the work, thus allowing the tool to cut; if there were no clearance the tool would not cut but would only rub the work. The rake gives the tool nose a more wedge-like form and is varied according to the metal being cut. Soft and ductile metals are cut with more rake than hard and brittle ones. The rake we have shown is the slope back from the front of the tool and is called *front* or *top rake*. Sometimes a tool nose is sloped sideways as well, the slope in this direction being called *side rake* (Fig. 109(d)).

The fundamental cutting form we have just described occurs in principle, on every type of cutting edge used for metal cutting. Sometimes it is not very obvious at first, but upon more detailed examination the fundamental rake (or rakes) and clearance may be picked out.

Chip formation. The type of chip produced when cutting metal depends on the material being machined and the cutting conditions at the time. These conditions may involve the tool, the rate of cutting, the type or condition of the machine and the use or absence of a cutting lubricant.

The three common types of chip from a single point tool are shown at

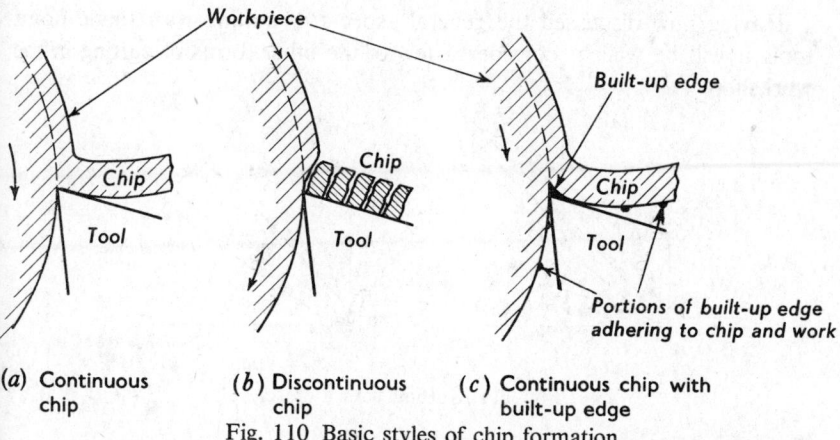

(a) Continuous chip   **(b)** Discontinuous chip   **(c)** Continuous chip with built-up edge

Fig. 110 Basic styles of chip formation

Fig. 110. The *continuous* chip leaves the tool as a long ribbon and is common when cutting most ductile materials such as mild steel, copper and aluminium. This chip is also promoted by increasing the tool rate angle and the cutting speed, decreasing the cut and using a cutting lubricant.

The *discontinuous* chip leaves the tool as small segments of metal and results from cutting brittle metals such as cast-iron and cast brass with tools having small rake angles. Cutting other metals with small rake angles at low speeds and with heavy cuts may also result in this type of chip.

The *built-up edge*, associated with the continuous chip, is the result of small particles from the workpiece being built up and welded to the tool face under the heavy pressure and the heat generated at the tip of the tool face. The condition often involves the continuous formation of the built-up materials, and its being swept away adhering to the work, and the underside of the chip. The phenomenon results in a poor finish from the cut and is caused by unduly high friction and cutting pressure, and when there is any metallurgical affinity between the tool and work materials. It can be minimised or prevented by using light cuts at higher speeds and using an efficient cutting lubricant.

Chip Breaker. When large amounts of metal are being removed with a continuous chip the resulting mass of tangle ribbons of metal constitutes a serious problem in its control and disposal, as well as a danger hazard. Under these conditions the chip may be broken up by grinding a groove on the tool face a little way back from the cutting edge.

Having now discussed the general aspects of cutting with single point tools it will be well to consider a few of the other forms of cutting in the workshop.

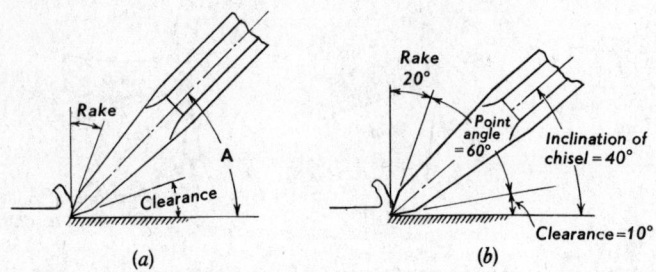

Fig. 111 Cutting with a chisel

The flat chisel. A diagram of a chisel point in the action of cutting is shown at Fig. 111(a), where the angles of rake and clearance are indicated. If, as is usual, the point is ground symmetrical, the rake and clearance will depend upon the angle of inclination (A) between the chisel and the work. A clearance of about 10° would be suitable, so that for a 60° point angle it would be necessary to hold the chisel with angle $A = 10° + 30° = 40°$ and the rake would then be $90° - (40° + 30°) = 20°$. This is shown at (b). Since cutting hard and brittle metals calls for less rake than soft metals, the chisel angle must be less for the softer metals, and the following table gives suitable values for the angles:

Table 13. Chisel angles

Metal being Cut	Angle of Chisel Point
Cast steel	65°
Cast iron . . \. . .	60°
Mild steel	55°
Brass	50°
Copper	45°
Aluminium	30°

As an exercise, the reader should calculate the angle (A) at which each of the above chisels should be sloped to give a clearance of 10°, and then find the cutting rake.

Hacksawing. An enlarged view of the tooth of a 300 mm 'Eclipse' saw blade, having a tooth pitch of 2 mm, is shown at Fig. 112, together with a

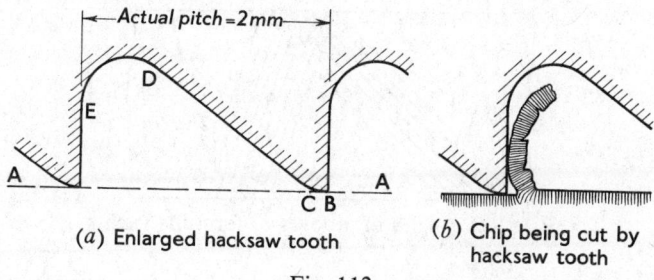

(*a*) Enlarged hacksaw tooth   (*b*) Chip being cut by hacksaw tooth

Fig. 112

diagrammatic view of a chip in the action of being cut. As the tooth is cutting relative to the line AA, it will be seen that this tooth is cutting with little or no rake. The clearance face is the portion BC of the tooth, and then its profile is sloped more sharply to give the form BCDE. This is necessary because if the tooth were formed by continuing line BC to the next tooth, the space between the teeth would be too shallow, and would become clogged with chips of the metal being cut. On the other hand, if the portion BC were omitted, and the back of the tooth formed by taking it from the sharp edge down the slope CD, the tooth would lack strength. In the same way the tooth would be weakened by adding much rake. The part BC is probably less than $\frac{1}{4}$ mm long, but it is very influential in giving strength to the back of the tooth. The radiussed portion DE also adds strength.

Saw-blade teeth are very small and thin, and as they are generally used on every kind of metal, the makers of them do their best to design the most durable tooth possible under the circumstances, rather than try to suit any particular metal.

The file. File making is a highly specialised trade, the cutting of the teeth calling for much skill and experience. The teeth on a file are cut with a sharp chisel having a point angle of 55° to 60°, and the operation may be carried out either by hand or machine methods. The most usual form of file is that with *cross-cut* teeth, the grooves in the face of the file running in two directions and dividing it up into small diamond-shaped teeth. In the *single-cut* file there is only one series of grooves. The operation of cutting files consists first of cutting a series of grooves sloping at an angle of about 45° to the file edge, and marked '1st cut' on Fig. 113. The extreme tip of the teeth is then filed off and a second series of grooves cut at an angle of 70° to the edge (marked '2nd cut'). The result of the two sets of cuts is to raise teeth in the form of small pyramids with edged tops turned back towards the

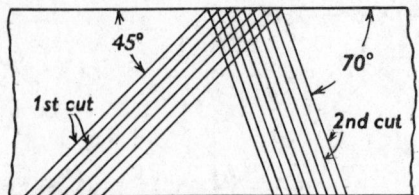

Fig. 113 Angles of grooves to form file teeth

direction in which the file must cut. When one side of the file has been cut it is turned over, embedded in a block of soft metal, and the process repeated on the other side. An enlarged photograph of the teeth of a file is shown at Fig. 114(a), and a section through one of the grooves at (b). It will be seen from (b) that the normal cutting point formation is again in evidence, but obviously, with such small teeth, we cannot expect to see precise rake and clearance faces. It will be noticed that the grooves being cut at an angle, instead of perpendicular to the length of the file, gives the teeth side rake when the file is moved in the direction of its length.

Samuel Peace and Sons, Ltd

(*a*) Enlarged photograph of file teeth.

(*b*) Enlarged form of grooves cut to form file teeth.

Fig. 114 File teeth

For many years all file teeth were hand cut and these were found to be greatly superior to the early machine-cut files. This was found to be due to the fact that the irregularity of hand-cut teeth gave them a higher cutting efficiency than the perfectly spaced teeth of machine cutting. In later developments of machine cutting the teeth were purposely made with irregular spacing.

The twist drill. It is difficult, on first looking at the point of a twist drill, to see where the rake and clearance enter into its action. We know, however, that even if a tool will cut without rake, it cannot do so without clearance, so let us first examine a flat drill, a tool used for boring holes long before the twist drill. One of these is shown at Fig. 115(a), and each of the two cutting edges is bevelled off to give clearance as shown. If the front face

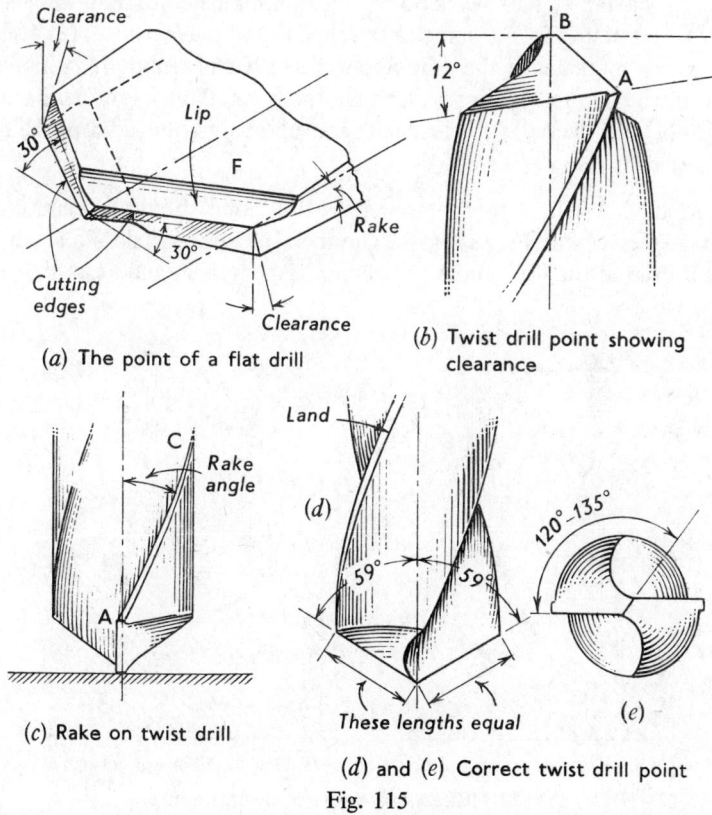

(a) The point of a flat drill

(b) Twist drill point showing clearance

(c) Rake on twist drill

(d) and (e) Correct twist drill point

Fig. 115

(marked F) is left quite plain, the tool has no rake, but if it is ground to a lip as shown, then the rake angle is as indicated. It would cut without rake, but for soft metals the addition of rake improves the action.

A side view of a twist drill is shown at Fig. 115(b), and here, as before, the metal behind the cutting edge AB slopes away. The correct angle for this slope is 12° at the outside diameter of the drill. In Fig. 115(c) a sketch is shown looking level with the cutting edge AB, and it will be seen that the effect of the twist formation of the drill (AC) is to give cutting rake. The rake angle is that indicated, and it is altered by making the 'twist' quick or slow; if the flutes were straight, then there would be no rake; if they were left-handed instead of a right-handed helix the rake would be negative. As well as the cutting clearance on the point of a twist drill, there are two clearances on its body: (a) the body of the drill is made slightly less in diameter, leaving a narrow 'land' of the full diameter running down the front of each flute. This is sometimes called the *body-clearance*; (b) a slight taper on the diameter of the drill, about 1 in 1000 of length, from point to shank, the shank end being smaller than the point. This allows all parts of the drill behind the point to clear and not rub against the sides of the hole being drilled.

Drill grinding. When properly ground the drill point should be central and the lip angles equal (Fig. 115(d)). The metal behind the lip should be as Fig. 115(b) and the line across the centre of the web should be as shown at (e).

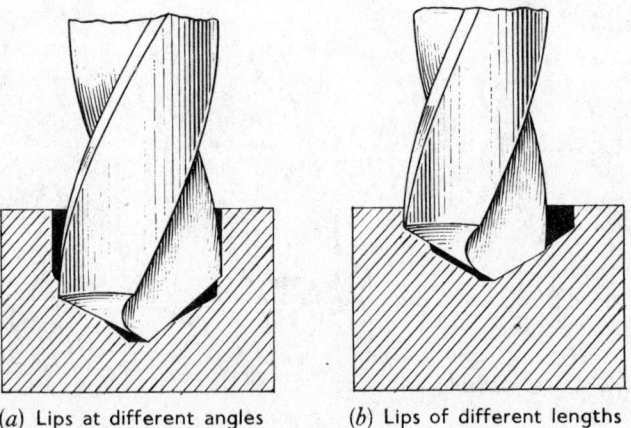

(a) Lips at different angles   (b) Lips of different lengths

Fig. 116 Effects of errors in drill grinding

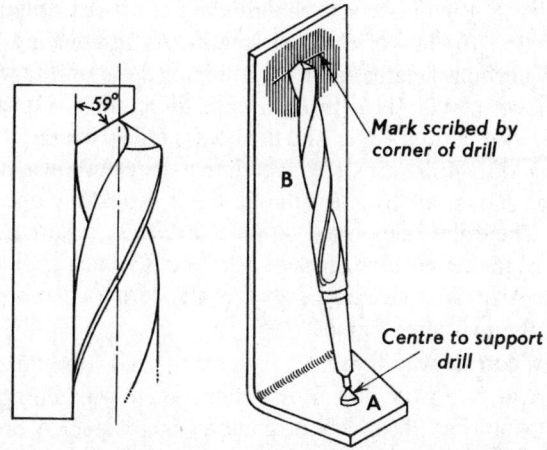

(a) Gauge for lip angle   (b) Gauge for checking drill point corners

Fig. 117 Drill grinding gauges

T. S. Harrison and Sons, Ltd

Fig. 118 Pedestal grinder with enlarged diagram of drill grinding attachment

The errors that may occur in the grinding of a twist drill are: (a) lips at unequal angles, (b) lips of unequal lengths, (c) lips having both unequal angles and unequal lengths, and the effect of these on the working of the drill are shown at Fig. 116. In each case the cutting is unequally shared between the two cutting edges and the hole is drilled oversize. For checking the grinding of a drill point the two simple gauges shown at Fig. 117 are useful. That shown at (a) is for the lip angle, whilst the one at (b) checks the points. The drill is supported on its centre at A, whilst a mark is made with each of the lip corners on the back face B, which should be smeared with chalk. Any difference in the level of the corners can be seen immediately from the two marks made.

The only correct way to obtain the complicated form necessary for the clearance faces of a twist drill is by grinding them with a proper drill grinding attachment (Fig. 118). Unfortunately, many people do not see the wisdom of this, so that grinding attachments are not as common as they

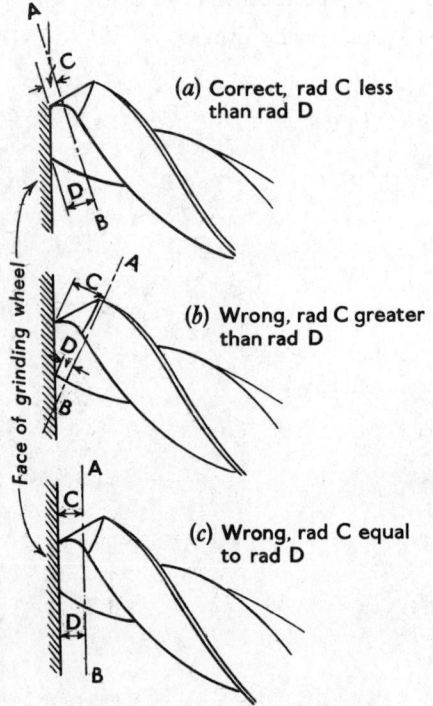

(a) Correct, rad C less than rad D

(b) Wrong, rad C greater than rad D

(c) Wrong, rad C equal to rad D

Face of grinding wheel

AB = *Axis upon which drill is rocked*   Fig. 119 Drill grinding should be, and it is likely that the reader may have to learn how to grind drills by hand. In order to obtain the correct clearance it must be ground in a path similar to that which it follows when being fed into the work, i.e. points nearer the centre travel in shorter paths than those further out. To attain this the drill must be rocked about an axis AB (Fig. 119(a)) such that radius *C* is *less* than *D*. The conditions shown at (b) and (c) are incorrect.

Point thinning. The metal at the centre of a drill (called the web) tapers and gets thicker towards the shank. This causes the centre of the drill point to get thicker as its length is reduced by grinding, and to prevent this thick edge from reducing the efficiency of the drill it should be ground thinner. A diagram showing this is given at Fig. 120.

Miscellaneous drilling hints

(*a*) For soft metals use a drill having a quick twist to its flutes, and vice versa for hard metals. For chilled iron a flat drill gives best results.

(*b*) Cut with soluble oil for steel and malleable iron, kerosene or turpentine for very hard steel. Cast-iron or brass should be drilled dry, or with jet of compressed air.

(*c*) If the corners wear away rapidly, the speed is too high.

(*d*) If cutting edges chip, reduce feed or grind with less clearance.

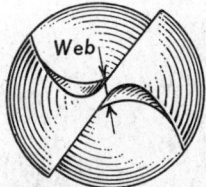

Fig. 120 A point-thinned drill

(*e*) If drill will not start drilling there is no clearance on lips.

(*f*) Examine relative sizes of turnings issuing from each flute. They should be approximately the same, and if not, the drill is wrongly ground with one lip doing more cutting than the other.

(*g*) Drill breakage may be caused by point wrongly ground; feed too great; not easing drill at 'break through'; binding in hole due to lands being worn away; drill choked in a long hole.

(*h*) The blueing of a high-speed steel drill is not detrimental but it is fatal to a carbon steel drill.

(*k*) A hard spot encountered may be removed by reducing speed and using turpentine.

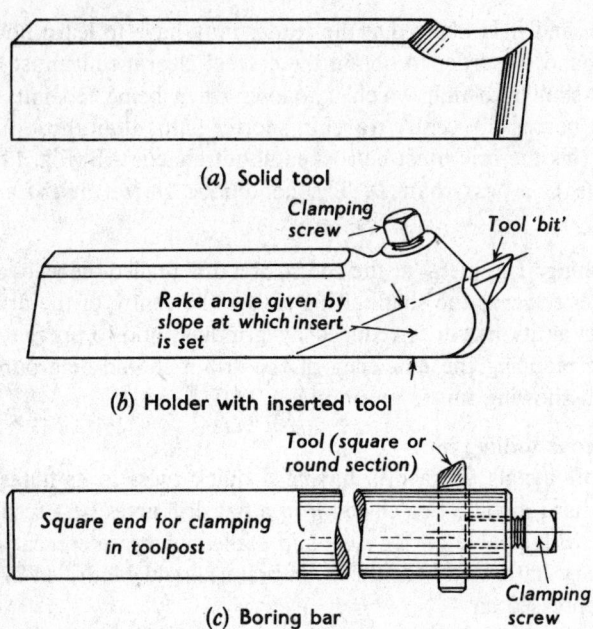

(a) Solid tool

(b) Holder with inserted tool

(c) Boring bar

English Steel Corporation

(d) Ceramic tool holder and insert

Wm. Jessop and J. J. Saville Ltd

(e) Carbide tipped tool

Fig. 121 Single-point tools

(*l*) When drilling deep holes release feed occasionally and withdraw the drill. Remove chips from bottom of hole with an old round file that has been magnetised.

(*m*) When opening out holes only slightly less in diameter than the drill be careful of points digging-in. For large holes use a 3 or 4-flute drill.

Turning, boring, shaping and planing tools

The single-point tools used for these processes may have a solid shank as shown at Fig. 121(a), be made of a tool bit held in a holder as at (b), consist of a small tool carried in a bar as at (c) (chiefly for boring), or consist of a shank to which a tip of the cutting material is brazed or clamped (d) and (e). Before we proceed to the shapes and applications of these tools it will be well for us to examine the disposition of the cutting angles as they may be influenced by the type of cutting being done.

Rake. Let us consider a tool such as that shown at Fig. 122(a) having 15° front rake and 15° side rake. The line of greatest slope on the face of this tool will be the line AE and it will be inclined at an angle of about 21° with the horizontal. If, now, this tool is made to cut by feeding parallel to its length it will cut with a top rake of 15°; if fed perpendicular to its length it will cut with a rop rake of 15°, and if fed at 45° to its length the top rake will be 21°. Then conditions are shown at (b), (c) and (d), and we see that the disposition under which a tool must cut influences the effective values of its rake angles. For example, at (b) the side rake has very little effect on the tool action and could be omitted, whilst at (c) the tool would cut almost the same without any top rake. These considerations are important when a tool such as a side tool is used for more than one pur-

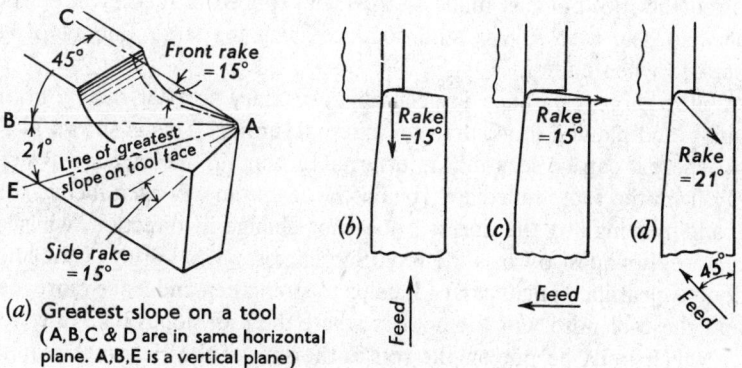

(*a*) Greatest slope on a tool
(A,B,C & D are in same horizontal plane. A,B,E is a vertical plane)

Fig. 122 Effect of cut disposition on rake value

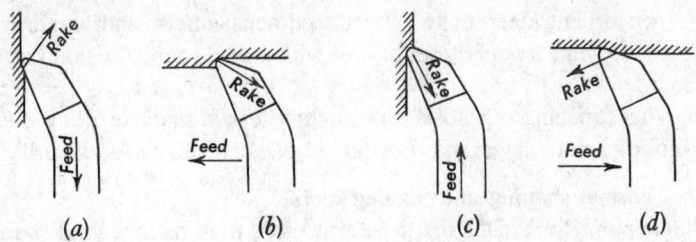

Fig. 123 Approximate direction of rake necessary for different cutting conditions

pose. In Fig. 123 (a), (b), (c) and (d) are shown four possible methods of cutting with the same tool and the best cutting conditions will be obtained when the line of greatest slope on the cutting face is approximately according to the arrow shown. When tool bits are used in holders the bit is generally set with an initial back slope as shown at Fig. 121(b). This gives to a flat-topped tool a front rake equal to the slope angle of the tool and should be taken into account when grinding the tool bit.

Clearance. The remarks we have just made about rake apply in general terms to clearance. The tool must be provided with sufficient clearance *relative to the surface being machined*, and if this is not achieved the tool will be prevented from a free cutting action by its front rubbing against the metal. In Fig. 123 the clearance for the four examples shown must have its general direction inclined towards the arrow, and clearances should be ground in planes parallel and perpendicular to the surface being machined. The amount of clearance should be no more than is necessary to permit the tool to cut cleanly, and 5° to 10° is usually sufficient. If more than this is allowed the tool point is made sharper and robbed of metal which would otherwise help it to survive and conduct away the large amount of heat generated when cutting.

A little more clearance is generally necessary for flat facing, boring, shaping and planing tools, than for external turning. This is shown at Fig. 124, where it can be seen that in external turning (a) the surface AB slopes away from the tool immediately below the centre line. For flat facing, shaping and planing (b) the surface does not change in direction, whilst for boring (c) it slopes towards the tool. Sometimes, when boring a small hole, a double clearance angle as at (d) helps to strengthen and leave more metal round the tool nose. On toolholders where the tool slopes back the clearance which must be put on the tool is the sum of the clearance required, and the slope angle of the tool (see Fig. 121(b)).

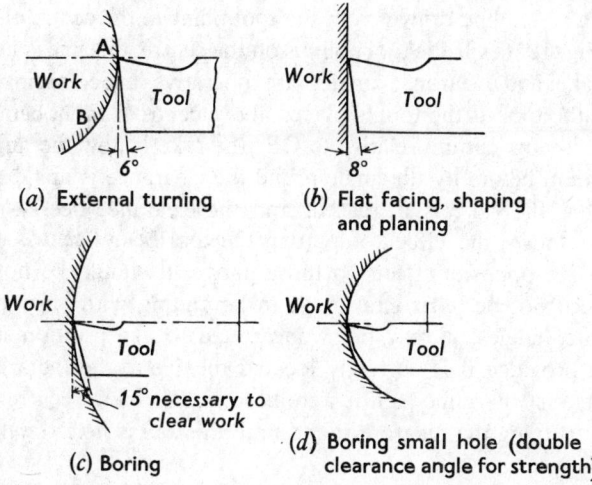

Fig. 124 Effect of cutting conditions on tool clearance

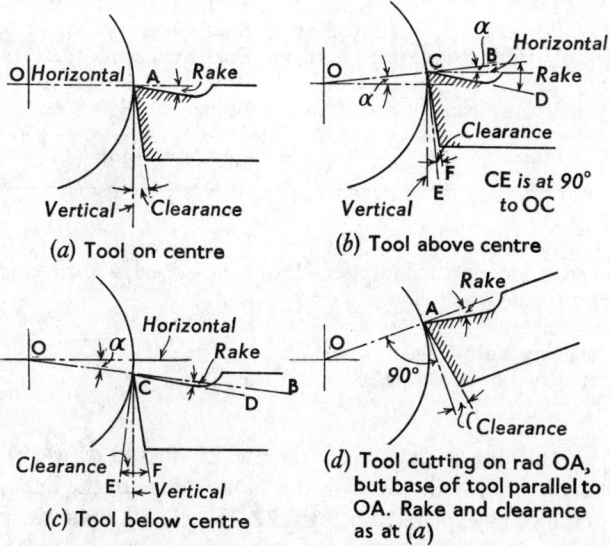

Fig. 125 Effect of tool setting on angles

Turning off centre. When a bar is being turned the cutting takes place relative to a radial line drawn from the tool point to the centre of the work as OA in Fig. 125(a). If the tool point is on the centre, this line is horizontal, and the rake and clearance angles are operative under their values as ground on the tool. If the tool is placed above centre without being rotated as at (b), it is now cutting relative to OB; the rake is now the angle BCD, or greater than before by the angle α and the clearance is angle ECF. We thus see that the rake is increased, and the clearance decreased, by α. Fig. 125(c) shows the effect of putting the tool below centre where the reader will see opposite effects to those above. It should be noticed that the tool need not be horizontal for it to be cutting with its normal rake and clearance angles. It may be swung round to any position such as at (d), where, provided it is correctly located relative to the radial line to its nose, it cuts just the same as if it were horizontally on the centre. This can be seen by turning the diagram round until the tool is horizontal.

Table 14. Rake angles for cutting various metals

Angle given is that of the true rake on the tool (see Fig. 126)

Metal being Cut	Hard Brass, Bronze and Cast-iron	Hard Steel, Medium Cast Iron, Brass and Bronze	Medium Steel, Soft Cast Iron, Brass and Bronze	Mild and Soft Steel	Aluminium and light Alloys
Rake	0°	8°	14°	20° to 27°	40°

CLEARANCE

This should be no more than is necessary to allow the tool to cut efficiently and may be approximately as follows:

External turning, 6° to 10°.

Facing, shaping and planing, 8° to 17°.

Boring, sufficient to allow heel of tool to clear.

Table 14 will give an idea of the rake which should be used for cutting various metals. The angle given in the table is that of the line of greatest slope on the tool as explained in Fig. 122. Depending upon the conditions of cutting (see Fig. 126), this angle may be obtained by a combination of front and side rake (for ordinary surfacing, e.g. Fig. 122(a); by front rake only, e.g. for a parting tool; or by side rake only, e.g. for a knife tool).

Plan profiles of tools

The top profiles of the more usual tools are shown at Fig. 126, and an indication of their uses is as follows:

(*a*) Straight Rougher: General surfacing work.

(*b*) Side Tool: Surfacing up to a corner facing down a side.

(*c*) Knife Tool: Facing ends of bars in lathe. Finish facing generally.

(*d*) { Parting Tool: Parting off in lathe, cutting grooves.
Cranked Parting Tool: Parting off near chuck; grooving near another face.

(*e*) Broad Finishing Tool: (Planer and shaper only.) Surface finishing on cast iron with broad feed.

(*f*) Boring Tool: Boring in the lathe.

The tools for lathes, planers, etc., should have a good solid shank and should be well supported by not protruding from the toolholder any more than is necessary.

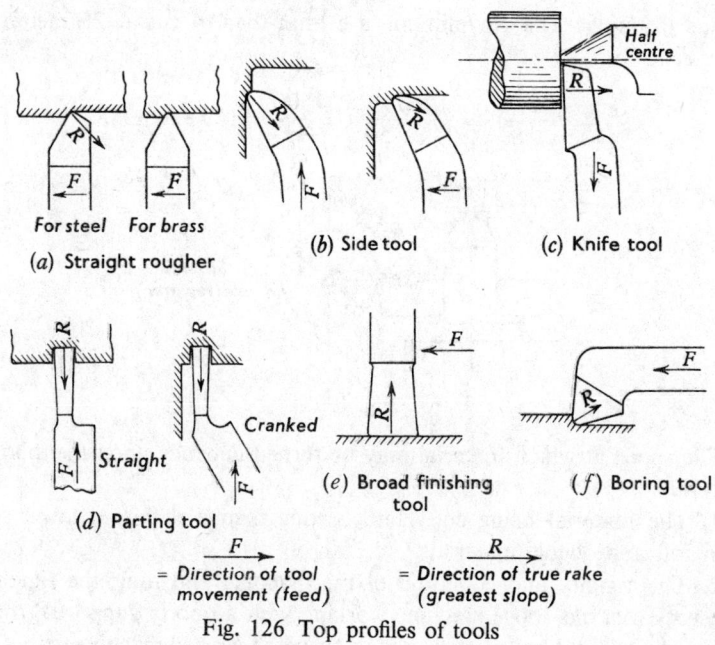

Fig. 126 Top profiles of tools

Cutting speed

The cutting speed for a cutting operation is the speed at which the cutting edge passes over the material, and it is usually expressed in *metres per minute*.

Cutting speed for turning. For turning, the cutting speed is the surface speed of the bar being turned. To find it we must multiply the circumference of the bar (in metres) by the speed in rev/min. Thus if d = diameter of work (mm) and N = speed (rev/min), Circumference (metres) = $\dfrac{\pi d}{1000}$,

Cutting speed (metres per minute) $S = \dfrac{\pi d N}{1000}$. (See Fig. 127)

We often require to find the rev/min when we are given the cutting speed (S) and the work diameter (d), and by changing the above formula we get

$$N = \frac{1000S}{\pi d}$$

For example: The rev/min for a 50-mm bar to cut at 25 metres per minute,

$$N = \frac{1000S}{\pi d} = \frac{1000 \times 25}{\pi \times 50} = \frac{500}{\pi} = 159 \text{ rev/min.}$$

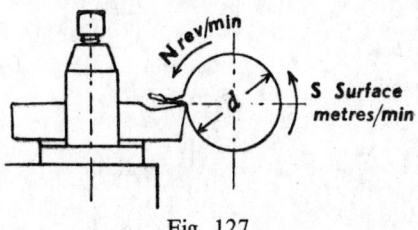

Fig. 127

The speed at which material may be turned depends upon the following factors:

1. The material being cut. Hard, strong metals require a lower speed than soft and ductile materials.

2. The rigidity and condition of the machine and tool; the rigidity of the work. An old, loose machine working with a poorly supported tool on a thin bar will not cut at such a high speed as a good machine with a rigid tool operating on a well-supported bar of reasonable dimensions.

3. The material of which the tool is made. Special cutting alloy and high-speed tools will cut at much higher speeds than low tungsten or carbon steel tools.

4. The depth of cut and the feed. A light finishing cut with a fine feed may be run at a higher speed than a heavy roughing cut.

The following speeds may be taken as a guide for the initial setting of the machine.

Table 15. Cutting speeds for turning
(The speeds given are for average cuts with high speed steel tools)

Material being Cut	Cutting Speed (metres per minute)
Mild steel	20 to 28
Cast iron	18 to 25
High carbon steel . . .	12 to 18
Brass	45 to 90
Bronze.	15 to 21
Aluminium	Up to 300

(It may be possible to exceed these speeds for light finishing cuts. For heavy roughing cuts they should be reduced.) One formula for cutting speed in which the cut and feed are taken into account is as follows:

$$\text{Cutting Speed } (S) \text{ metres per minute} = \frac{C}{\sqrt[3]{A}}$$

Where $A =$ Area of cut = feed $\times$ depth of cut
$C =$ a constant

Values of C:

Material being Cut	Soft Steel	Hard Steel	Cast-iron	Bronze	Brass
C	24	13	18	21	48

Example: To find the cutting speed for mild steel working on a cut of 2 mm with a feed of 0·75 mm, using the above formula and taking $C = 24$

We have $S = \dfrac{C}{\sqrt[3]{A}}$. Where $C = 24$ and $A = 2 \times 0 \cdot 75 = 1 \cdot 5$.

Hence
$$S = \frac{24}{\sqrt[3]{1 \cdot 5}} = \frac{24}{1 \cdot 145} = 21 \text{ metre/min}$$

Cutting speed for shaping and planing

In turning, the material passes over the tool at a uniform rate, but in shaping and planing, for each double stroke of the machine the following takes place:

Cutting Stroke. The tool starts from rest, attains a maximum cutting speed and slows down to rest again.

Return Stroke. The same as for cutting except in a shorter total time.

This renders the estimation of a machine speed rather difficult, but the following method may be used as a guide in setting the machine:

Divide twice the length of the stroke (in metres) into the lowest speed given in Table 15. This will give the number of double strokes per minute.

Thus, for planing cast-iron with a table travel of 500 mm, the number of double table strokes per minute will be:

$$\frac{18}{2 \times \frac{1}{2}} = 18 \text{ double strokes of the table per minute.}$$

Cutting speed for drilling. The formula given on page 184 for the cutting speed in turning will apply to drilling if d is made to represent the drill diameter instead of the work diameter.

Thus for a 14 mm drill to cut at 15 metres per minute

$$\text{Speed of drill (rev/min)} = \frac{1000S}{\pi d} = \frac{1000 \times 15}{\frac{22}{7} \times 14} = \frac{1000 \times 105}{22 \times 14}$$
$$= 381 \text{ rev/min}$$

For drilling speeds use about $\frac{2}{3}$ of those given in Table 15.

Cutting speed for hacksawing. The conditions under which a hack-saw operates are much the same as for shaping and planing except that the saw teeth are much more delicate than a solid tool. Suitable speeds for sawing operations are as follows:

Table 16. Sawing speeds

Operation	Light Cutting Dry		Medium Cutting with Cutting Lubricant	
Metal being cut	Hard steel	Mild steel Soft metals	Hard steel	Mild steel Soft metals
Double strokes per minute	40 to 50	50 to 60	60 to 80	100 to 110

Beginners at hand-sawing nearly always operate the saw much too fast. This results in undue wear of blades.

Cutting speed for filing. The file, wielded by human energy, cannot be fitted in to hard-and-fast rules of cutting speeds. There are so many variable factors to be taken into account, such as material being filed, type of work, rate at which material is being removed, strength and endurance of the worker, and so on. It has been found that the range of speeds over which experienced users of the file operate varies from about 45 to 85 strokes per minute, with an average of about 65 strokes per minute.

Cutting feeds

A metal-cutting operation consists of causing the tool to travel relative to the work and feeding it in such a way that it cuts the metal. In shaping, for example, the tool reciprocates backwards and forwards, and just before each cutting stroke the work is fed a small distance across. We have already discussed the speed at which the work must travel past the tool and must now study the rate at which the tool is fed.

Turning. In turning, the tool is fed along the bar for turning a cylinder, and across for a flat face. This gives the effect of removing the metal by turning a very fine thread as shown in Fig. 128. The *feed* of the tool is the distance it moves along for each revolution of the work (e.g. $\frac{3}{4}$ mm per rev as shown). The feed that should be used for turning or shaping depends on the following:

(*a*) The smoothness of the finish required. A coarse feed will give wider and deeper machining marks, and an inferior finish to a fine feed. The

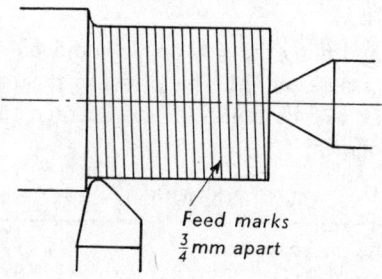

Feed marks
$\frac{3}{4}$ mm apart

Fig. 128 Turning with a feed of $\frac{3}{4}$ mm

shape of the tool enters into this as well because a blunt-nosed tool will give a better finish than a sharp tool for the same feed.

(b) The power available, the condition of the machine and its drive. The product of the speed, feed and depth of cut gives us the amount of metal that is being removed, and hence the power necessary. A coarse feed on a poor or badly driven machine may be too much for the machine or tool, or cause the drive or belt to slip.

These two considerations must guide the reader in his choice of a feed. As a general rule, when roughing a job down, give the coarsest feed that the machine will stand because finish is unimportant. When finishing adjust the feed so that it is fine enough to give the class of finish required.

Shaping and planing. The remarks we have just made for turning apply in a general way to these two methods of machining.

Drilling. The feed for a drill is expressed in millimetres per revolution of the drill (e.g. 0·10 mm per revolution, or ten hundredths per revolution). For a drill, the feed depends on the drill for the small sizes, and upon the machine for large drills. To make this a little clearer the reader will appreciate that if we overfeed a 6 mm drill it will break, but if we overfeed a 35-mm drill we may either break the machine, or cause the belt to slip off; it is unlikely that we shall break the drill. The dividing line between these two cases occurs for about a 20-mm drill; below that the drill is the weaker member and, above it, the machine. The following table gives a guide as to twist drill feeds:

Table 17. Twist drill feeds

Diameter of Drill (mm)	3	6	8	10	11	13	16	19	22 to 26	32
Feed millimetre per revolution ($\frac{1}{100}$ mm)	7	12	15	18	20	23	25	28	30	37

Sawing. In sawing and filing the saw or file is moved over the work and is fed into it by the pressure put on. This pressure, then, really determines the rate at which metal will be removed and average suitable pressures are approximately as follows:

SAWING

Blade thickness (mm) . .	$\frac{3}{4}$	1	$1\frac{1}{2}$
Pressure (kgf)	5·5	9	30

Filing. As we have stated, filing is such an individual and variable operation that we cannot state hard-and-fast rules, but can only give figures, more for the reader's interest than anything else.

Filing involves a vertical pressure on the file and a thrust to cause it to move over the metal. These will not only vary according to the file, and the material being filed, but also according to the physique and habits of the workman. For example, a strong man will exert more force than a weaker one, and a filer who habitually files large articles will press harder than one who is always working on thin, fragile parts. Another factor in filing is that the pressure varies throughout the stroke, being greatest just over half-way. Tests on six experienced filers showed that the maximum pressure they exerted (sum of both hands) varied from 135 N to 220 N, and the average pressure from 95 N to 170 N. The thrust required to push the file forward will itself depend on the pressure, on the material, file condition, etc. For example, in a series of tests with new files, the average values of the ratio: $\dfrac{\text{Forward thrust}}{\text{Vertical pressure}}$ were found to be as follows:

Metal being filed	Steel	Cast-iron	Copper	Brass
Average of thrust pressure . .	1·3	1	1·6	1

The lubrication of cutting tools

The machining of most metals is greatly improved by the use of a cutting lubricant. Cast iron should always be machined dry; brass, bronze and aluminium may be cut dry or wet, whilst steel should always be machined with a lubricant. The improvements effected by using a lubricant are as follows:

(*a*) The tool and work are cooled and higher cutting speeds may be used.

(*b*) The cutting fluid helps in lubricating the severe rubbing action taking place between the chip and the top face of the tool. This effects some saving in power, prolongs the life of the tool and promotes a better finish. (many readers are probably aware what an improvement a drop of whale oil will effect to the finish of a thread.)

(*c*) A heavy flow of lubricant helps to wash away the chips and keep the cutting point clear.

Cutting lubricants may consist of a pure oil, a mixture of two or more oils—or a mixture of oil and water. Oils are generally divided into what are termed the 'fixed' oils, and the mineral oils. When these are compared it is found that the fixed oils have a greater 'oiliness' than the mineral oils, but they are not so stable and tend to become gummy and decompose when heated. The 'fixed' oils consist of animal, fish, and vegetable oils, and the chief of these used for cutting are lard oil (animal), sperm and whale oil (fish), olive, cottonseed and linseed oil (vegetable). Turpentine, which is distilled from vegetable oil, is also used. The mineral oils come from crude petroleum oil pumped at the oilfields, and paraffin is a cutting lubricant belonging to this class.

In order to combine the advantages of the stability of the mineral oils with the good lubricating qualities of the fixed oils they are often mixed, and sometimes sulphur is added to give the property of 'wetting' the metal with a highly adhesive oil film. Such oils are known as compounded oils and sulphonated oils.

The most common type of lubricant used for cutting is a soluble oil, which, when mixed with water, forms a white solution known as 'suds' or 'slurry'. This has better cooling properties than oil, but does not lubricate as well. The oil part of it is generally a mineral oil mixed with a soap solution, sometimes with an EP agent added. There are many forms of soluble oil on the market and the supplier's instructions should be followed as regards the proportions of oil and water required.

Oils with an extreme pressure (EP) additive

The arduous cutting conditions imposed by heat-resistant and stainless steels, nimonic alloys, etc., have led to the introduction of EP additives to cutting lubricants. These are based on chlorine and sulphur and under extreme conditions of temperature and pressure form solid films of iron chloride or iron sulphide between the chip and the tool. These films are easily sheared and have high melting point so that they assist in preventing portions of the chip welding themselves to the nose of the tool. (The reader will no doubt be aware of the use of EP oils for coping with the severe conditions imposed on the crown wheel and pinion in a car rear axle.)

Application of cutting lubricant. The cutting lubricant may be applied to the tool or cutter by hand (using a brush), by means of a small drip tank attached to the machine or by a pump which is driven by the machine and pumps the fluid from a drain sump. For heavy and continuous cutting the pump system is the only one which can be said to effect proper cooling of the tool. Some attention has recently been given to forcing the lubricant upward between the tool and the work for turning operations and it is claimed that this method is more efficient than feeding from above. Generally an overhead flow is used as well to reduce fumes.

Surface finish

Within recent years a great deal of attention has been given to the investigation of the surface texture produced by various machining and finishing operations, and, although we have not yet reached the stage when surface finish requirements can be specified and produced in the same way that is possible for a given dimension, or fit, information is being built up which may eventually bring this about.

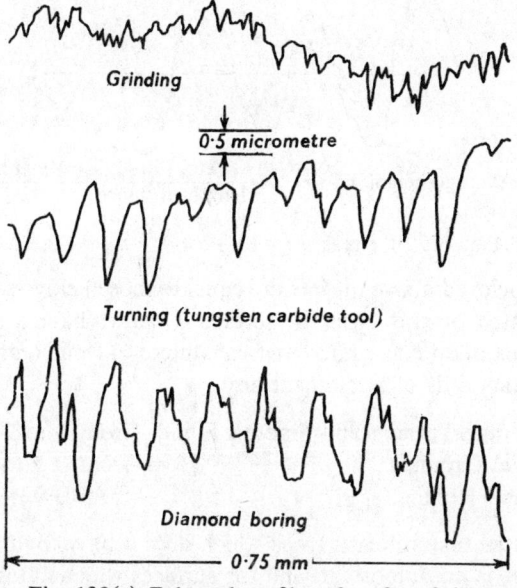

Fig. 129(a) Enlarged profiles of surface finishes

If a machined surface is reproduced to a large enough magnification it will be found to have a wavy profile, the character and shape of the waves varying for different machining methods. Equipment is now available which will produce an enlarged contour of a surface by traversing a needle-like probe over it, the instrument meanwhile plotting on a chart a greatly magnified reproduction of the up-and-down movement of the probe with its longitudinal traverse. Examples of such profiles are shown at Fig. 129(a), but in studying them the reader should allow for the fact that the vertical magnification is about 8000 whilst the horizontal enlargement is only about 100, so that in reality the spiky peaks and hollows are quite gentle curves. (If the reader is in doubt as to the reason for such differing scales of magnification he should estimate how long the diagrams would be if the horizontal enlargement were the same as the vertical.)

Measure and influence of finish

The present method of expressing a measure of surface finish is to find the average height of the undulations of its contour with reference to a horizontal straight line drawn through the contour in such a position that the

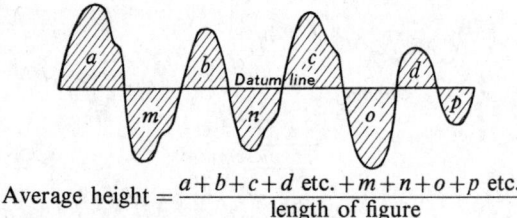

$$\text{Average height} = \frac{a+b+c+d \text{ etc.} + m+n+o+p \text{ etc.}}{\text{length of figure}}$$

Fig. 129(b) Areas $a+b+c+d = m+n+o+p$ etc.

loop areas enclosed above the line are equal to those below, see Fig. 129(b). When expressed on this basis the surface is said to have a Ra (roughness average) Index of so many micrometres* (micro = 1 millionth). In practice, reasonable standards of attainment are:

Good Average Turning and Smooth Filing—from 1 to 2 Ra
Commercial Grinding   —from 0·5 to 1 Ra
Lapping and Honing   —from 0·025 to 0·2 Ra

In service the texture of surfaces which do not have to fit and mate with other surfaces is not important, and for surfaces which fit together without any relative movement a reasonably smooth finish is good enough to provide the support and alignment necessary. When, however, two surfaces both fit and *run* together their finish is of the utmost importance, since

---

* The reader might prefer to think of this in terms of millimetres, i.e. $\frac{1}{1000}$ mm.

where there are 'hills' and 'hollows' the first thing that happens in the early periods of their rubbing together is that the 'hills' wear down.

When these have become levelled off the surfaces settle down to a comparatively long period of contact without much further wear taking place. Let us suppose that the hills are $\frac{1}{400}$ mm high on a shaft and in the bearing to which it has been fitted. The initial wear which takes place rapidly removes these, resulting in the shaft wearing to a diameter $\frac{1}{200}$ mm less, and the hole $\frac{1}{200}$ mm more in size, a total difference of $\frac{1}{100}$ mm. If the reader does not realise what this means let him bore a hole about 25 to 30 mm diameter and turn a plug to be a nice smooth fit in it (or bore a hole to a standard plug gauge). Now make another plug about $\frac{1}{100}$ mm less in diameter and try it in the hole.

It is this problem of the initial wear taking off the hills and hollows which makes workshop engineers so particular in obtaining the smoothest possible finish on motor and aeroplane cylinders and pistons, on high speed bearings, and so on.

7 The checking and measurement of surfaces

The actual removal of metal is not in itself a very difficult task, but the initial setting for it, and the judgment of the correct point at which to stop it, often call for high skill. If we examine a number of finished articles chosen at random, we shall find that the majority of the finished surfaces on them are either flat or cylindrical in form as shown in Fig. 130. Now the drawings of these parts would give us all the information we required regarding the shape and sizes of the various faces of the components. There are many other things, however, which the drawing would not tell us, but which we should be expected to produce correctly. For example, it would not tell us that the centre hole in the disc had to be exactly on the same centre as the outside diameter, nor that the edge of the disc had to be square with its face. In the bracket on the right, it would be taken for granted that we should bore the large hole with its centre line parallel to the base, that the centre lines of the two small lug holes were co-axial (i.e. on one and the same centre line), and perpendicular with the sides of the slot, and so on. The aim of the workshop engineer, then, should be to give careful thought to each job and try to fulfil the obligations which have been left unwritten on the drawing, but which are as important as the most accurate size relationship.

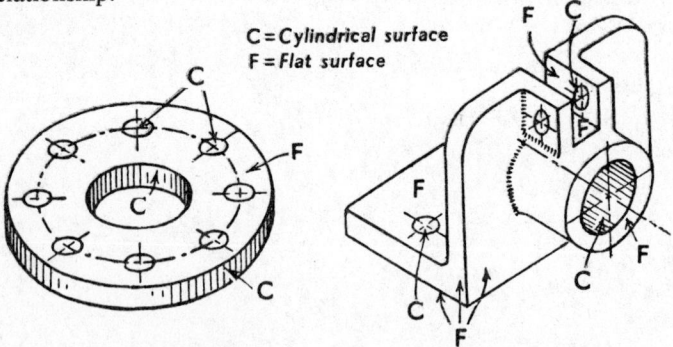

C = Cylindrical surface
F = Flat surface

F = Flat surface.　　　C = Cylindrical surface.

Fig. 130 Flat and cylindrical surfaces on components

Even though we may infer from a drawing that two faces have to be square, or perhaps parallel, it does not follow that we must always spend time in getting the relationship accurate to a hundredth part of a millimetre. It may be that in some cases the job will be satisfactory if within half a millimetre, and the reader will be justified in asking, 'How am I to know which parts must be very accurate and which not so accurate?' In reply to this we can only say that, in time, his experience, study and instincts will be his guide, and he should work to acquire the knowledge in the shortest possible time. His ruling principle always should be to so arrange the production of a job that, when finished, it will be *right*. This involves qualities of mind and character which we have already discussed and differentiates the skilled artist from the 'just ordinary' worker.

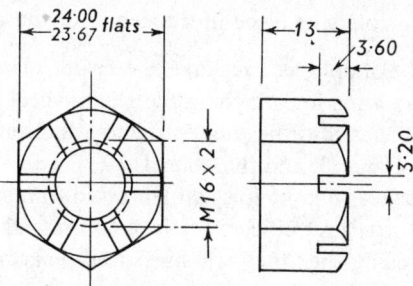

Fig. 131 Slotted hexagon nut

In the light of the above remarks, we will discuss the unwritten accuracy relationships on two simple parts. At Fig. 131 is shown a slotted hexagon nut just as it would appear on the drawing. The outside dimension for the flats of the nut is given as limits with a tolerance of 0·33 mm so its interpretation is clear, but it would be less serious for the flats to be a little below the minimum size than over the maximum, for in the latter case they might be too large for the spanner to fit over them. The three sets of flats must of course be uniform, although only one set is dimensioned. Nothing is said about the angle between the flats, but we know, if the nut is truly hexagonal, this must be 120°. The threaded hole in the nut, stated as M16 × 2, is beyond our control in its dimensions since it would be finished with a tap, but the position of this hole is very much under our control. In the first place, it must be in the centre of the nut; secondly, the hole should be square with the bottom face of the nut. If it is not so, when the nut is tightened down it will not sit flat on the mating surface and the bolt will be bent in the tightening process. The bottom face should be flat (or very slightly saucer-shaped) and, as just stated, square with the thread; the top face is unimportant since it does nothing, and the nut thickness is relatively unimportant. The slots across the top of the nut which accommodate the split pin will have been dimensioned with sufficient clearance for the pin, so that extreme accuracy is not necessary for their width or depth, but if they do not pass through the centre of the nut it may not possible to get the split pin through.

We can thus summarise our discussion as follows:

1. The flats should be uniform and should not be wider than the maximum.
2. The bottom face of the nut should be square with the thread, and should be flat, or slightly saucer-shaped, but not rounded (why?).
3. The split pin slots should be in the centre of the nut.

As our second example we will take the crank of a bicycle, choosing the type which has arms for attaching the chain wheel (Fig. 132). For the purposes of its application, the only surfaces which need machining are those marked A, B and C, and the holes D, E, F and G. Provided the remainder of the surfaces can be forged or pressed to a good smooth finish, and can be made to sizes sufficiently close to bring the crank within the limits set for its weight, then they will only need polishing so as to provide a suitable foundation for the plated or enamelled finish. If the primary forging process is not able to give a sufficiently good finish, it may be necessary to machine these surfaces, but for the purpose of our discussion we will ignore them.

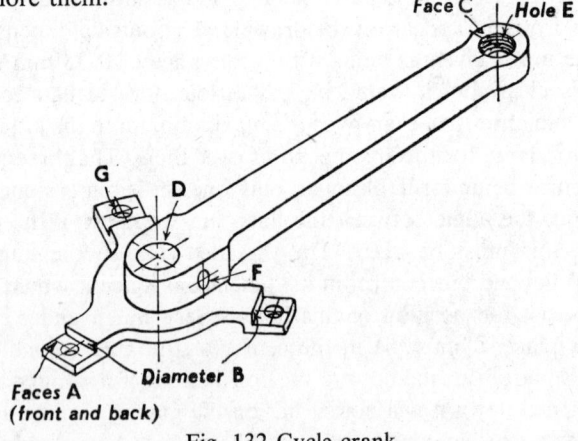

Fig. 132 Cycle crank

When the chain wheel is assembled on the crank, the screws pull it up against the front faces A, and a hole in the wheel fits on the diameter formed by the three portions B. This gives us conditions for the accuracy requirements of A, B and G.

(*i*) The diameter of which the portions B form a part must be held to close limits so that any chain wheel will fit any crank.

(*ii*) The above diameter must be true with hole D; if it is not, the chain wheel will not run true.

(*iii*) Front faces A must run true when the crank is rotated on the hole D. If they are not true, then the chain wheel will wobble out of truth sideways.

(*iv*) When the chain wheel is assembled, the holes G must correspond with similar holes in the chain wheel; otherwise the screws will not enter. These holes should also be a good fit for the screws which pass through them. The back face of A must be cleaned up to provide a seating for the screw heads.

Now for holes D and E.

(*i*) The diameter of D must be to close limits to ensure a good fit on the axle of the bicycle.

(*ii*) Hole E must be screwed so as to be a good firm fit for the pedal when it is screwed in.

(*iii*) The distance between the centres of D and E is not important to $\frac{1}{2}$ mm, as no rider would notice such a small variation. He would notice, however, if D and E were not in the centre of the circle of metal containing them.

(*iv*) The centre lines of D and E must be parallel in all directions. If they are not so, then the pedal will appear as though it has been knocked on one side. The face C, against which the pedal spindle screws, should be faced up true with hole E.

The relation between the cotter pin hole F and hole D is important. Its distance from the centre of D must be such that the usual form of cotter pin enters for the correct distance, locking up the crank. Its distance from the face of the crank fixes how far the crank goes on to the axle and both these positions should be held to less than $\frac{1}{2}$ mm. The diameter of the cotter pin hole must allow a standard cotter pin to a nice sliding fit.

It will be seen that simple though our two examples have been, they contain many problems that must be solved satisfactorily if production is to be successful, and every piece of work that is attempted deserves similar consideration, however simple it may appear.

Tests for the form and relationships of surfaces

We now realise that in checking a piece of work for accuracy, we have not only to verify its dimensional correctness, but also the accuracy and relationships of its surfaces. We will discuss surfaces first.

Flatness

A flat surface is one of the fundamentals of workshop engineering, and although most of the surfaces we produce are flat enough for their purpose, most of them would be far from the precision engineer's standard of flatness.

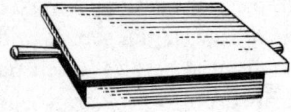

Fig. 133 A surface plate

The methods which are most convenient for verifying flatness are either to test the surface against another surface which is known to be flat or testing with a straightedge. A *surface plate** (Fig. 133) has a surface of proved flatness, and when testing, the top of the plate should first be rubbed with a thin smear of a paste made up of red lead and oil. The face to be tested should be wiped clean, and then placed in contact with the surface plate and moved about. If it is reasonably flat, upon examination after this treatment, spots of the red lead will be visible all over it. Another method using the surface plate is to rest the face to be tested on the plate with a single thickness of cigarette paper under each corner, and if necessary, at other points as well. Pull at each of the papers and if they are all tight the surface is resting on all, and may be assumed flat. This method fails, however, on a surface (e.g. a round one) which is concave, because although the edges would grip the cigarette papers, the centre would not be in contact with the surface plate. If the work is larger than the surface plate, then the plate may be rested on the work, instead of the work on the plate.

A *straightedge*† may be in the form of a steel strip (Fig. 134(a)), or as a stiff casting with the edge straight, as at (b). For lengths greater than 300 mm the ribbed cast-iron pattern is to be recommended. The surface to be tested should be compared with the straightedge in several directions. Red lead or cigarette papers spaced along it can be used with the cast-iron

---

* See BS 817.　　　　　　　　　　　　　　　　　† BS 818 and BS 852.

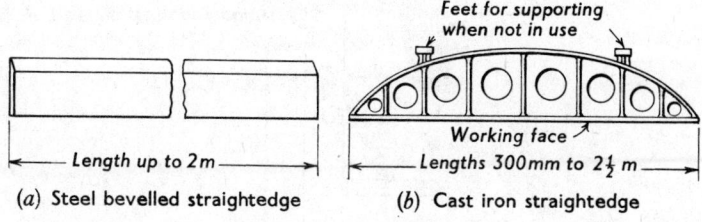

(a) Steel bevelled straightedge   (b) Cast iron straightedge

Fig. 134 Straightedges

straightedge which has an edge of appreciable width, but the steel pattern, which has a knife-edge, must be used either with a number of cigarette papers, or by the appearance of 'daylight'. If an edge is placed against a surface and then held up to the light, any small discrepancy can be detected by the appearance of the light between the two, and our eyes are so sensitive that a gap of $\frac{1}{200}$ of a millimetre can be seen. If we can guarantee our straightedge, this method then gives us a good test for flatness.

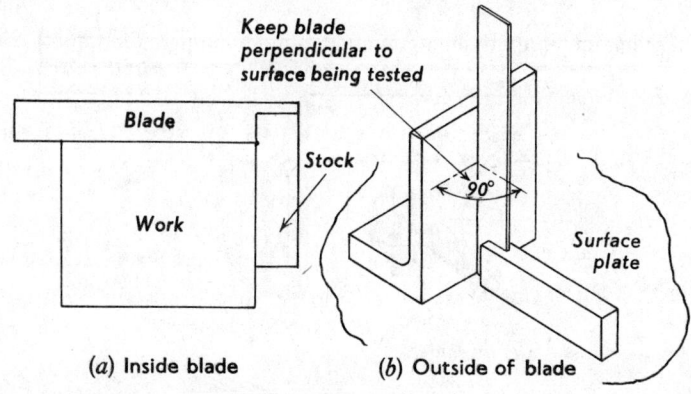

(a) Inside blade   (b) Outside of blade

Fig. 135 Using the try square

Squareness

The *try square** is the most common tool for testing squareness, and diagrams of the use of this for internal and external testing are shown at Fig. 135. When using the square care should be taken to ensure that its blade is held perpendicular to the surface being tested or errors may occur

* BS 939.

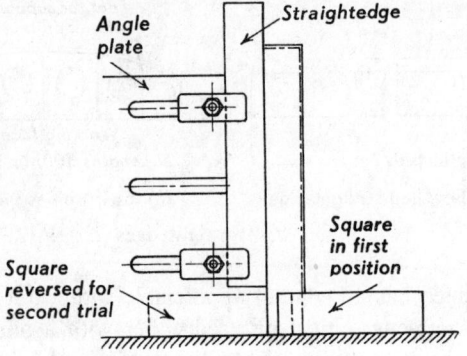

Fig. 136 Testing a square

(see Fig. 135(b)). A good square is valuable, and should be treated with as much care as a watch, but as it is a delusion to put one's confidence in a square which is not accurate, its accuracy should be tested occasionally. One method of doing this is shown at Fig. 136, where a straightedge is

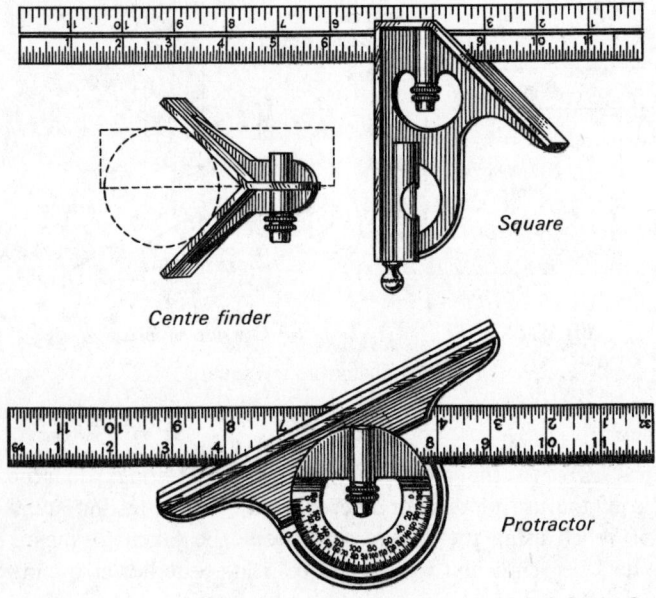

Centre finder

Square

Protractor

L. S. Starrett Co

Fig. 137 Combination square

clamped to an angle plate, and set so as to correspond accurately with the *outside* of the square blade. The square is reversed, and if it is correct, the *inside* will also check with the straightedge. This assumes, of course, that the blade of the square is parallel; if it is not, the square is useless anyhow.

The *combination square* consists of a blade which may be used in conjunction with any one of three heads. The various heads enable the tool to be used as a square, a protractor or a gauge for marking lines passing through the centre of round bars (Fig. 137). The complete set forms a very useful workshop accessory, and fulfils many needs, but when a high degree of accuracy is required for squareness, we recommend that an accurate solid try square be used in preference to the square of the combination set.

Parallelism

Depending upon the character of the work, it will be necessary to employ a variety of methods for testing the parallelism of two surfaces. If the dimensions of the part are small enough to be within the capacity of calipers, or of a micrometer, its parallelism may be tested by one of these instruments. For larger work, and for testing the alignment of holes with

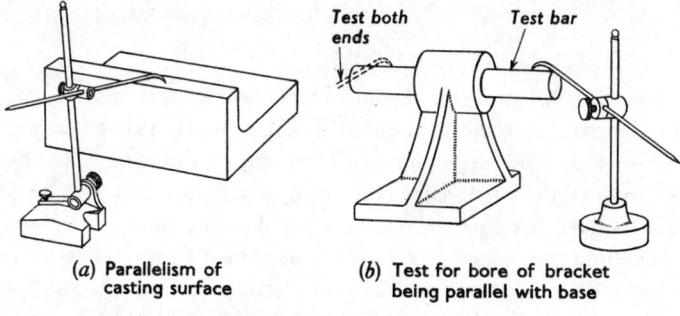

(a) Parallelism of casting surface   (b) Test for bore of bracket being parallel with base

Fig. 138 Test with scribing block

flat faces, the *surface gauge* or *scribing block* (Fig. 138) used in connection with a surface table, is useful. The scriber of the surface gauge has one end bent at right-angles and this may be used as shown at (a). When a hole is being checked as shown at (b), the longer the bar that is put through the hole, the more accurate will be the test, but the bar itself must be parallel. Fig. 139 shows how the ends of a motor engine connecting rod may be checked for parallelism in both directions. Before this test is made, the Vee blocks supporting the bottom bar should be checked by trying the scriber along the top of the bar supported in them.

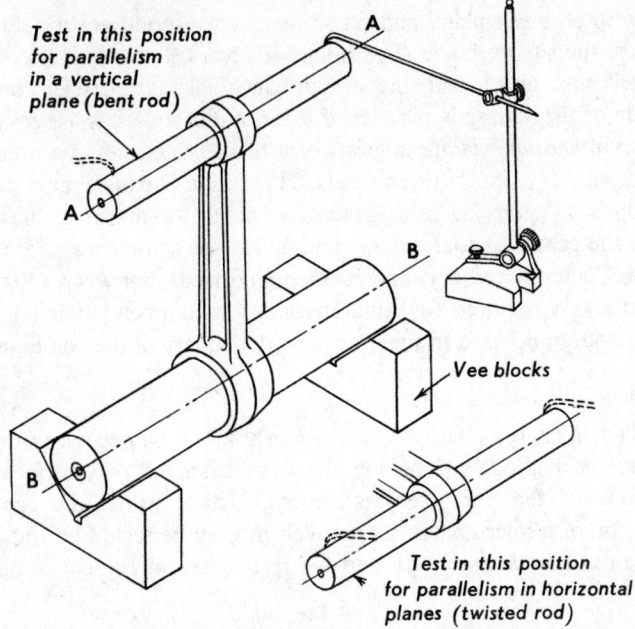

Fig. 139 Alignment tests on a connecting rod

To test parallelism (and distance) between a step and an edge, the combination square may be used as in Fig. 140(a). The blade should be adjusted so that when it is slid over one end of the step, with the stock pressed against the other edge, the edge of the step can just be felt. By repeating at the other end of the step, a check on the parallelism is obtained. By this method the square is also being used as a depth gauge to measure the distance between the two faces. Yet another method which is useful for setting up work on machine tables is shown in Fig. 140(b). This makes use of the little pegs found in the base of a surface gauge which, if pushed through so as to project beyond its base, enable the gauge to move parallel to the edge of the table.

The Dial Gauge.* Two drawbacks to the use of the scribing block for tests of parallelism are: (1) the accuracy depends upon the sensitiveness of our 'feel' with the bent end of the scriber on the work. (2) If the heights differ at each of the faces being tested, our test does not give an accurate measure of the difference.

* See BS 907.

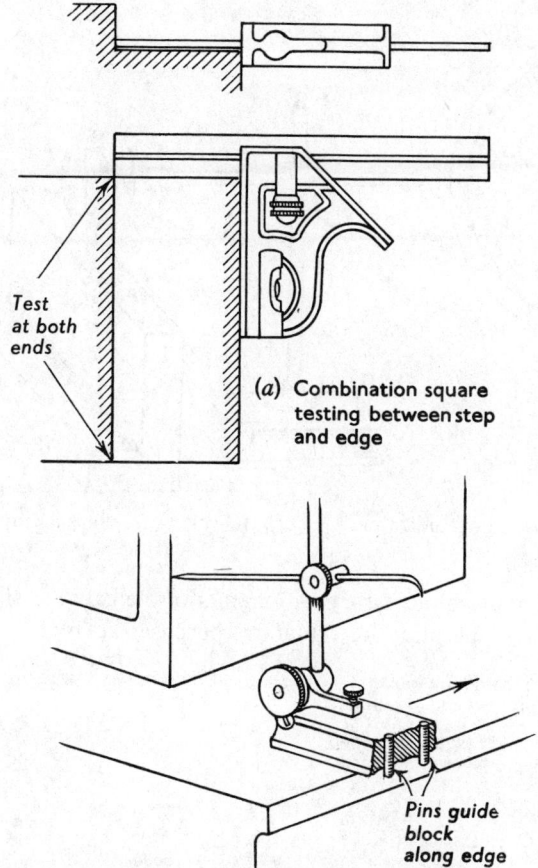

Test
at both
ends

(a) Combination square
testing between step
and edge

Pins guide
block
along edge

(b) Using pins in scribing block base to set
surface parallel to edge of table

Fig. 140 Testing for parallelism

These objections are overcome by the use of a dial gauge, the essential part of which is like a small clock with a plunger projecting at the bottom (it is often called a 'clock indicator'). Very slight upward pressure on the plunger moves it upwards and the movement is indicated by the dial finger which is generally arranged to read in $\frac{1}{100}$ mm of movement. For very accurate work, gauges reading in $\frac{1}{1000}$ths may be obtained. The head is supported on a base and upright very much like that of a scribing block, and for testing parallelism it is used in much the same way. For some purposes

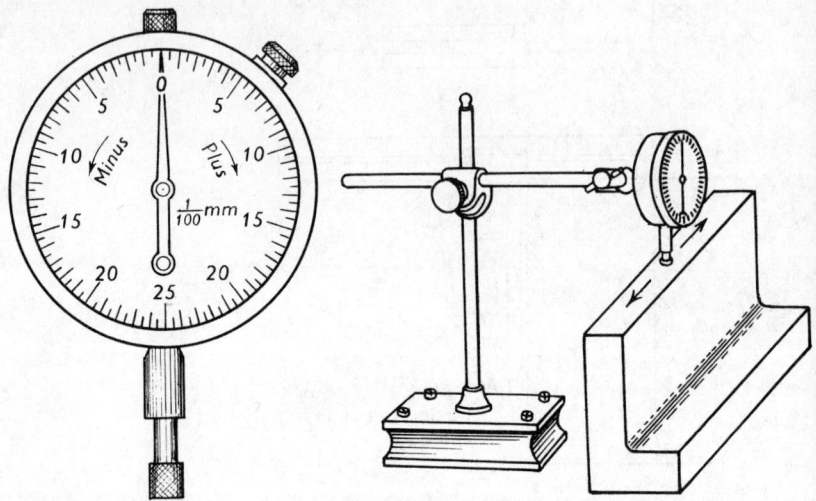

(a) Dial Gauge or Clock Indicator   (b) Checking for Parallelism

Fig. 141

a magnetised base is a useful accessory. A diagram of the gauge is shown at Fig. 141(a), and its application for the above purposes at (b).

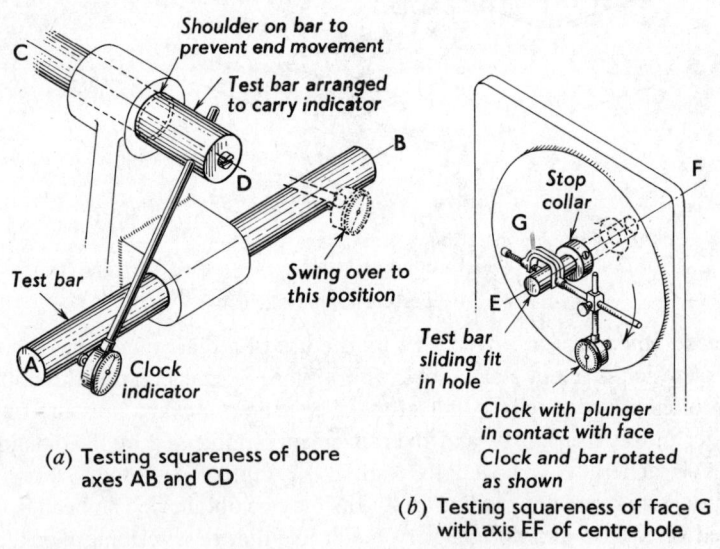

(a) Testing squareness of bore axes AB and CD

(b) Testing squareness of face G with axis EF of centre hole

Fig. 142 Uses of dial indicator

Use of dial gauge for squareness of hole axes. A useful application of the dial gauge is in testing the squareness of two holes or between a hole and a flat face. If a dial gauge is attached to, and rotated with, a shaft which fits into a hole, it rotates on the hole axis and thus may be arranged to indicate the squareness of this axis with a bar in another hole or with a flat face. An indication of how these tests may be carried out is shown in Fig. 142(a) and (b). At (a) the position of gauge is adjusted until a reading is shown when it is in contact with the left-hand end of the bar as shown on the diagram. It is now swung over until it contacts the other end of the bar and the two bars are perpendicular if both clock readings are the same. In the test shown at (b) the reading must remain constant whilst the gauge and bar are rotated a complete turn. The reader may, at first, be rather doubtful as to why this test shows up an error in perpendicularity. If he is, let him imagine how the clock would swing if there were a *large* error, and he will then realise that the method is really very accurate.

Tests for roundness and concentricity

Due to faults in machines and tools, distortion and spring in work and so on, parts which should be round are very often not so. The easiest and most obvious way to verify roundness is to use a micrometer or calipers and test the diameter of the part in several different positions in the same plane. If these dimensions vary, the part is not round. If the test shows the diameter to be the same at every position, the part can be assumed to be round. We should like to show, however, that it is possible to have a shape

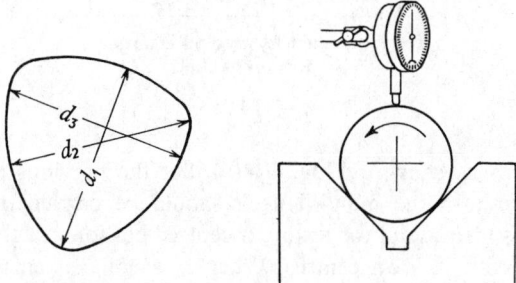

(*a*) 'Constant Diameter' figure which would appear to be round if measured across diameters

(*b*) Test for roundness with Vee blocks and clock indicator

Fig. 143

which, when measured across different extremities, will appear to be round, although it is not so. Such a shape is shown exaggerated in Fig. 143(a), and similar shapes to this can be and are produced on some classes of machine if conditions are not correct. In the shape shown, the diameters d_1, d_2, d_3, etc., are all equal, so that a micrometer test would give a false impression of roundness, and the only true test is to rotate the figure in a Vee block under a dial gauge as shown at (b). This will serve to impress the reader that in the workshop things are not always what they seem.

Concentricity. On many turned parts produced on the lathe, it is important that various diameter shall be true with one another or have the same centre. This is called concentricity. The clock gauge is again useful for this test, and if the part can be pushed on to a bar or mandril, it may be tested

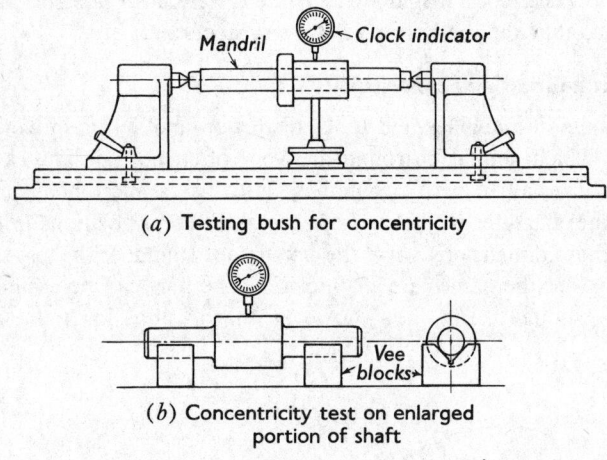

(*a*) Testing bush for concentricity

(*b*) Concentricity test on enlarged portion of shaft

Fig. 144

between centres as shown at Fig. 144(a), but the previous precaution of testing the truth of the mandril itself should be carried out first. This arrangement is also useful for testing a centred bar for straightness and for concentricity with its own centres. When it is not convenient to put the article between centres it may be rotated whilst resting in a Vee block as shown at (b). For these tests, as well as for certain marking out operations, a pair of centres mounted on a base-plate, and lined up to a central slot, form a very useful piece of workshop equipment. This is indicated in Fig. 144(a).

Angular testing

Angular testing

When two surfaces are at any angle other than 90°, the angle between them must be tested with some form of protractor. Instruments for this purpose may have a scale of degrees, enabling the angle to be read off, or they may consist of a gauge which must be set to the angle before use. The *bevel* (Fig. 145) is an example of this second variety of gauge, and must be set to the correct angle before use.

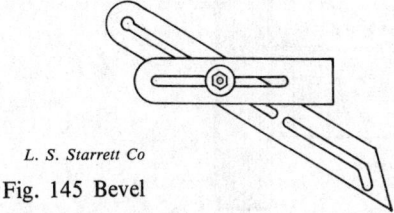

L. S. Starrett Co

Fig. 145 Bevel

The protractor head of the combination square has its own scale of degrees marked on it, and will be found capable of dealing with most of the angular testing encountered. For very accurate work, however, this protractor is not suitable because the scale is very fine, and the degrees are not subdivided. Accurate angular testing should be carried out with the *vernier protractor*. A diagram of this is shown at Fig. 146, together with an enlarged diagram of the vernier scale. The main scale on the protractor is divided up into degrees from 0 to 90 each way. The vernier scale is divided up so that 12 of its divisions occupy the same space as 23° on the main scale (Fig. 146(b)), so that one vernier division $= \dfrac{23}{12} = 1\frac{11}{12}°$, i.e. $\frac{1}{12}$ degree or 5 minutes less than 2°.

The instrument therefore allows settings to 5 minutes of angle to be obtained.

Reading the vernier protractor

1. Read off directly from the scale the number of whole degrees between 0 on the main scale and 0 on the vernier scale.

2. Count *in the same direction* the number of divisions from the 0 on the vernier to the first line on it which is level with a line on the main scale. As each division on the vernier represents 5 minutes, the number of these divisions multiplied by 5 will be the number of minutes to be added to the

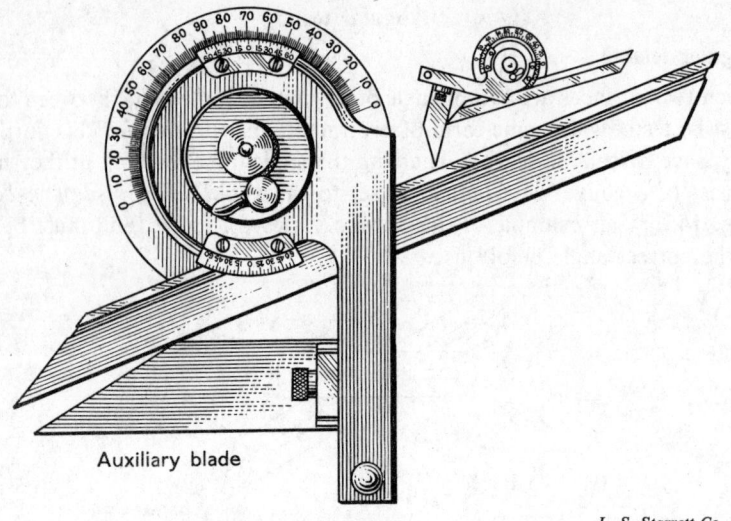

L. S. Starrett Co

(a) Protractor

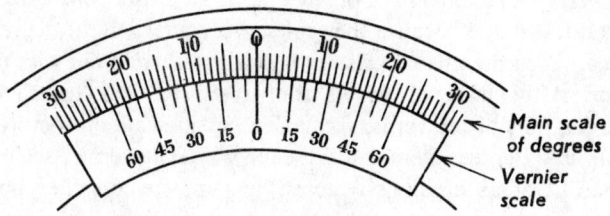

Main scale of degrees

Vernier scale

(b) Vernier scale showing 12 divisions on vernier scale equal to 23° on main scale

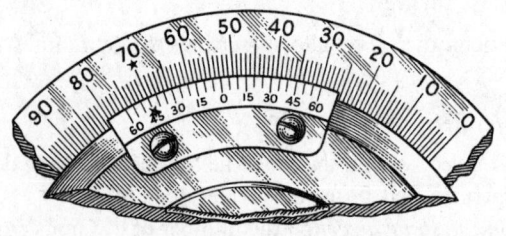

L. S. Starrett Co

(c) Reading 52° 45′

Fig. 146 The vernier protractor

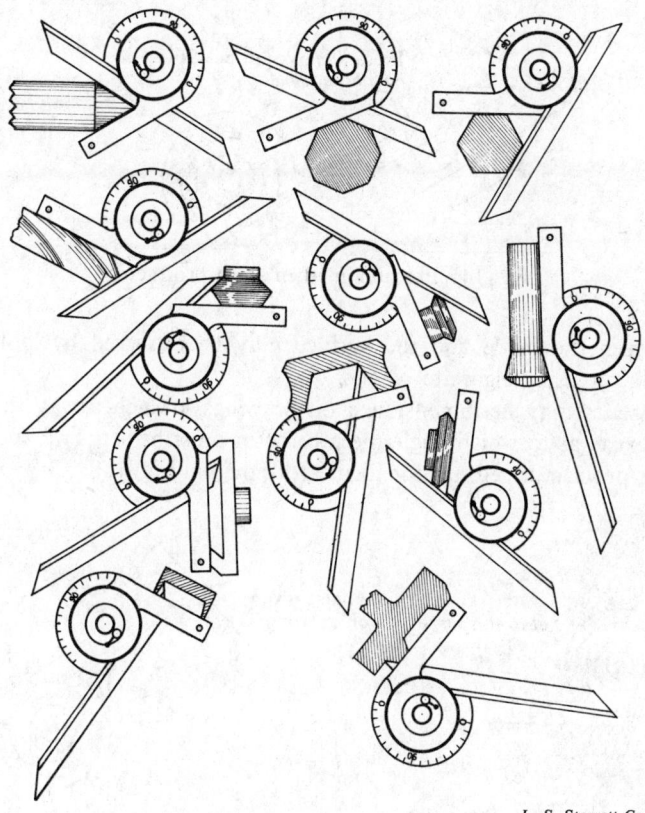

L. S. Starrett Co

(*d*) Applications of vernier protractor.
Fig. 146—*contd.*

whole number of degrees. Actually the multiplication is not necessary, as it has been done on the scale.

Thus in Fig. 146(c), the number of whole degrees is 52, and the mark on the vernier representing 45 minutes is level. The reading is therefore 52° 45′. Diagrams showing some uses of the protractor are shown at Fig. 146(d).

Hints regarding the measurement of angles

1. *The protractor blade should always be perpendicular to the surface being tested and should lie on a line of greatest slope* (see Fig. 147). The

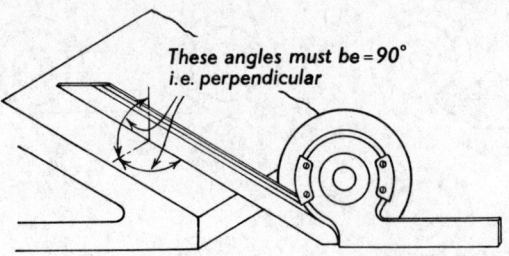

Fig. 147 Precaution when using protractor

fit between the blade and the surface may be observed by looking for 'daylight' or using cigarette papers.

2. Angles may be dimensioned on drawings in such a way that some doubt exists as to which angle the protractor must be set. Two such cases and the protractor settings for them are shown in Fig. 148.

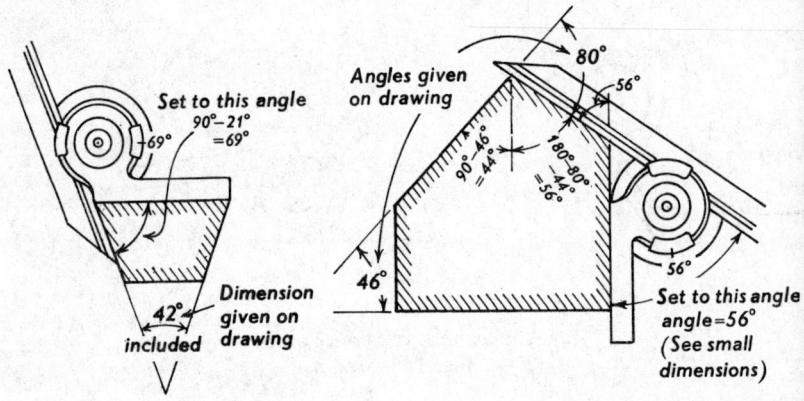

Fig. 148 Protractor setting for obscure angles

Although, as a general rule, the distances between lines on a drawing should *never* be measured with a rule, the reader will find that angles shown on drawings are fairly consistent in their accuracy. If, therefore, the reader is in doubt whether he should set the protractor to the angle shown, or to its difference from 90°, he will be fairly safe in putting the protractor up against the drawing.

3. Always check the reading of the protractor *after* the clamping screw has been locked up, as the setting may have moved.

Measurement

So far we have dealt with checking the accuracy of our work without much regard to its actual sizes. We must now consider the methods employed for the actual measurement of it.

'Line' and 'end' measurement

A length may be expressed as the distance between two lines (line measurement) or as the distance between two faces (end measurement). The most common example of line measurement is the rule with its divisions shown as lines marked on it, whilst we use end measurement when we employ calipers, micrometer, solid length bars, etc., to obtain our size. One of the most difficult problems in precision work is to transfer a dimension from its 'line' to its 'end' form, and this has probably been brought home to the reader in the setting of his calipers. We can set calipers much more accurately to a given size from the faces of a block than we can from the markings on a rule.

Actually when we set the calipers from our rule we do get an 'end' effect on one leg of them if we hold it against the end of the rule.

For this reason, in the shop, we employ end methods as far as possible when we wish to obtain an important dimension; the process of comparing anything with lines for the purpose of its measurement really requires the assistance of a microscope if any degree of accuracy is to be obtained.

When drawings are in SI metric units they are generally dimensioned in millimetres only; lengths will merely be put down as 260; 55; 18·63, etc., generally without the 'mm' qualification. It is not usual to employ any fractions except the $\frac{1}{2}$ and in the shop it is general to refer to the decimal part of a millimetre as so many hundredths. Thus 0·63 would be referred to as 63 hundredths.

The rule. When a dimension is given on a drawing without any limits or covering note as to a particular type of fit (e.g. press fit, or to suit such and such a part), or without stating any particular method of machining (e.g. in the case of a hole, '20 mm *reamer*'), it may generally be assumed that the dimension may be made with a rule. If calipers and rule are used it is still a rule dimension. The degree of accuracy to which work may be produced when measurements are made by a rule depends on the quality of the rule, and on the skill of the user. The marks on a good-class rule vary from 0·12 mm to 0·18 mm wide, so that we cannot expect to obtain a degree of accuracy much closer than within 0·012 mm, but a good workman should be able to work as closely as this. An important factor is that of possessing a good rule and getting used to it, and the reader is advised to buy the best rule he can obtain and accustom himself to its markings.

The most useful and convenient markings are millimetres on one face of the rule, and English units on the other.* A good example is the Rabone Chesterman No. 64, the English side of which is the lower rule shown in

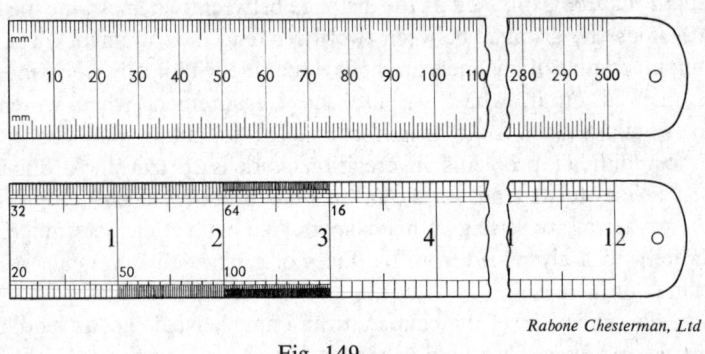

Rabone Chesterman, Ltd

Fig. 149

Fig. 149. On one side this is graduated in millimetres on one edge and $\frac{1}{2}$ millimetres on the other. The reverse side has $\frac{1}{8}$ inches and their subdivisions along one edge and the subdivisions of $\frac{1}{10}$ths along the other. If the length of the graduation lines for small divisions were reduced to about $\frac{1}{3}$ of the length they are generally made, the eye would be better able to make use of them. This applies also to the $\frac{1}{2}$ mm and $\frac{1}{64}$-in rulings on the average rule, and the reader will find it easier to measure, say, $22\frac{1}{2}$ mm or the full scale of mm, than on the half scale. The upper rule shown in Fig. 149 is an alternative rule (No. 47) marked in millimetres throughout and if the reader prefers a 150-mm (6-in) rule, both these patterns are available in that length.

A good rule is worth looking after and should neither be used as a feeler gauge under shaper tools or milling cutters, nor for scraping Tee slots and

machine tables. The end of the rule particularly should be preserved from wear, as it generally forms the basis for one end of the dimension. Rusting of the rule should be avoided by oiling it during weekends and when it is

* The retention of English scales of measurement is desirable until the SI system has completely displaced the Imperial measure.

not in use. The thickness of a rule may often be put to good use, and the rules mentioned above will be found 1 mm and $\frac{1}{2}$ mm thick respectively.

When taking measurements with a rule, it should be so held that the graduation lines are as close as possible (preferably touching) to the faces being measured.* The eye which is observing the reading should be as near as possible opposite to the mark being read. This avoids what is called 'parallax', and the reader may try for himself the errors which parallax may cause if he reads the large hand of a clock when it is vertical, by looking at it not from the front, but from a very acute side angle.

As far as is possible, the end of the rule should be held flush with one of the faces to be measured, the reading then being taken somewhere along the rule.

Calipers. To measure the diameter of a circular part involves straddling across it to obtain the length of its greatest dimension. Even if the end of

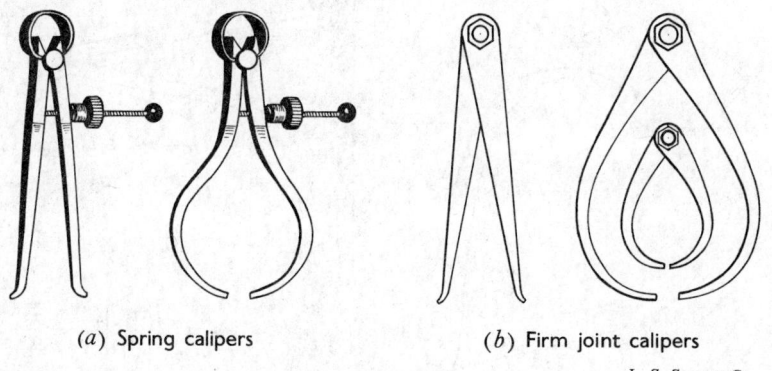

(*a*) Spring calipers   (*b*) Firm joint calipers

L. S. *Starrett Co*

Fig. 150 Calipers

the bar is available, a rule alone is not a convenient method of taking its size, so that for round work calipers are used. The shape of calipers varies according to whether they are for external or internal use, and they may be stiff jointed at the hinge of the legs, their opening being maintained by the friction at the joint, or the joint may be free and spring controlled, the opening being adjusted and maintained by a nut working on a screw. Screw-controlled spring calipers are more easily adjusted, but just as good work

---

* To assist in bringing the graduations close to the work, rules may be obtained with bevelled edges, but great care is necessary that the edges do not get knocked about, as they are easily damaged.

may be done with a good pair of stiff jointed ones which the reader can make for himself.

Examples of calipers are shown at Fig. 150.

When working with outside calipers they should be adjusted by tapping one leg (stiff joint), or by the adjusting screw, until when the work is straddled by the legs, it is just possible to feel the contact between the calipers and the work. The contact should not be too heavy, otherwise the legs may be slightly sprung and a false reading obtained. When a nice feel has been obtained on the job the size should be read on a rule by resting the end of one leg on the end of the rule and taking the reading at the other (Fig. 151(a)). To set outside calipers to a fairly particular size they should be set from a block or gauge of the given dimension.

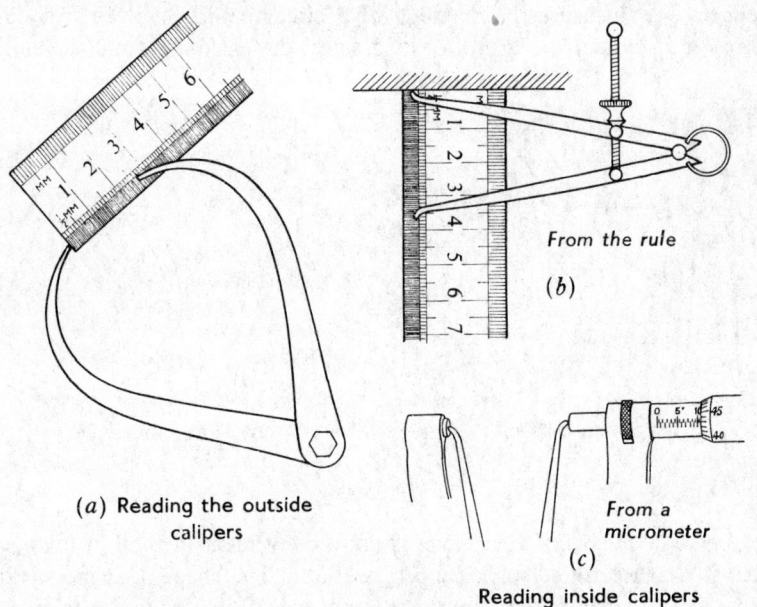

(a) Reading the outside
calipers

From the rule

(b)

From a
micrometer

(c)

Reading inside calipers

Fig. 151

Inside calipers are adjusted in the same way, but their manipulation to obtain the size of the hole they are measuring requires a little more skill than with outside conditions. They should be adjusted until they are at the *largest* size at which their legs can just be felt contacting the extremities of a diameter of the hole, and to find this, the joint should be held by the

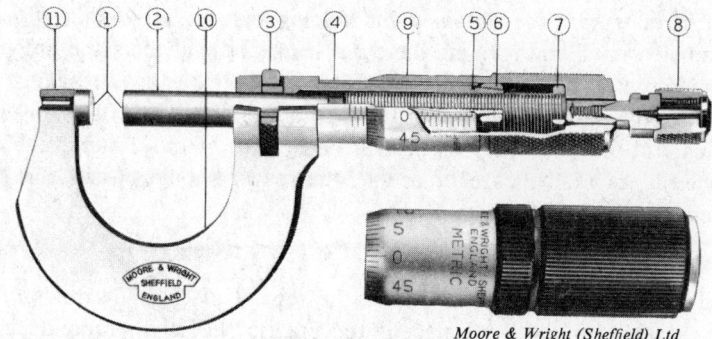

Moore & Wright (Sheffield) Ltd

Optional friction thimble.

Index to parts
1. Anvils
2. Spindle
3. Locknut
4. Sleeve
5. Main nut

6. Adjusting nut for main nut
7. Thimble adjusting nut
8. Ratchet stop
9. Thimble
10. Frame

(a) Micrometer showing interior mechanism.

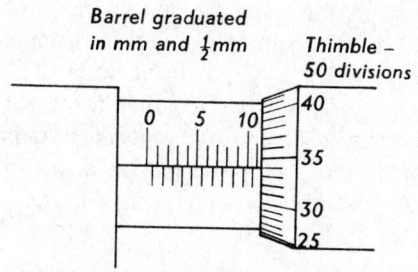

(b) Divisions on micrometer. (Reading 11·34 mm).

Fig. 152.

thumb and first finger, one leg held stationary in contact with the inside of the hole and the other leg rocked about in a small circle. After a demonstration by a skilled user, and a little practice, the reader will soon learn the manipulation necessary for the operation. The opening of inside calipers may be checked by a rule or micrometer. If the rule is used, hold the end of the rule and one leg of the calipers pressed against a vertical flat surface (e.g. edge of a lathe bed) and read the other leg (Fig. 151(b)). The micrometer reading may be obtained by holding one leg in contact with the micrometer anvil, and rotating the other leg slightly, at the same time that the micrometer is screwed up, until the 'feel' of the top leg against the spindle of the micrometer is eventually obtained (Fig. 151(c)). In this way, with careful use, it is easy to measure a hole to within a hundredth of a millimetre. As in the case of a rule, the reader should possess his own calipers and look after them.

The micrometer*

When a part has to be measured to the second place of decimals in the metric system, or the third place in the English, we need a more accurate method of measurement than can be obtained with a rule, and the micrometer is commonly used. A micrometer consists of a semi-circular frame having a cylindrical extension (the barrel) at its right end, and hardened anvil inside, at the left end. The bore of the barrel is screwed $\frac{1}{2}$ mm pitch and the spindle, to which is attached the thimble, screws through. Adjustment is provided for the longitudinal position of the spindle, and for the tightness of the screw thread. The barrel is graduated in mm and $\frac{1}{2}$ mm for a length of 25 mm and the rim of the thimble is divided into 50 equal divisions. See Fig. 152(a) and (b). The measurement is taken between the face of the anvil and the end of the spindle, and the range of the micrometer is 25 mm, so that if we wish to measure up to 150 mm, we must have six micrometers; 0 to 25, 25 to 50 and so on, with 125 to 150 mm as the largest size. As a compromise, it is possible to obtain adjustable micrometers which cover a range of sizes by producing a frame to suit the largest in the range required, and a set of detachable anvils to accommodate the smaller intervals. An example of such an instrument for measuring from 0 to 50 mm is shown in Fig. 153.

Reading the micrometer

The screw in this micrometer has a pitch of $\frac{1}{2}$ mm, so that the jaws open $\frac{1}{2}$ mm for each turn of the thimble. The rim of the thimble is divided into 50 parts, which gives a reading of $\frac{1}{2} \div 50 = \frac{1}{100}$ mm. The barrel is marked in millimetres and $\frac{1}{2}$ mm divisions, so that to take a reading we add the number of hundredths indicated on the thimble to the millimetre and $\frac{1}{2}$ mm uncovered on the barrel. In Fig. 154 there are $11\frac{1}{2}$ mm uncovered, and the thimble reading is 33 hundredths. The reading, therefore, is $11 \cdot 5 + 0 \cdot 33 = 11 \cdot 83$ mm.

* BS 870.

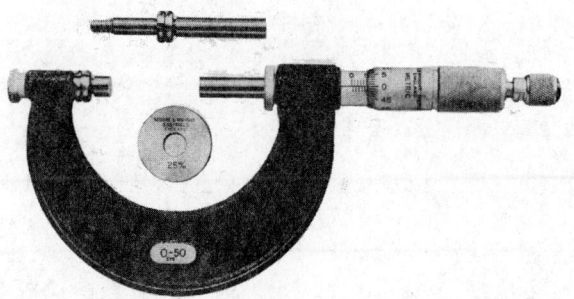

Moore & Wright (Sheffield) Ltd

Fig. 153 0–50-mm Micrometer.

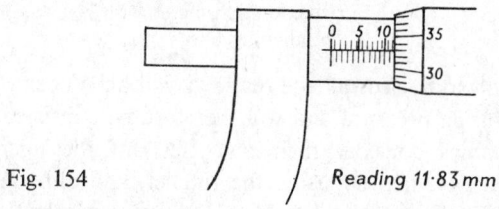

Fig. 154

Reading 11·83 mm

The inside micrometer.* For the accurate measurement of the bore of holes an inside micrometer may be used. This consists of a head similar to the barrel and thimble of the ordinary micrometer, and a set of lenghtening bars to increase the range. The movement obtained on the head is 13 mm and the bars are inserted up to the size required. The manipulation of this instrument is similar to that for inside calipers, and the reader should seek the help of an experienced mechanic to show him when using for the first time (Fig. 155).

The accuracy of the micrometer. The accuracy with which measurements can be made with a micrometer depends on the skill of the user, the accuracy of the micrometer screw, the number of times the instrument has been dropped and various other factors. The pressure with which it is screwed up to the work will vary the reading and a 'heavy' user will obtain a smaller reading than one who uses the instrument lightly. Micrometers can be obtained with a ratchet on the end of the thimble (Fig. 152(a)), and if the ratchet sleeve instead of the thimble is held when screwing up, the ratchet slips when the pressure on the screw reaches a certain amount. If

*BS 959.

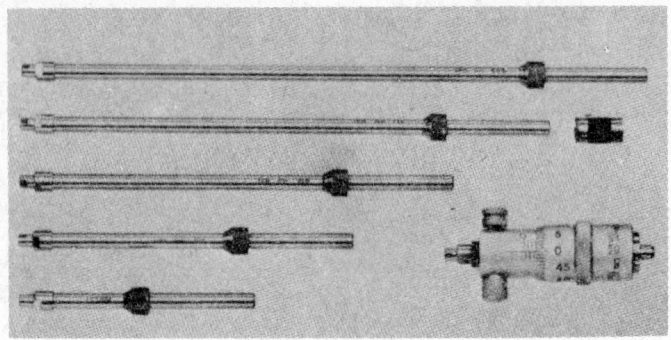

The L. S. Starrett Co Ltd

Fig. 155 Inside micrometer with
lengthening bars

the ratchet is always used, all the readings will have been obtained under the same measuring pressure and will therefore be consistent.

Some mechanics consider their own 'feel' of the micrometer to be superior to that obtained by using the ratchet, and although this may or may not be true, the ratchet does give consistent results for *any* user of the instrument. Results as between different mechanics with their individual 'feels' are bound to vary; in fact the pressure exerted by one individual will not always be the same but will vary with personal factors, e.g. health, temper, tiredness, etc. However, as probably more people use the micrometer without the rachet than otherwise, the reader should learn a proper conception of micrometer-feel from an experienced mechanic, as it is one of the things he cannot learn from books. There is now a micrometer constructed with a friction connection incorporated in the thimble, so that readings *have* to be taken through the thimble (see Fig. 152).

The accuracy of the micrometer screw will depend on the price paid for the instrument. It cannot be assumed that if the reading is 0 when the jaws are closed, it will be correct when they are open at (say) 18 mm for example; the screw may vary along its length in such a way that at 10 mm the opening is a little *less* than the reading, and when screwed out to 20 mm it might be a fraction of a hundredth *more* than that indicated. The safest way, if a trustworthy reading at any particular size is desired, is to check the instrument against a master gauge at or near the size required. Rough usage and dropping will soon impair the accuracy of any micrometer, and it should be treated with as much care as any other valuable and useful instrument.

The English micrometer

In external appearance, the English micrometer is similar to the instrument shown in Fig. 152(a) and the method of handling it is the same. The only difference on the English instrument is that the screw on the spindle and in the barrel is 40 threads per inch, and the thimble and barrel markings differ accordingly. These graduations are shown at Fig. 156(a) where it will be seen that the barrel is divided up into $\frac{1}{10}$ths and $\frac{1}{40}$ths of an inch for a length of 1 inch and the thimble is marked round into 25 equal divisions.

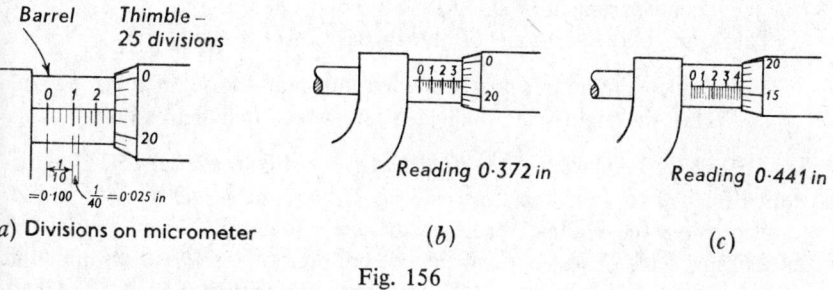

(a) Divisions on micrometer   (b) Reading 0.372 in   (c) Reading 0.441 in

Fig. 156

Reading the English micrometer

Since the spindle screw has 40 threads per inch, 1 turn of the thimble opens the jaws $\frac{1}{40}$ of an inch. Now $\frac{1}{40}$ in corresponds to $\frac{1000}{40} = 25$ thousandths or 0·025 in, and as the rim of the thimble is divided into 25 parts round its circumference, one of these divisions corresponds to $\frac{1}{1000}$ in (0.001) of movement of the spindle. The barrel is divided into $\frac{1}{10}$ths (numbered 1, 2, 3, etc.), and each tenth is divided into four parts ($\frac{1}{40}$ths). Each of these small divisions corresponds to the movement caused by one turn of the thimble, so that each time we turn the thimble one of the small divisions is uncovered.

To take a micrometer reading add the tenths, the number of 0·025's showing beyond the last tenth, and the number of odd thousandths on the thimble.

Thus in Fig. 156(b) the reading is $0·3 + (2 \times 0·025) + 0·022$

$$= 0·3 + 0·05 + 0·022$$
$$= 0·372 \text{ in}$$

and in Fig. 156(c) it is:

$$0·4 + 0·025 + 0·016$$
$$= 0·441 \text{ in}$$

The Vernier Calipers*

With a micrometer we take an 'end' measured between two jaws, the opening of which is controlled by a very accurate screw. The vernier also gives an 'end' measurement, but the position of the jaws is controlled from a 'line' scale, accurate transference being made possible by a vernier scale. A vernier scale is the name given to any scale making use of the difference between two scales which are nearly, but not quite alike, for obtaining small differences. We have already seen how such a scale is applied to a protractor for reading small divisions of a degree; its principle for length measurement is along similar lines.

There are two variations of the metric Vernier scale:

1. In which the main scale is divided into mm and $\frac{1}{2}$ mm;
2. Where the main scale is divided into whole millimetres only.

An enlarged diagram of the first of these is shown at Fig. 157. The distance from 0 to 1 on the top scale is 10 mm, and it will be seen that 25 divisions on the sliding vernier scale are equal to 12 mm on the top measuring scale. The length of the bottom divisions is $\frac{12}{25} = 0.48$ mm, and since the top divisions are $\frac{1}{2}$ (0.5) mm, the difference is $0.5 - 0.48 = 0.02$ mm, which represents the accuracy to which readings may be taken. To

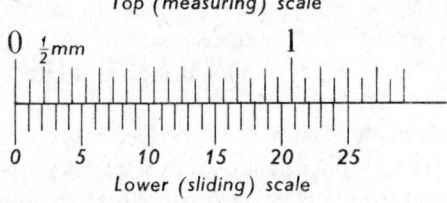

Fig. 157 Metric vernier scale

read the vernier, count up the total length indicated as far as the 0 on the sliding scale, and note the mark on the sliding scale which is level with a mark on the top scale. This latter amount will represent the number of 0.02 mm which must be added to the first reading. Hence, opening of jaws = amount indicated up to the 0 + *twice* the indication where the marks are level. Fig. 158 shows metric verniers reading (*a*) 32.32 mm and (*b*) 18.6 mm.

The form of metric vernier giving readings to 0.02 mm on a scale of full

* BS 887.

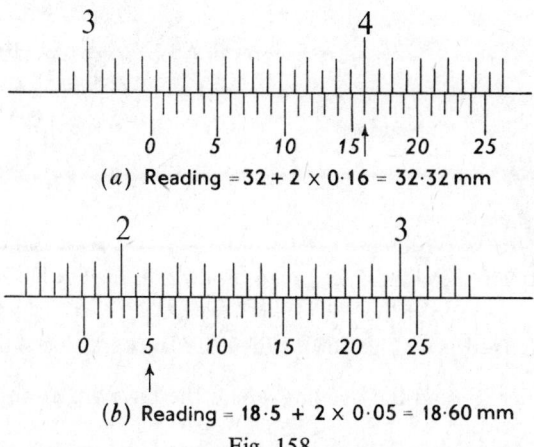

(a) Reading = 32 + 2 × 0·16 = 32·32 mm

(b) Reading = 18·5 + 2 × 0·05 = 18·60 mm

Fig. 158

mm is obtained by dividing 49 mm on the main scale into 50 divisions on the vernier scale. One main scale division is now 1 mm, and a vernier scale division is $\frac{49}{50}$ mm, the difference being $1 - \frac{49}{50} = \frac{1}{50} = 0·02$ mm. A diagram of this scale is shown in Fig. 159(a) and a combined instrument, metric/English, each having its sliding scales divided into 50 parts is shown at Fig. 159(b). On the English side of this the small divisions are $\frac{1}{20}$ inch and the readings, as set, are metric 27·42 mm; English 1·079 in.

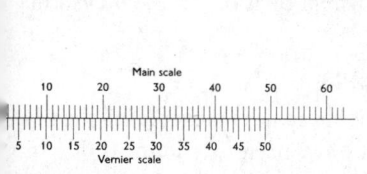

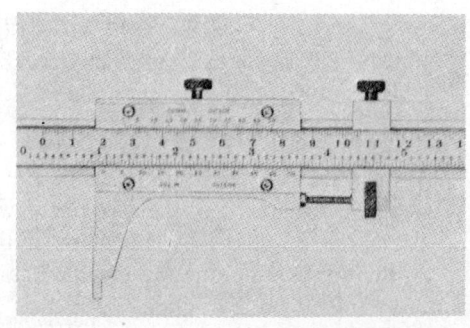

L. S. Starrett Co Ltd

Fig. 159

) Vernier scale with 49 mm divided into 50 parts

(b) This scale applied to a metric/English Vernier calipers

In the vernier calipers the lower scale in the diagrams above is part of the sliding jaw and the upper scale forms the body of the instrument which is

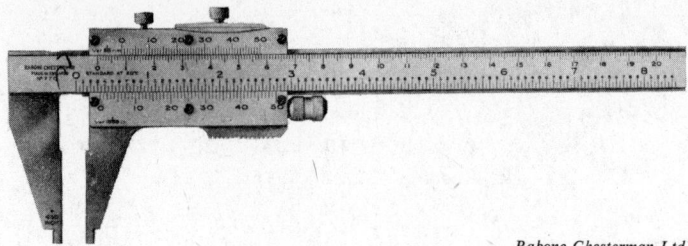

Rabone Chesterman Ltd

Fig. 160 770 Vernier caliper gauge. 50 division English and metric
(Conforming to BS 887: 1950)

solid with the fixed jaw (Fig. 160). When the jaws are closed the reading is
$\frac{0}{0}$ + and in order to read the opening where the jaws are at any setting, we
count up the total amount on the measuring scale, to the left of the 0 on the
sliding scale and add it to the amount indicated on the vernier scale.

English vernier scale. The principle of the English length vernier will be
gathered from Fig. 161. On the upper (measuring) scale, 1 in is divided into
10 parts (numbered 1, 2, 3, 4, etc.) and each $\frac{1}{10}$th is divided into 4, giving

$$\tfrac{1}{4} \text{ of } \tfrac{1}{10}\text{th} = \tfrac{1}{40} \text{ in} = \tfrac{25}{1000} \text{ in.}$$

On the lower (sliding) scale, $\frac{6}{10}$ in is divided into 25 parts. Each of these
will have a length of

$$\tfrac{1}{25} \text{ of } \tfrac{6}{10} = \tfrac{6}{250} = \tfrac{24}{1000} \text{ in,}$$

and the difference in length between one small division on the measuring
scale and one on the sliding scale is

$$\tfrac{25}{1000} - \tfrac{24}{1000} = \tfrac{1}{1000} \text{ in.}$$

If, then, the two zeros are level and the sliding scale is moved until the first
small marks are level, the movement will have been $\frac{1}{1000}$ in, or 0·001 in;
if it is moved until the 15th mark on the lower scale is level with a mark
on the upper scale, the lower scale has moved 0·015 in, and so on.

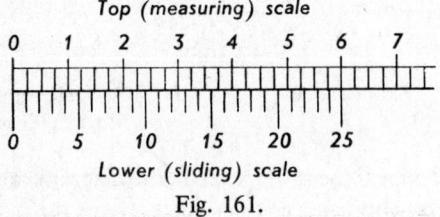

Top (measuring) scale

Lower (sliding) scale

Fig. 161.

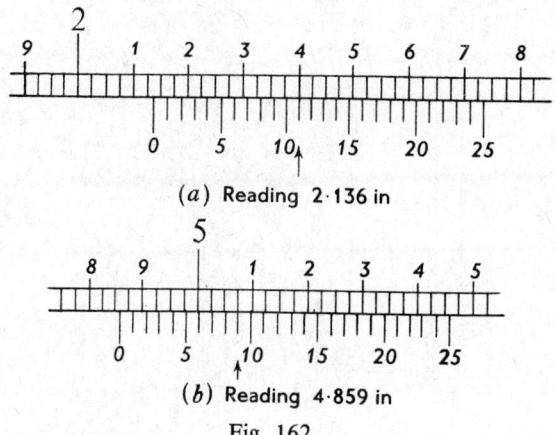

(a) Reading 2·136 in

(b) Reading 4·859 in

Fig. 162

Fig. 162 (a) and (b) show two settings of this vernier, the readings being as follows:

(a) To the left of the 0 on the sliding scale the reading is 2·125. The 11 mark on the vernier is level.

Hence, total reading = 2·125 + 0·011 = 2·136 in

(b) The reading to the left of the sliding scale zero is 4·850 and the 9 on the vernier is level.

Hence, total reading = 4·850 + 0·009
= 4·859 in

On an alternative English vernier scale the main scale is divided into $\frac{1}{10}$ths and $\frac{1}{20}$ths inch, and on the vernier scale 49 divisions (2·450 in) are divided into 50 parts. This enables differences of 0·001 in. to be read. This scale is shown at Fig. 159(b).

Use of the vernier calipers. The vernier is made in various sizes from 150 mm upwards, a good size being one capable of working up to 300 mm. It is not used for straddling round bars in the same way as a micrometer, but may be employed for measuring large diameters on their ends, or large bores (Fig. 163). When it is used for a bore, or any other inside measurement, the reading will give the *inside* jaw opening, and allowance must be made for the total thickness of the jaw ends. This may be obtained by closing the jaws and finding their combined thickness with a micrometer. For setting the vernier to the work or to a reading, the extra slide shown to the right of the main slide is used. If this is locked, and the main slide left

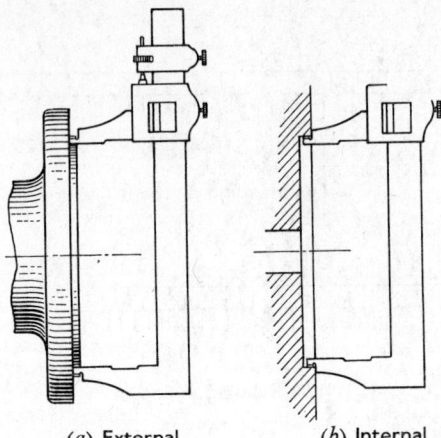

(*a*) External   (*b*) Internal

Fig. 163 Use of vernier to measure spigot and recess

free, turning the knurled finger nut will move the screw, and the main slide attached to it. Without this it would be difficult to obtain a fine movement. When setting the vernier this way, lock up the small slide but do not leave the screw on the main slide quite free, but give it the slightest bit of tension. After adjusting the instrument to the work with the finger nut, clamp up the main slide and check up again on the work.

If the vernier is being set to a particular dimension, check the reading after the main slide has been clamped, and *always clamp the main slide before using.* When using the vernier, balance it lightly in the hands and *under no circumstances use the slightest force to gauge the work.* The construction of the instrument is such that the jaws are easily sprung open, and the instrument ruined.

As in the case of the micrometer, the reader will be well advised to seek the advice of a skilled mechanic on the manipulation of the vernier.

The use of gauges.

The academic distinction between measurement and gauging is that, in measurement, we use some instrument to *measure* the dimension of the part, whilst in gauging we check the accuracy of the work by *comparing* it with a gauge, without being particularly interested in its size.

The variety and scope of the gauges employed in engineering workshops is so great that it would occupy more than this book to describe them all, but the reader ought to be acquainted with one or two of the more common types.

Hole gauging.—Plug gauges. When a mating hole and shaft have to be produced it is always necessary to make some form of fit, depending on the conditions. If the combination is a working bearing, the shaft will have to be a running (easy) fit; if the shaft has to be a fixture in the hole, such as for a tightly fitting pulley; the connection between the hole and shaft will have to be such that some force is necessary to assemble them. The pulley might be driven on with a hammer (driving fit) or the shaft forced in under a press (force fit). Naturally, different kinds of fit will require different allowances to be made between the hole and shaft sizes; for a running fit the shaft will be slightly smaller than the hole, whilst for a force fit it will have to be a little larger. Whatever these variations might be, the hole is always made to the standard (nominal) size and the shaft varied to suit. When boring a hole, therefore, into which another part has to fit, the size of the hole must conform to the exact standard size, and to check this a plug gauge is used. A skilled turner could easily work to the exact bore required by using his inside calipers and a micrometer, but he will always prefer to have the gauge as a final check. A *standard plug gauge* (Fig. 164

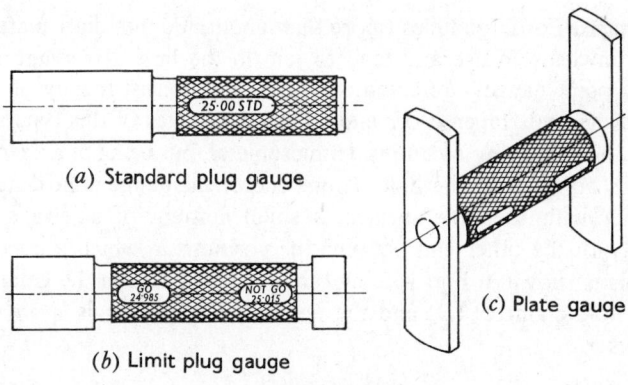

(*a*) Standard plug gauge

(*b*) Limit plug gauge

(*c*) Plate gauge

Fig. 164 Hole gauges

(a)) has its diameter finished to the standard size (e.g. 25·00 mm for a 25 mm hole) and is used in general engineering workshops, toolrooms, etc., where production quantities are not great, and accuracy is more important. In production workshops, where large quantities have to be produced with less skilled operators, a *tolerance* is allowed on the size of the holes which are dimensioned with *limits*. For example, a 25-mm hole might be dimensioned $25 \begin{smallmatrix} +0 \cdot 030 \\ -0 \cdot 000 \end{smallmatrix}$ and provided it measures anything within these sizes, it would be suitable. The $25 + 0 \cdot 030$ dimension $= 25 \cdot 03$ mm is the *high limit*, and the $25 - 0 \cdot 000 = 25 \cdot 00$ mm is the *low limit*. The difference between them (in this case $0 \cdot 030$), is called the *tolerance*, and represents a safety margin for errors in workmanship, wear of tools, etc. Holes of this type are gauged with *limit plug gauges* which have 'GO' and 'NOT GO' ends. The 'GO' end of the gauge must enter the hole, being made the size of the lower limit, whilst the 'NOT GO' end, which is the size of the upper limit, must not enter. Usually the 'GO' end is made longer than the 'NOT GO' as shown at Fig. 164(b). This is because it is continually being subjected to frictional contact with the sides of the hole and would wear rapidly if short; also if short it would tend to jam in the hole. The 'NOT GO', which should never enter a hole, need only have sufficient length to carry its size.

For gauging large holes a solid plug gauge would be unduly heavy, and various methods are utilised to reduce weight. The plate gauge (Fig. 164(c)) is such an example and has the extremities of its end plates ground to the required diameter.

Point gauges. For large holes (more than about 300 mm dia.), plate gauges become unwieldy in use and tend to jam in the hole. To gauge holes of this type point gauges are commonly used and consist merely of a bar of steel with its ends tapered off and rounded. Gauges of this type are used in much the same way as an inside micrometer, but being of a fixed length, cannot be adjusted to the hole. If one end of the gauge is held stationary in contact with the hole surface, a small amount of sideways rock is possible with the other end between the positions in which it contacts the hole. This is shown in Fig. 165, and the relation between the gauge length (L), the hole diameter (D) and the amount of rock ($2w$) is approximately as follows:*

$$D = \frac{w^2}{2L} + L$$

Shaft gauging. Shafts and male components are usually checked with *caliper gauges*, an example of which is shown at Fig. 166(a). In their smaller sizes, these gauges are generally called *snap gauges*, and are made with the 'GO' jaws at one end and the 'NOT GO' at the other as shown at (b). The use of caliper and snap gauges is almost entirely restricted to pro-

* The proof of this is given in the Author's *Senior Workshop Calculations*.

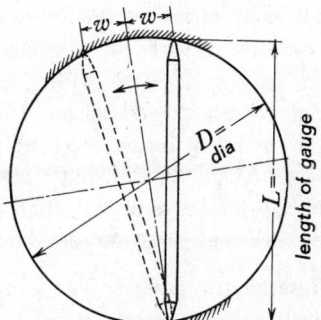

Fig. 165 Gauging large bores with point gauge

duction work where large quantities of similar articles are made. In general engineering shops and toolrooms shafts are brought to size by means of the micrometer. In production work, shafts are made to limits in the same way as holes, and these limits are put on the caliper gauges. Caliper gauges may be made with adjustable anvils as shown at (c). These are useful in workshops carrying out the production of medium quantities of different parts, as they may be set to different sizes and limits, thus saving time and expense of making new gauges.

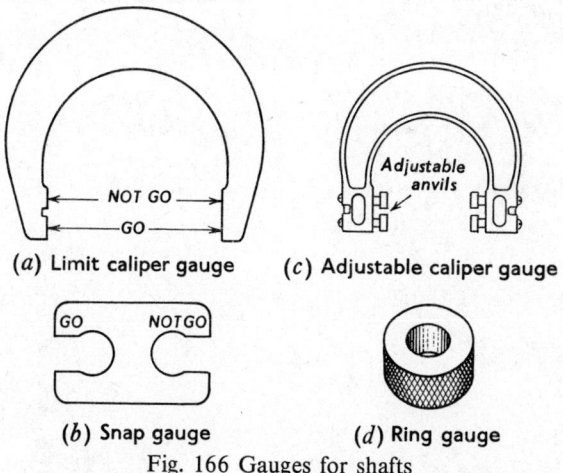

(a) Limit caliper gauge

(c) Adjustable caliper gauge

(b) Snap gauge

(d) Ring gauge

Fig. 166 Gauges for shafts

The *ring gauge* (Fig. 166(d)) would seem to be an obvious method of gauging a shaft, but in practice it is rarely used. This is because a gauge of this type, which entirely envelops the surface being gauged, places an extremely stringent test on its accuracy, and prohibitive time and expense would be incurred in working to ring gauges. Furthermore, a ring gauge can only be used from the shaft end, and must be threaded all along the shaft to test its parallelism. This condition is not always convenient when a shaft is being turned between centres. Lastly, a ring gauge either does, or does not, go on the shaft; it tells us nothing about a shaft which may be slightly too large or too small for it, and we should require two ring gauges for 'GO' and 'NOT GO' conditions.

Material for gauges. The gauging surfaces of gauges must be extremely hard and resistant to wear as otherwise they would very soon lose their size. For caliper and snap gauges, the most common material employed is cast steel which is hardened on the gauging faces only. Plug gauges present a more difficult problem because of the distortion and cracking which is liable to occur when cast steel is quenched. If this material is used, extreme care is necessary in hardening to avoid this risk. The heat treatment of gauges is greatly facilitated, and the risk of spoilt gauges eliminated, by using one of the alloy tool steels which harden by oil quenching. Plug gauges of case-hardened mild steel are quite satisfactory as they possess the glass hard skin and tough core necessary for good wear combined with serviceable strength.

8 The bench. Flat surfaces —Filing, chipping and scraping

The work carried out at the fitting bench of a workshop will depend upon the class of work for which the shop exists. The benchwork in a toolroom, or general engineering shop working on individual pieces of equipment, will be of a highly skilled nature, calling for extensive knowledge and experience. Here the work will include marking out, and the preparation of work for machining; the adjustment of details after machining; the building up of jigs, mechanisms and other assemblies after the machined details have been made; the making and finishing of tools, gauges and other small details; the repair and adjustment of machine tools, and so on. A competent fitter at such a bench needs to have a knowledge of the chief methods of machining, with the ability to carry them out himself, a knowledge of materials, measurement and testing. In addition he must be skilled in the various aspects of his own trade as a fitter.

At the other extreme, in a shop working on a mass-produced article, that part of the work called 'fitting' might be the mere operation of assembling parts together, a job which, in a simple case, could be learned completely in a few months. We advise the reader to give such an occupation a wide berth, as although he might earn higher wages than if he were learnin the trade of a true fitter, he may realise too late that he has been following a 'blind alley' occupation and have his whole life ruined thereby.

The bench and vise

The work of a fitter will probably take him to various places, but the bench is his headquarters. This should be a rigid structure provided with racks or shelves at the back for storage of articles and tools, so that the bench top may be kept tidy and clear. Lock-up drawers should be available for the reception of each fitter's private tools. A rigid bench is essential, as nothing is more aggravating, and conducive to inaccurate work, than a bench which wobbles about as one applies force to a job supported on it. For holding work whilst operating on it the fitter has a *vise* (or vice) attached to the bench. This consists of an iron or steel cast body into which is fitted a square section slide formed to a jaw at its outer end. The corresponding fixed jaw is incorporated on the body of the vise, and the two jaws are faced with hardened steel jaw-pieces, screwed to the jaws and cut with teeth to help grip the work. The sliding jaw is operated by a screw and nut, and in the quick-operating types the nut only embraces the lower portion of the screw, being capable of disengagement from it by the operation of a trigger at the front of the slide (Fig. 167). The height of the bench should

(*a*) Elevation

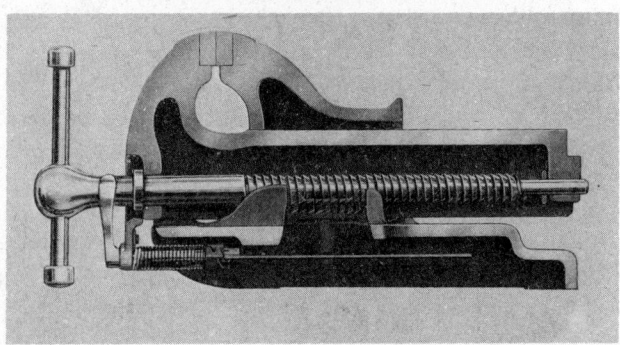

(*b*) Section of quick operating type

Fig. 167 Fitter's vise

be such that the top of the vise-jaws is at about the same height as the fitter's elbow when he stands normally at the bench with his upper arm hanging vertically, and forearm bent horizontally.

To avoid damage to the surface of finished work by the hardened jaw-pieces, it is usual to employ clamps (or clams) made of copper, brass, lead or soft steel.

The marking-out table. This is an essential item of equipment in a fitting shop, and consists of a cast-iron table of fairly substantial dimensions supported horizontally on rigid legs. The top of the table is planed flat, and its edges should be machined square. The smallest practicable size for a marking-out table is about a metre square, and we shall refer later to its uses (Fig. 168).

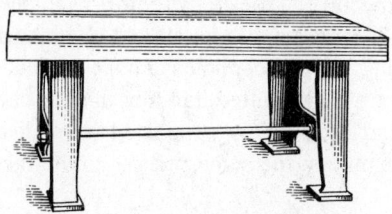

Fig. 168 Marking-out table

The hand-hammer*

The hammer is probably the most used of all the tools possessed by the fitter, and is an item of his equipment upon which he sets great value. The most common shapes of engineers' hammers are shown at Fig. 169

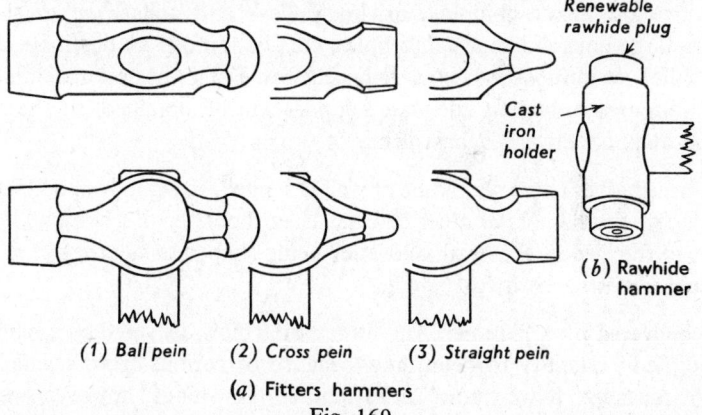

Renewable rawhide plug

Cast iron holder

(*b*) Rawhide hammer

(*1*) Ball pein   (*2*) Cross pein   (*3*) Straight pein

(*a*) Fitters hammers
Fig. 169

and they are specified by the shape of the end opposite the striking face, called the *pein*. The ball pein is the most common of the three, the ball end being used mostly for riveting over the ends of pins and rivets. Cross and straight peins are also useful for riveting in awkward places; another use for these ends is that of *peening*. This consists of striking the surface to be

* BS 876

peened a number of light, sharp blows with the pein of the hammer, the effect being to stretch the metal and elongate it. Thus if the concave side of a bent bar is peened and elongated, it is often possible by this means to straighten the bar.

Hammer heads are made from a plain steel of about 0.6% carbon and are shaped by stamping or forging. The two ends must be hardened and tempered, the centre of the head, with the eye, being left soft. When correctly hardened the face and pein should be able to pass the following tests without showing any perceptible damage or indentation: (a) a blow from the ball pein of a $\frac{1}{4}$ kg shafted hammer of correct hardness, (b) twelve full blows on the end of a rigidly supported normalised bar of the same material and approximately the same section as the face of the head being tested.

The principal consideration in choosing a hammer for any particular purpose is that of its weight. Engineers' hammers are made in weights varying from $\frac{1}{8}$ kg to $1\frac{1}{2}$ kg, and of this range the most commonly used are hammers ranging from $\frac{1}{4}$ to $\frac{3}{4}$ or 1 kg. Light hammers are used for lighter and more delicate operations such as light riveting, striking centre punches and small chisels, driving in small pins, etc. Heavy heads are necessary on heavy chipping, driving pulleys and collars on to shafts, driving large pins and shafts into holes, etc. To remove an obstinate shaft, pin, pulley or any other detail requiring force, it is useless to employ a light hammer; providing adequate support can be obtained, the hammer will be most effective if it has plenty of weight.

Hammer shafts. These should be of well-seasoned, straight-grained hickory or ash, free from knots or other defects. They should be of a size and length suited to the size of the head, and after being shaped to suit the eye should be secured with a hardwood wedge.

Blow delivered by a hammer. The force of the blow delivered by a hammer is a difficult quantity to determine, because of various factors which we cannot measure. If we know, or assume certain things, however, we can form a rough estimate of it.

When a body of mass m kg is moving at a speed v, metre/sec its kinetic energy is $\dfrac{mv^2}{2}$ joules

If we can estimate the speed of a hammer head, we can thus calculate the energy it possesses at the moment it strikes its objective, and this energy is available for doing work on the object being struck.

Now Work = force × distance moved, so that if we know the distance moved in bringing the hammer to rest we can estimate the *average* force of the blow.

Example: A hammer of mass $\frac{1}{2}$ kg moving at 1 metre per second strikes a pin and drives it a distance of 3 mm. Estimate the average force of the blow.

Kinetic energy of hammer $= \frac{1}{2}mv^2 = \frac{1}{2} \times \frac{1}{2} \times 1^2 = \frac{1}{4}$ joule.

This is work stored in the hammer head, and if the head is brought to rest in 3 mm ($\frac{3}{1000}$ m) we have:

Work = force × distance

$$\tfrac{1}{4} \text{ joule} = \text{Force} \times \frac{3}{1000}$$

From which force $= F = \dfrac{1000}{3 \times 4} = \dfrac{1000}{1^2} = 83 \cdot 3$ newtons.

(The reader will probably get a better mental appreciation of this force if we express it in kilogramme units. To do this we must divide its value in newtons by the acceleration due to gravity, i.e. $9 \cdot 81$ m/s^2. Hence F $= 83 \cdot 3$ newtons

$$= \frac{83 \cdot 3}{9 \cdot 81} = 8 \cdot 5 \text{ kgf}$$

Soft hammers. For hitting finished surfaces which would be bruised or damaged by the hardened face of a hammer some form of soft hammer should be used. The chief soft hammers used have heads of rawhide, copper or lead. The rawhide hammer has a head of cast-iron into which plugs of rawhide are inserted, renewals of these being made when the old ones are worn out (Fig. 169(b)). The lead hammer has a lead head cast on to a length of steel tube, whilst the copper hammer has a copper head fastened in some way to a shaft made of tubing.

Files*

The chief cutting tool used by the fitter is the file. We have already discussed the file tooth, but we must now consider the chief types of file and their uses. The chief files are as follows (see also Fig. 170):

1. *Flat file*. This file is parallel for about two-thirds of its length and then tapers in width and thickness. It is cut on both faces (double cut) and both edges (single cut).

2. *Hand file*. The width of this file is parallel throughout, but its thick-

*See BS 498.

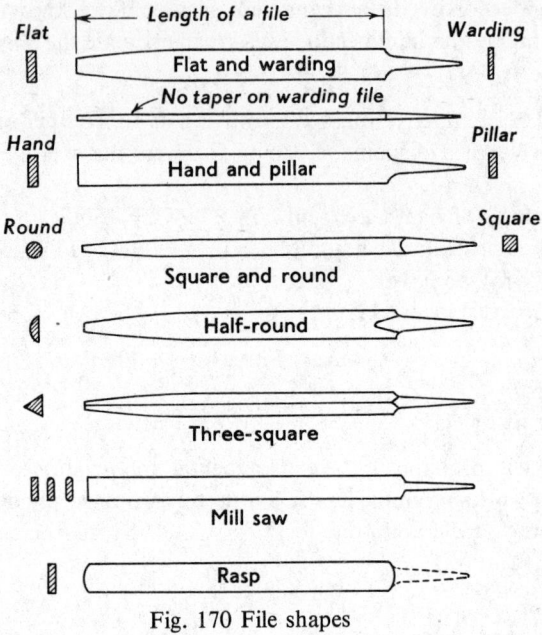

Fig. 170 File shapes

ness tapers similarly to the flat file. Both faces are double cut, and one edge single cut. The uncut edge is called the 'safe' edge and prevents cutting into one face of a square corner whilst the other face is being filed. Both these files are used for general surfacing work, the hand file being used more particularly when filing up to a step which must be straight and square.

3. *Square file.* This is parallel for two-thirds of its length and then tapers off. It is double cut on all sides and is used for filing corners and slots where the hand file could not be entered.

4. *Round file.* Tapers similar to square file. Used for opening out holes, producing rounded corners, round-ended slots, etc. Round files are usually double cut on the rough and bastard qualities over 150 mm long, whilst the rough and bastard under 150 mm, together with the second-cut and smooth, are single cut.

5. *Half round.* the rounded side is not a true half-circle, but only a portion of a circle. This side of the file is useful for many purposes involving the formation of a radius. The flat side of this file is always double cut. The second-cut and smooth are only single cut on the curved side.

6. *Three square.* Used for corners less than 90°, and in positions where awkward corners have to be taken out. Double cut on all faces.

Other and less common files are as follows:

7. *Warding file.* Similar to the flat file but thinner and parallel on its thickness. Useful for getting out narrow slots.

8. *Pillar file.* Nearly the same as the hand file but narrower. Useful in narrow apertures which the hand file would not enter.

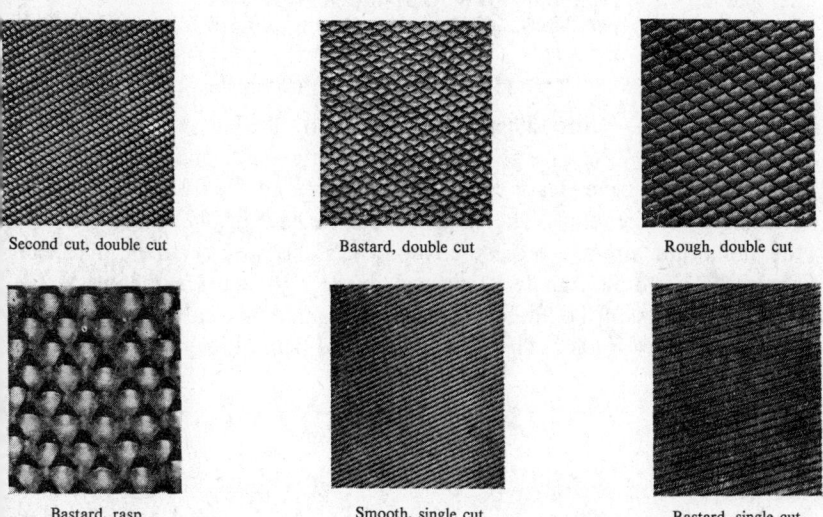

| Second cut, double cut | Bastard, double cut | Rough, double cut |
| Bastard, rasp | Smooth, single cut | Bastard, single cut |

Samuel Osborn & Co., Ltd

Fig. 171 File cuts (300 mm file)

9. *Mill file.* Similar to the flat file but parallel on both width and thickness. The *mill saw* file may have one or both edges rounded for forming the radius on saw teeth and in slots.

10. *The rasp.* The Horse Rasp is useful for filing soft metals, wood and soft non-metallic materials. The rasp tooth is cut with a pointed punch which gives the tooth form shown (Fig. 171). Generally the faces of a horse rasp are cut partly with rasp teeth, the ramainder being ordinary single or cross-cut file teeth.

Grades of cut. Files may be cut with teeth of the following grades: Rough; Bastard; Second Cut; Smooth and Dead Smooth. The rough grade is not greatly used and the relative tooth sizes for the other grades may be gained from the following table and Fig. 171, which shows a selection of the cuts with which 300 mm files may be obtained.

Table 17. Recommended practice for cuts of files
(*Applying to Flat; Hand, Half Round (Flat Face) and 3 Square*)

Length of file (mm)		100	150	200	250	300	350	400	450	500
Bastard	Pitch of	0·65	0·8	1	1	1·2	1·35	1·4	1·5	1·6
Second cut	teeth (mm)	0·6	0·67	0·8	0·9	1	1	1·15	1·3	1·4
Smooth	parallel to	0·43	0·5	0·58	0·6	0·65	0·67	0·7	0·75	0·8
Dead smooth	file length	0·3	0·3	0·32	0·34	0·35	0·37	0·4	0·42	0·45

Use of the file.—Cross-filing. The production of a flat surface by filing is a difficult task. Whilst we give the following brief description of *cross-filing*, the reader will never learn to file without some advice from a skilled filer, and plenty of practice. The work should be held firmly in the vise with the minimum amount projecting, and with the surface to be filed truly horizontal. The file handle is grasped in the right hand as shown at Fig. 172(a), the end of the file-handle pressing against the palm of the hand in line with the wrist-joint. The left hand should be used to apply pressure at

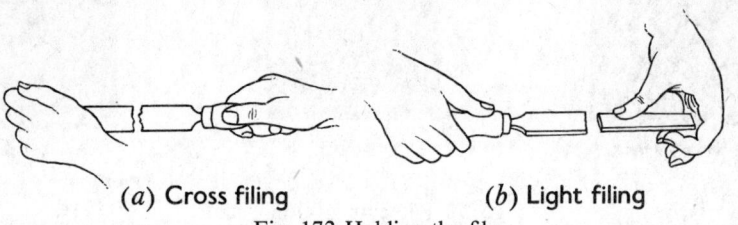

(*a*) Cross filing   (*b*) Light filing
Fig. 172 Holding the file

the end of the blade as shown. For lighter filing a rather lighter grip is used with the right hand and the end pressure is applied with finger and thumb (Fig. 172(b)). A position should be taken up on the left side of the vise and the feet firmly planted, slightly apart. A stroke should be made by a slight movement of the right arm from the shoulder, and by a sway of the body towards the work, each of these movements being about equal. As the file moves over the work it does not move parallel to its length, but in an oblique direction from left to right. The file must remain horizontal throughout the stroke, which should be long, slow and steady, with pressure only applied on the forward motion. On the return stroke, although the pressure is relieved from it, the file remains in contact with the work.

Success in filing flat is dependent on keeping the file horizontal throughout its stroke, and this is controlled by the distribution of pressure as between the two hands. The fault with beginners is to apply too much downward pressure with the right hand at the beginning, and with the left hand at the end of the stroke, causing the file to rock, and produce a round instead of a flat surface. Careful practice in gradually shifting the pressure from the left to the right hand will ultimately bring success in the production of a flat surface. But even with the best regulated filing there is always a tendency to rock the file slightly, and the curved surface of a tapered file tends to make a hollow surface, thus counteracting any slight rocking.

To test the surface of work during filing a straightedge should be placed on it occasionally, and the line of contact viewed for 'day-light'. When any considerable amount of metal has to be removed, the bulk of it should be removed by a rough or bastard-cut file and the surface progressively brought to a finish by second-cut and smooth files.

Draw filing. File marks may be removed and a good finish imparted by draw-filing (Fig. 173). For this purpose a fine-cut file with a flat face should be used (e.g. a mill file).

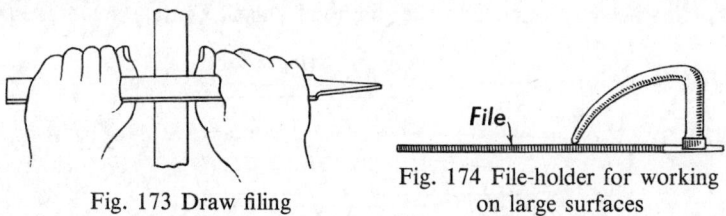

Fig. 173 Draw filing

Fig. 174 File-holder for working on large surfaces

Filing broad surfaces. When a surface is so wide that the handle would prevent the full stroke of the file, a surface file-holder must be used. This clamps on the tang of the file (Fig. 174). Surfaces produced in this way are generally finished flat by scraping (see later), so that small errors in flatness may be corrected during that process.

The pinning of files. Soft metals when filed tend to clog the file teeth with minute lumps of metal ('pins'), and if the file is not cleared this accumulation will not only stop the efficient cutting action, but also scratch the work. If the pins are not too firmly wedged they may be removed with a *file card*, which is a brush made from a strip of webbing having thin, hard wire bristles, and nailed on to a piece of wood. Tightly-wedged pins must be scraped out with the point of a scriber. Pinning may be partly prevented by chalking the file or applying turpentine, but these methods should not be used when filing cast-iron or brass as they are only applicable to ductile metals and cause the surfaces of brittle metals to glaze under the file.

The care of files. The teeth of files are very brittle and easily broken. For this reason files should never be heaped together, nor mixed carelessly with other tools, but should always be kept each in its separate rack. For similar reasons a new file should not be used on a surface of unknown hardness such as the surface of a casting or welded joint. Brittle metals such as cast iron and brass are not readily filed with a worn file, so that new files should be started on these metals, later being transferred to working on mild steel which can easily be cut with a file having teeth slightly worn. When filing a cast-iron surface with edges in the black, scaly condition, the edges should be chamfered slightly with the *edge* of the file to prevent damage to the file teeth on the hard scale.

Chisels and chipping

The cold chisel is an important cutting tool used by the fitter, and engineers' chisels are distinguished from other chisels by the fact of their not having a wooden handle. The four most important types of chisel are the flat, the cross-cut, the half-round and the diamond chisel. Chisels are made from

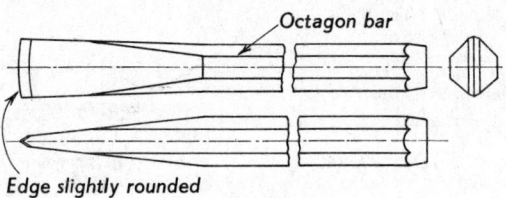

Octagon bar

Edge slightly rounded

Fig. 175 A flat chisel

cast tool steel of octagonal cross-section, the heaviest being made from about 20 mm material and the smaller sizes from lighter section steel down to about 10 mm. Alloy tool steels may also be used for chisels, with improved results (see p. 71). *Centre punches* and *drifts* are modifications of the chisel.

The flat chisel. This is the general-purpose chisel of the fitter and in its heaviest form is made about 200 mm long when new, the lighter sizes being proportionately shorter. It is tapered and flattened for about one-third of its length to the cutting edge, which should be about 2 mm to 3 mm thick on the large chisel, and less in proportion for smaller chisels.

The cutting edge is ground to an angle suited to the material being worked upon, as we have already discussed on page 170, and it should not be exactly straight, but given a slight curvature, as shown in Fig. 175. After forging to shape and roughly grinding, the chisel edge should be hardened and tempered, as explained on pages 45 to 47. Consequent grinding after finishing should be carried out with care, as otherwise the edge will be over-heated and softened. If this occurs it will be necessary to harden and temper the chisel all over again. On no account should the head of the chisel be hard where it is struck by the hammer.

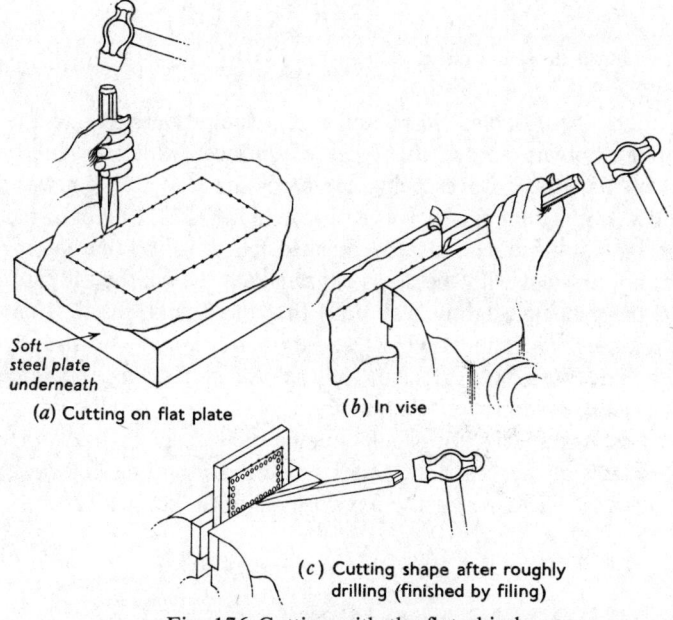

(a) Cutting on flat plate

Soft steel plate underneath

(b) In vise

(c) Cutting shape after roughly drilling (finished by filing)

Fig. 176 Cutting with the flat chisel

The flat chisel is used for cutting sheet and plate material, as shown at Fig. 176(a) and (b), for cutting out slots after their outline has been previously drilled, as shown at (c), for surfacing work as explained later, and for miscellaneous cutting jobs as they arise.

The cross-cut chisel. The cutting end of this chisel is forged to the shape shown at Fig. 177, the cutting edge AB being 6 mm to 9 mm wide. From the edge, the metal thickness tapers off slightly, being slightly thinner at CD. This is to permit the body of the chisel to clear when a groove is being cut.

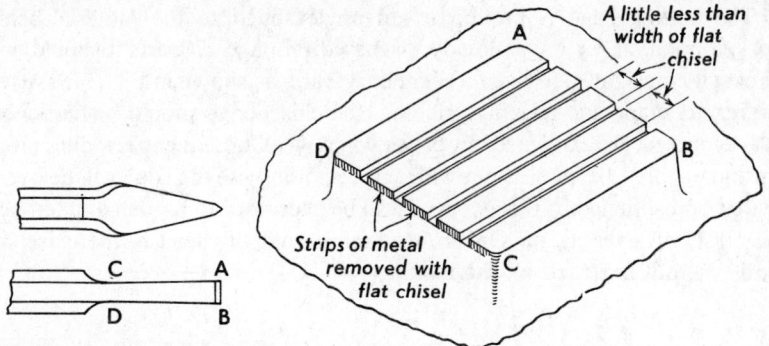

Fig. 177 Point of cross-cut chisel   Fig. 178 Grooves cut by cross-cut chisel

The original use of this chisel, and from which it derived its name, was for cutting grooves across the faces of surfaces which required to be finished off flat, the grooves being the preliminary to the removal of the main metal, with a flat chisel. This is shown in Fig. 178, where the face A B C D in the rough state must be finished flat down to the line marked. Grooves are first cut with the cross-cut chisel as shown, the width of metal between them being slightly less than that of the flat chisel. This metal is then removed with the flat chisel, and the surface finally finished by filing and scraping. With the extension of the use of shaping, planning and milling machines, however, this method of surface production is only employed in cases where it would not be convenient to use a machine. The chief uses of the cross-cut chisel now are for cutting keyways, slots, and in entering places where the flat chisel is not a convenient tool.

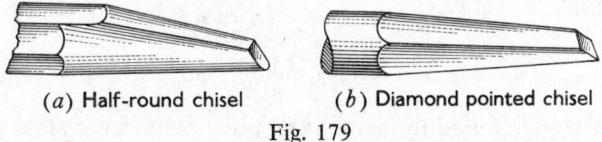

(a) Half-round chisel   (b) Diamond pointed chisel

Fig. 179

The *half-round* and *diamond* chisels are shown at Fig. 179. The half-round pattern is useful for cutting grooves such as oil grooves in bearings and similar work. The diamond chisel is often used for cutting holes in plates such as boiler plates, and for grooving the start of a drilled hole to correct for an error in the starting of the drill.

General points on chipping. The chisel should be held about midway between the head and edge and at the correct inclination, as if it is at too great an angle the edge will cut too deep, whilst if the angle is not enough the chisel will cease to cut. Within these limitations, the smaller the angle the more effective are the hammer blows, and hence the greater the efficiency. The edge should be kept well up against the shoulder formed by the cut and chip, and particles of metal should be kept away. When the chisel approaches the edge of a surface, particularly if the metal is cast iron, it should be reversed, or the cut taken at right angles to the previous one. If this precaution is neglected the edge of the metal is likely to be broken away.

Chipping ribs. When a surface, such as the base of a bracket, has to be bedded on and screwed to the surface of a rough casting, it would be a difficult and tedious operation to mate the two surfaces together if the surface to be fitted consisted of solid and continuous metal. The bedding together is greatly facilitated, however, by casting the bracket with its base in the form of a ribbed pattern, and this is bedded on to the rough surface by chipping and filing these ribs until they contact at all points with the surface to which their faces must be fastened. Such ribs are called chipping ribs, and their use facilitates the attachment of extra brackets to the body or legs of machine tools (Fig. 180).

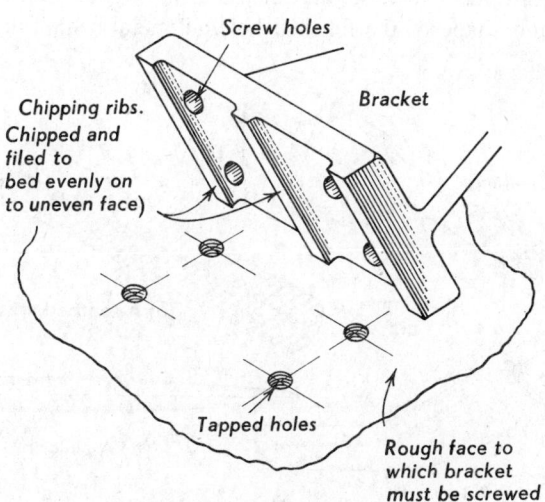

Fig. 180 Chipping ribs

The centre punch

This tool, which is shown at Fig. 181, is used when circular dot marks are required. When a job has been marked out it is usual to follow along the lines with small dot marks in case the lines themselves become obliterated. The surface which is so marked is then finished to the centre of the dot

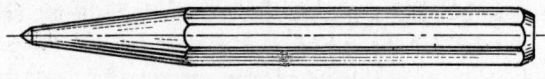

Fig. 181 Centre punch

mark. Centre punches are also used for marking the centre point of drilled holes for the purpose of giving a start to the drill; marking the centre of circles to provide a point for placing one leg of the dividers to scribe the circle, and so on. Punches are usually made of cast steel, being hardened and tempered in the same way as a chisel.

Drifts

A drift is a tool used to finish off small non-circular holes to shape. Holes which are square, hexagonal, rectangular, etc., in shape, particularly if their dimensions are small, constitute very difficult problems in finishing to size. They are generally finished very near to size and shape with chisel, file or any other convenient method, and then a drift is made with its end to the exact size and shape of the finished hole, all metal behind this being filed

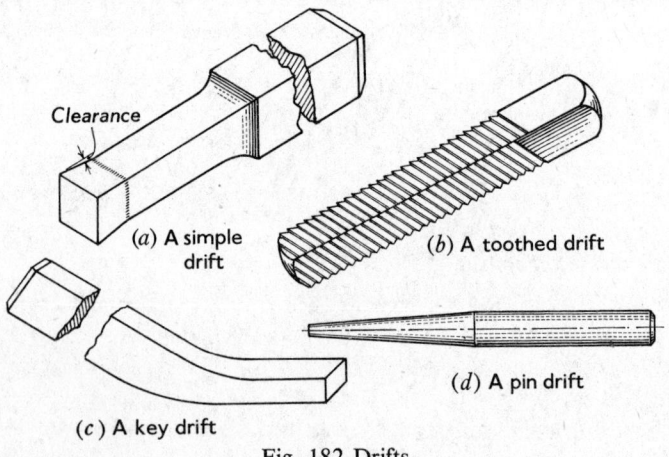

Fig. 182 Drifts

away for clearance. After hardening and tempering, the drift is hammered, or pressed through the hole, and if the previous work has been done properly the drift will just clean up the sides of the hole to size and shape. A simple drift is shown at Fig. 182(a) and a more elaborate one made with a number of cutting edges at (b). On the drift shown at (b) each tooth removes a little bit more metal than the one in front of it, the final two or three teeth being made to the finished size and shape of the hole. Drifts of this type are generally forced through under steady pressure as from an Arbor press (Fig. 183). A *Key-drift* (Fig. 182(c)) is a tool for driving out keys from pulleys and gears (keys such as that in Fig. 188). The *Pin-drift* is a round tapering punch for drawing out pins such as are used to hold crank handles, levers, etc., on to shafts. This is similar to a centre punch except that it has a flat point and a longer taper (Fig. 182(d)). Neither the key-drift nor the pin-drift do any cutting, but merely act as connections between the hammer and the object to be moved.

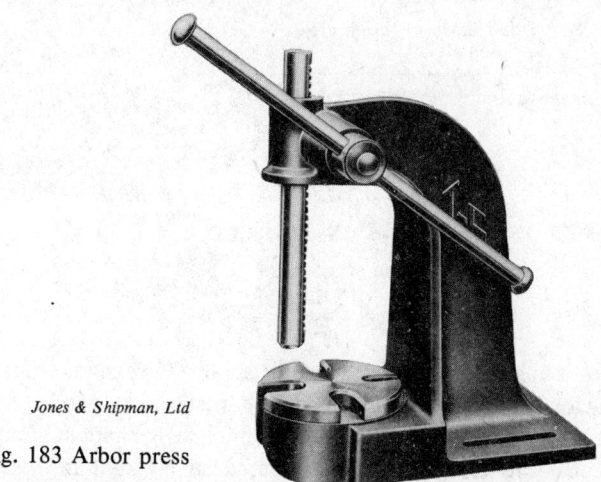

Jones & Shipman, Ltd

Fig. 183 Arbor press

Scrapers and scraping

The shaper, planer and milling machine have robbed the file and chisel of a large amount of the metal removing they did in the past and the use of surface grinding methods has supplanted a great deal of scraping for the finishing of machine surfaces and slides. It is not likely, however, that scraping will ever be completely replaced as there are still many occasions where the type of work renders grinding impossible (e.g. Fig. 185) or where its quality demands the individual attention associated with the scraping process.

It will be appreciated that a surface produced by chipping and filing cannot be finished accurately flat by these methods. Furthermore a machined surface, although being nearer to flatness than one that has been filed, will probably still need slight correction due to warping and distortion taking place after machining.

The flat scraper. The flat scraper is a tool for working up the surface of a flat face for the purpose of removing errors in flatness, and leaving a smooth finish. From Fig. 184(a) it will be seen that the flat scraper

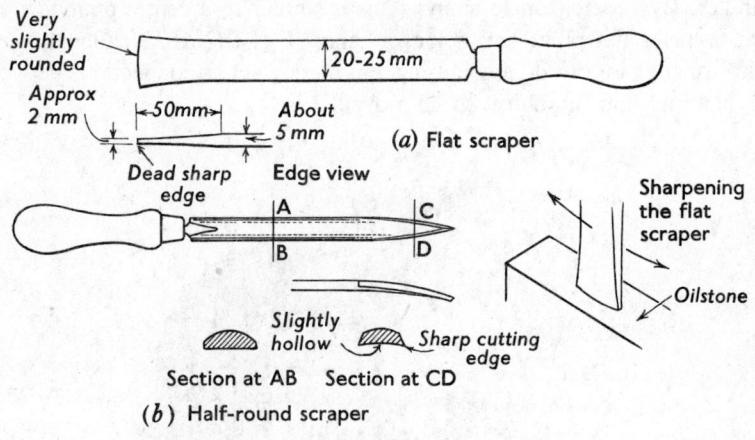

Fig. 184 Scrapers

resembles a file, and in many cases flat scrapers are made from old files by flattening out the end, grinding and re-hardening. A good scraper should be made from the best steel procurable and its edge kept in good condition. As the cutting force on a scraper is comparatively light and without shock, its edge need only be given a slight degree of temper, a fact which is favourable to its successful working since the fine scraping cut which it takes tends to dull the edge rapidly. After the edge has had a preliminary grinding, taking care not to draw the temper, it should be finished off by rubbing on an oilstone, by holding the scraper close to the end and rubbing the end on the stone with a rocking motion in a plane containing the edge of the tool (see diagram). This should be repeated, in addition to stoning the sides of the cutting edges, until the edges are dead sharp. During use, it will be necessary to bring up the edge from time to time by end rubbing on the oil-stone as just explained.

Using the scraper. The purpose of the scraper is to correct slight irregularities from flatness, and if these are very great, the surface should be brought nearer to perfection by some method other than scraping, as the process is rather laborious, and not intended for the removal of much metal. We will assume that a surface has been planed reasonably flat, and is to be scraped up to a flat finish. The first thing necessary is a standard of flatness with which to compare the surface being scraped, and for this a surface plate (Fig. 133) is necessary. If the surface to be worked on is smaller than the surface plate it can be rubbed on the plate, otherwise the surface plate will have to be reversed and rubbed on the surface.

After thoroughly cleaning the plate and surface, the plate should be smeared with a thin layer of prussian blue, or red lead in oil, the other surface then placed on it and rubbed about slightly. If there is any deviation from flatness, the high parts will be smeared with some of the marking substance from off the surface plate. These portions must now be reduced in height by scraping, and this is done by holding the scraper as shown at Fig. 185, pressing on with the left hand, and making short strokes 12 mm

Fig. 185 Using the flat scraper

to 25 mm long. The scraper is left in contact with the surface for the return stroke but no pressure is applied to it. Having gone over the surface with strokes in one direction, the next set of strokes should be made at right-angles to the first, and this changing round procedure repeated until sufficient metal has been removed. From time to time, of course, a trial should be made by rubbing the work on the surface plate, then working at the parts where the marking compound shows contact to be taking place. Gradually the surface will be brought to such a condition that its whole area is covered by small areas of contact and it may be considered sufficiently accurate for average requirements when these spots are 3 mm to 5 mm apart. Much patience and practice is necessary for scraping, and when the reader has become proficient at producing a flat surface he may then try producing the handsome frosted effects seen on the beds and slides of machine tools. For successful cutting keep the scraper sharp, and do not hold it too high.

The half-round scraper. This scraper (Fig. 184(b)) is used for working up internal cylindrical surfaces and is also a very useful tool in the kit of a centre lathe turner. It is often made from an old half-round file. Our remarks regarding the condition and finish of the cutting edge of flat scrapers applies to this tool. The half-round scraper is useful for finishing the surfaces of cylindrical bearings for the purpose of bedding in the shaft. This operation is carried out as follows: We will assume that the bearings are in two halves, and that they have been bored as near as possible for the shaft to be a running fit; also that the shaft has been ground or otherwise smoothed on the parts where it takes its bearing, and any 'out of roundness' corrected.

The bottom halves of the bearings are fitted into position in the casting which supports them, and the bearing portions of the shaft are smeared with a thin layer of red lead paste or prussian blue. The shaft is placed in position, rotated for a few turns and then taken out, when the high spots on the bearing liners will be shown up by traces of the marking agent adhering to them. These areas must be scraped with the half-round scraper, and the operations of trying in the shaft and removing the high spots repeated until the whole area of both bearings shows a covering of contact points 3 mm to 5 mm apart. When the bottom liners have been completed the shaft may remain in position, the caps and top liners put on and the nuts pulled up until the shaft can just be rotated. Upon removal of the top liners the high places will again be visible and these must be scraped until contact over the whole area is obtained and the shaft is a good running fit with the bearing cap nuts tight. It may be necessary to file some metal off the flat joining faces of the two bearing halves or, alternatively, thin pieces of sheet copper or brass (shims) may have to be placed between them to open them out slightly. Subsequent wear up to a limited amount may later be taken up by removing the shims and re-bedding the shaft, or filing metal off the joint faces and rebedding. If the shaft is not re-bedded after such an operation it may only be making contact with the bearing liner at top and bottom.

The hacksaw. The hacksaw is the chief tool used by the fitter for cutting-off, and for making thin cuts preparatory to other chipping and filing operations. There are many sizes and types of hacksaw blades on the market with special recommendations for particular classes of work and in view of the variable factors governing the selection of a blade it is advisable to keep a stock of different blades to meet the different sawing conditions. A diagram

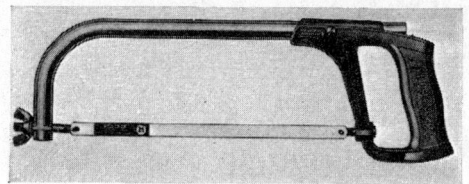

James Neill & Co., Ltd

Fig. 186 Hacksaw frame

of a hand-saw frame is shown at Fig. 186, and the blade fits over two pegs which project from the pins sliding in the ends of the frame. The wing nut at the front end of the frame on some types, and the screwed handle on others, is for tensioning the blade and adjustment is provided in the construction of the frame for blades of different lengths. The length of the blade is that between the outside edges of the holes which fit over the pins, and the most usual blade for hand work is one measuring 250 mm long by 12 mm wide. Blades may be of carbon or high-speed steel, and may be finished either with the cutting edge only hardened, or they may be hard right through. The soft-backed blades are tougher and less liable to snap than the all-hard blades, but are not as efficient. High-speed steel blades, although more expensive than those of carbon steel, will generally be found cheaper, if the cost is reckoned in terms of useful life.

Hacksaw teeth are specified by the *pitch* of the teeth (Fig. 187(a)). If the reader examines the teeth of a hacksaw blade he will notice that the teeth have a *set* to the sides. This causes the blade to cut a slit wider than itself, and prevents the body of the blade from rubbing or jamming in the saw-cut. Often alternate teeth are set to right and left, every third or fifth tooth being left straight to break up the chips and help the teeth clear themselves (Fig. 187(b)). Fine-toothed blades, for cutting thin metal, are sometimes made with a wavy set to minimise stripping of the teeth from the blade.

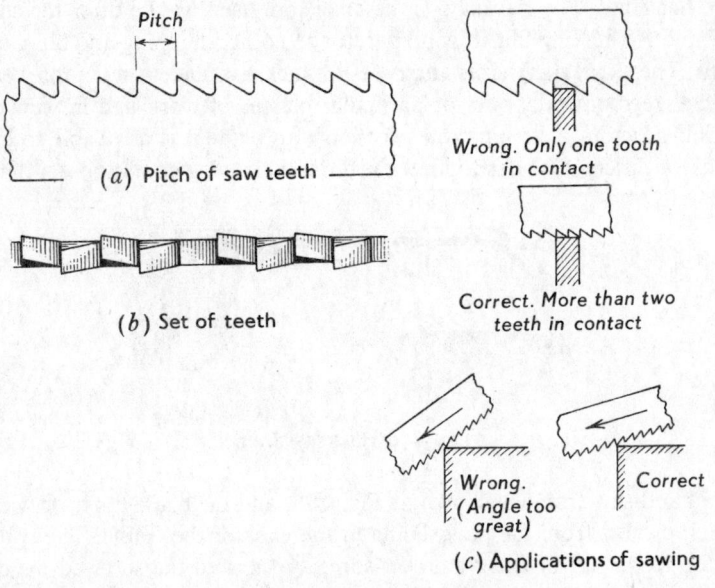

Fig. 187 Hacksaw teeth

The choice of a blade for any particular class of work is largely governed according to the pitch of the teeth; these should be as large as possible to give maximum clearance for the chips and avoid clogging, but at the same time at least two or three teeth should be in contact with the surface being sawn. If this is not attained the teeth will be stripped from the blade sawing too sharply over a corner will also result in teeth being torn off (see Fig. 187(c)).

The best all-round blade for hand use is one with about $1\frac{1}{2}$ mm pitch

For other and special classes of sawing the following blades should be used:

(*a*) Solid brass, copper and cast iron $1\frac{3}{4}$ mm pitch.

(*b*) Silver steel and thin cast steel rods; thin structural sections 1 mm pitch.

(*c*) Sheet metal and tubing (e.g. steel, copper and conduit tubing) $\frac{3}{4}$ mm to 1 mm pitch.

Blades must be strained tightly in the frame and slow, firm and steady strokes (50 per minute) should be used, lifting the blade slightly on the return stroke. Breakage of blades may be caused by the following: (*a*) rapid and erratic strokes, (*b*) too much pressure, (*c*) blade held too loosely in the frame, (*d*) binding of the blade from uneven cutting, (*e*) work not held firmly in the vise. Solid metals should be cut with a good pressure and thin sheets and tubes with light pressure. Insufficient pressure at the start of a cut may cause the teeth to glaze the work, and so rub their edges away.

Work on flat surfaces

We have now discussed the characteristics and operation of the tools used for cutting and finishing flat surfaces on the bench and we might, with advantage, discuss one or two cases in which the use of these tools will be necessary.

The fitter is the agent whose responsibility it is to finish and adjust the machined surfaces of parts so that the parts assemble together correctly and with satisfactory results. In the course of this work he may have to make certain small details for which it would not be worth setting up a machine.

Example 1. Fitting a key.

Let us consider the job of fitting and keying a detail such as a lever on to the end of a shaft, with a taper gib-headed key (Fig. 188). When the details arrive at the bench for fitting, the shaft will have had the keyway cut on a milling machine, the boss will have had its keyway cut by slotting or key-seating if such a machine is available and the key may have been roughly shaped up with metal left on all over for the fitter. The shaft will' have been turned so as to be the correct type of fit in the hole of the boss so that this aspect will not concern the fitter. If no machining method is

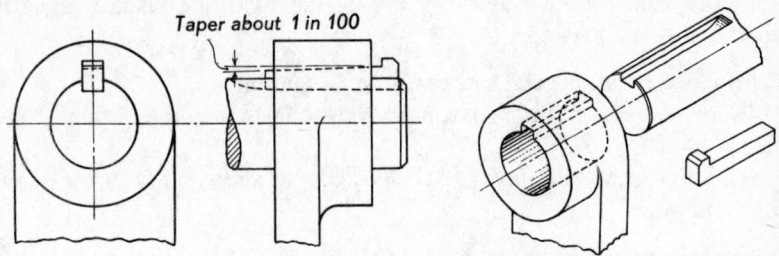

Fig. 188 Lever fitted to shaft with taper gib key

available for the keyway in the boss, the fitter will have to cut it with hand tools, and even if it has been cut he may have to file the taper on its bottom face. In this, and any other keying job, it is important that a good fit is obtained between the *sides* of the key and the keyways in shaft and boss. In this particular case of a taper key, the top tapered face of the key must also fit all along the bottom of the boss keyway, since this fit, when the key is driven home, secures the boss on to the shaft. The order in which the work must be done and the precautions necessary may be summarized as follows:

1. The width of the keyways in the shaft and in the boss must be exactly the same, and if they are not already so, the narrowest must be filed on its sides. They can be tested for similarity and parallelism with a block gauge or inside calipers.

2. If the key has been roughly shaped to size, it must now be filed up on its width so as to be parallel, and a nice push fit in the two keyways. If it is in the rough material form, it must be marked out and cut to shape by a combination of sawing, chiselling and filing according to circumstances. This preliminary shaping should leave about $\frac{1}{2}$ mm on the width and thickness, after which the sides may be finished as above.

3. The bottom of the keyway in the boss should now be inspected for taper. For example, if this is 1 in 100, the measurement between the far side of the hole and the bottom of the keyway should be more at the front end of the boss than at the other. If the boss is 50 mm long, the difference will be $\frac{50}{100}$ mm $= \frac{1}{2}$ mm. This is not particular to a few hundredths but should not vary too much from its correct amount. If the taper is not correct, or if none has been put on, it must be filed. At the same time check the depth of the keyways as being to the drawing.

4. The boss must now be assembled on the shaft and the key finished by filing its top and bottom faces flat, and with the correct taper on the top. To do this, file the bottom of the key flat and square with the sides, and then file the top to approximately the correct taper, until it can be entered a little way. A smear of red lead paste or prussian blue on the bottom faces of the boss keyway will now show whether the key taper conforms to that on the keyway. If adjustment is necessary, take the least possible amount of metal off in doing it as the key may enter too far. When the taper is correct, file the top down to it until the key enters with the head end about 12 mm from the boss face. Drive the key in fairly tight and drive it out again, examining its surface for polish where contact has taken place. Finally touch up to get perfect contact and until, when driven home hard, the underside of the head is 3 mm to 6 mm from the boss face. Cut off any surplus length from the tail end of the key, and finish the head by smoothing up and chamfering all edges and corners.

If the boss keyway has to be cut on the bench, this should be made the first operation. Mark off the keyway at each end, and if possible, run two sawcuts with their outer edges about 2 mm from the sides of the keyway and their depth $\frac{1}{2}$ mm less than the depth of the keyway. Remove the intervening metal with a cross-cut chisel and finish the keyway by filing with a parallel file, taking care that whilst you are watching for the size at one end, the other end of the file does not wander and overstep the mark, and test for parallelism with inside calipers from time to time.

Example 2. Squaring the head of a screw, and preparing a suitable hole in the end of a key.

The finished screw and key are shown at Fig. 189(a), and the screw, when it comes from turning, will have a solid head 18 mm diameter, with a line turned on it to mark the end of the squared portion. We will assume that the key has been forged to shape, with the end left blank.

(*i*) Using a safe-edge file, commence to file the screw for one of the flats of the square, terminating the flat about $\frac{1}{2}$ mm short of the scribed line. When the filing is about 2 mm deep, test for flatness, and with micrometer or calipers test for parallelism from the opposite turned edge (Fig. 189(b)). Correct for any errors in flatness or parallelism, and complete the filing of the flat until it measures 15·5 mm from the opposite side of the bolt head (i.e. one half head dia. (9 mm) + one half the square (6·5 mm)).

(*ii*) Turn over and file the opposite side of the square until it is flat, and parallel with the first one, finishing it off to a width of 13 mm (Fig. 189(c)).

(*iii*) Commence filing the third flat of the head and get it flat, square with one of the other faces, and parallel with the turned surface opposite to it. Continue to file it down, maintaining the above relationships, until it measures 15·5 mm from the opposite side.

(*iv*) Finish off by filing the fourth flat of the square, making it parallel with the opposite one and finishing to 13 mm thick. Finish the end shoulders of the flats uniformly to the scribed line and file about a ½ mm chamfer on the front edges and corners of the square.

(*v*) With odd-legs find centre of key-boss, and scribe its centre lines AB and CD. With dividers, scribe and dot punch a 13 mm diameter circle. Set dividers to one half the diagonal of a 13 mm square (approx. 9·2 mm) and, from the centre, mark this distance off on each centre line. Scribe and dot punch the 13 mm square (Fig. 189(d)).

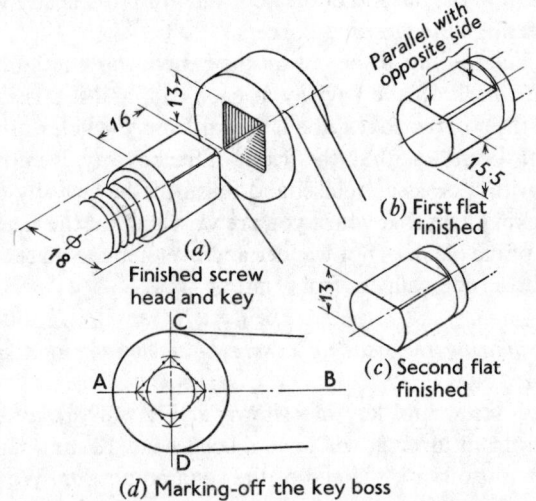

(*a*) Finished screw head and key

(*b*) First flat finished

(*c*) Second flat finished

(*d*) Marking-off the key boss

Fig. 189 Filing a square-head screw and key

(*vi*) Enlarge the punch mark in the centre, and drill out with a drill about 12 mm diameter.

(*vii*) Use a square file to open out the drilled hole to within about ½ mm of the marking out for the square hole. (Be careful that the file does not cut on its sides and bite in beyond the marking out.) Check the filing for being square with the face of the forging.

(*viii*) With a smooth file complete the filing of the squared hole. During this finishing, the work should be constantly tried with the square on the screw head and the hole carefully filed until the square will enter with a nice sliding fit in any of its four possible positions. Finish the key by taking off all rough corners and edges. The screw and key heads will be improved by giving them a final potash surface-hardening treatment.

Example 3. To fit the sliding block shown at Fig. 190(a).

When they come to the bench, the slot in the casting, the cast-iron block and the two steel retaining plates will have been shaped or milled, with a few hundredths left on for adjustment when fitting. The block has now to be let into its groove, and the retaining plates set and secured with screws. As the block is required to slide it will be advisable to finish scrape its bearing surfaces, i.e. top, bottom and sides.

(*i*) With a vernier, measure the width of the block and the sides of the slot to see how much material has been left on. At the same time check them for parallelism. There should not be much metal to remove, suitable conditions being such that the width of the block at its corners may just be forced a little way into the slot. The thickness of the block should be about the same as the depth of the slot.

(*ii*) If much material remains on the width it is advisable to remove most of it by filing. Verify from which part to remove it to bring the size to the drawing dimension and set to work. When filing, maintain flatness, parallelism and squareness as near as possible. Continue removing metal with the file until a corner of the block may just be forced into the slot.

(*iii*) Now set to work on the base of the block and scrape it up flat to a surface plate (see p. 198). When this is finished, scrape up the two sides flat, parallel and square with the base. Check occasionally with the slot to ensure that you are not removing too much metal (Fig. 190(b)).

(*iv*) The sides of the slot may now be scraped until the block will enter, and as soon as this occurs feel which are the tight places and scrape them down. When the block can be moved about, put a thin smear of red lead paste on it and find the high spots on the sides of the slot by pulling the block backwards and forwards. Continue scraping down the high spots until the block is a nice sliding fit. Chamfer its bottom edges if necessary to let it down to the bottom of the slot (Fig. 190(c)).

(*v*) When the sides are fitting satisfactorily, smear some marking paste on the bottom of the block and *without forcing it down* rub it along the bottom of the slot. Scrape down the high portions and continue the process until contact is taking place over the whole area of the base.

(*vi*) The conditions between the top faces where the retaining strips go

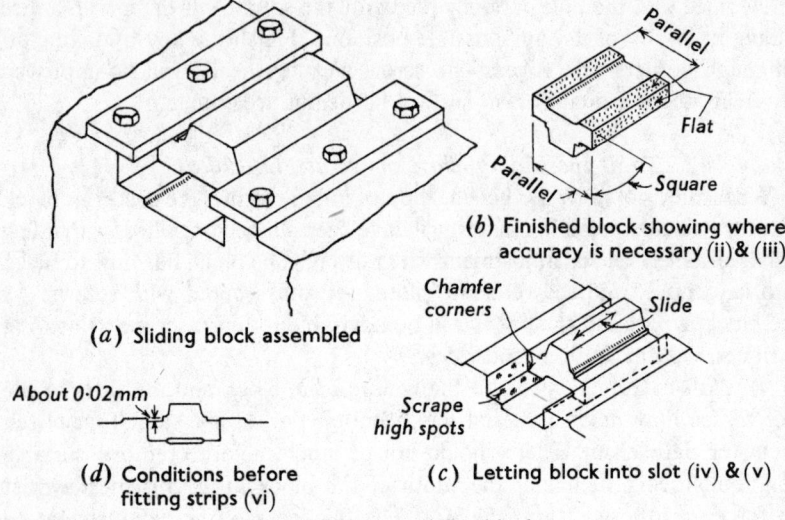

(a) Sliding block assembled

(b) Finished block showing where accuracy is necessary (ii) & (iii)

(d) Conditions before fitting strips (vi)

(c) Letting block into slot (iv) & (v)

Fig. 190 Fitting sliding block

may now be checked. With a micrometer measure the block to see if its top faces are parallel with the base, and the thicknesses equal at both sides; if not, file up until they are correct. When the block is now placed in the slot its top faces must be adjusted so that they are about 0·02 mm *below* the top faces of the slot (Fig. 190(d)). This will probably entail adjustment to one pair of faces, and when doing it, take care to preserve flatness and parallelism. (If material has to be taken off the slot faces, test across the top with a straightedge.) The above clearance should be obtained with the top faces of the block finished by scraping.

(*vii*) Now pick up the retaining strips and file or scrape their bottom surfaces perfectly flat. Polish the other faces (for appearance) and remove all sharp edges and corners having a near chamfer.

(*viii*) Mark out and drill these strips for the screws, then clamp them in position and spot the casting for the tapping holes. Drill and tap the casting, finally screwing on the plates to complete the job (see next Chapter for drilling, tapping, etc.).

9 The bench (*cont.*). Marking out, drilling, screwing

Unless work is of such a simple nature that its surfaces can be machined without any guidance other than the measuring facilities at the disposal of the machine operator, or unless the machine operator himself is very skilled and may be relied upon to obtain the correct surface relationships without any help, it is usual to indicate the position of finished surfaces by lines scribed on the component. These lines assist the machinist in setting up the work in his machine, and indicate to him the limit to which he may allow the cutting to proceed. The process of marking out is one of the shop activities for which the fitter is generally responsible, and is employed when the number of similar parts to be made is not great. For large quantities, marking out would be unduly wasteful of time and expense; in such cases it is eliminated by holding the component, whilst machining, in a jig or fixture, which locates it in the correct position, and provides some means for guiding the tool or cutter in the proper path.

Tools used for marking out

Before going on to the discussion of actual marking out examples, we must take note of a number of workshop tools and appliances used in connection with marking out as well as for other purposes, and with which we have not yet had occasion to deal. We have mentioned the *marking out table* which is an essential part of the equipment.

The **Angle Plate** (Fig. 191(a)) is for supporting a surface at right-angles to the surface of the table. It is provided with plenty of holes and slots for accommodating the bolts necessary to secure articles to it. An *adjustable* angle plate is useful for angular work.

Vee blocks (*b*) are used for supporting shafts and bushes and are generally made in pairs. It is important that a shaft, when resting in the blocks, shall be parallel with the surface table. This may be checked by trying a dial indicator over the shaft at each end. A shaft, clamped in a long vee block,

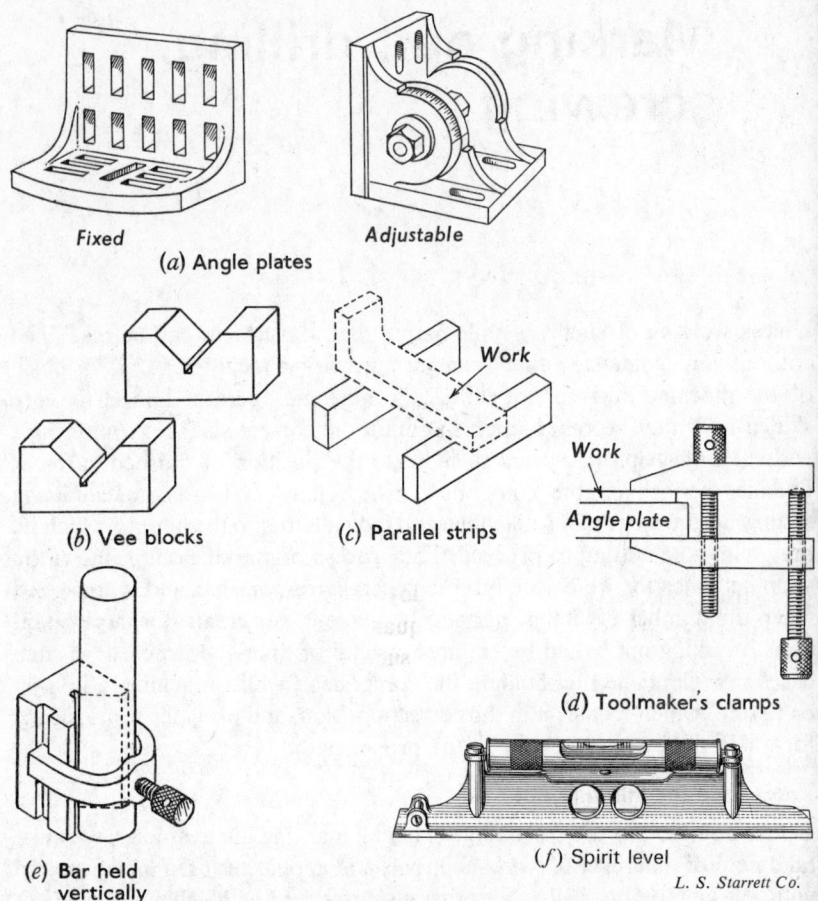

(a) Angle plates

Fixed   Adjustable

(b) Vee blocks   (c) Parallel strips

Work

Work

Angle plate

(d) Toolmaker's clamps

(e) Bar held vertically

(f) Spirit level

L. S. Starrett Co.

Fig. 191 Bench and marking-out accessories

may be supported perpendicular to the table if the end of the shaft is square with the vee surfaces (e). Vee blocks may be of cast-iron, or of mild steel hardened and ground, the latter generally being used for the smaller sizes.

Parallel strips* (c). These are useful for supporting work on the marking-off table, and may be of cast-iron smoothly machined, scraped or ground, or of steel which, when hardened and ground, makes very durable strips. The corresponding widths of a pair of strips must be of exactly similar dimensions, the strips must be perfectly parallel, and all their faces square; their

* BS 906.

lengths are relatively unimportant. Several sets of strips of varying sizes should be available.

Toolmaker's clamps (*d*) are an essential item of a fitter's kit, and two or three pairs should be available. They are handy for clamping work as shown, in addition to many other uses.

Spirit level (*f*). If the top of the marking-off table is level, as it should be, the spirit level is useful for setting surfaces parallel to it. The level is also necessary for levelling up machine beds and tables.

Dividers and trammels (Fig. 192 (a) and (b)) are used for scribing circles, marking off lengths, etc. Trammels are made to extend beyond the range of dividers, and those shown at (b) have a bar 350 mm long, but two or more of these bars may be joined together with sleeves of the form shown. The bent legs enable the trammels to be used in the same way as inside calipers.

Hermaphrodite calipers (*d*). In spite of its name, this is a useful tool, being half calipers and half dividers. When a line has to be scribed parallel to an edge, or an arc parallel to the rim of a circular disc or bar, the caliper end is moved along in contact with the edge and the line scribed with the point of the instrument. A better name to give this instrument in the workshop is *jennies* or *odd-legs*.

Feeler gauges* (*c*). These form a useful accessory not only on the bench, but also in connection with many other jobs in the shop. They consist of a series of blades or leaves, having thicknesses ranging from about 2 to 60 hundredths. Feelers may be used to gauge small distances (e.g. in flatness tests) and may be used in conjunction with other gauges. For example: the width of a slot $25 \cdot 25$ mm could be measured by using a 25 mm standard plug gauge together with a feeler of $0 \cdot 25$ thickness.

The scribing block, square, protractor, etc., which we have already described, will also be needed for marking off. In addition, a number of thin steel wedges, packings, jacks (Fig. 192(e)) and other oddments will be useful. These are gradually collected as different needs arise, and they should be kept in a box along with an assortment of bolts and nuts, clamps, etc.

Examples of marking out

We will now discuss the marking out of a few typical components, first of all reminding the reader that for a scribed line to be visible, it is first

***** BS 957.

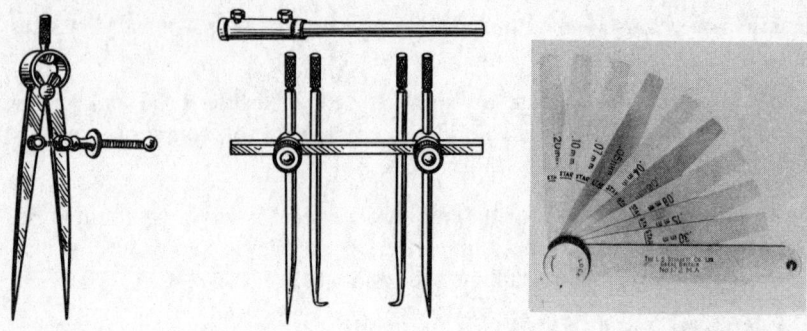

L. S. Starrett Co

(*a*) Dividers   (*b*) Trammels, with joining sleeve above   (*c*) Feeler gauges

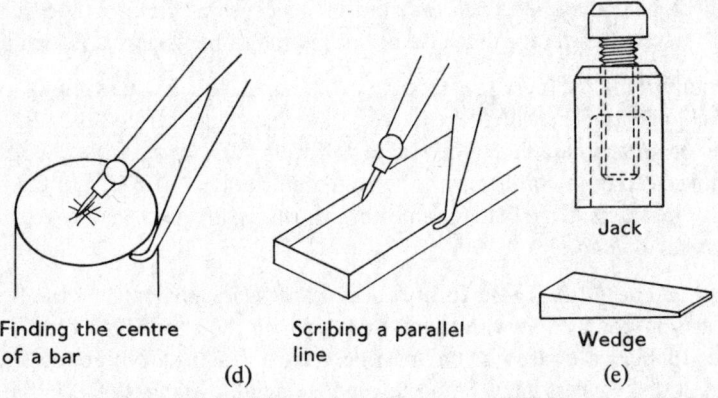

Finding the centre of a bar   Scribing a parallel line   Jack   Wedge

(d)   (e)

Fig. 192 Marking out and general accessories

necessary to prepare the surface which is to receive it. The rough faces of castings which have to be marked should be brushed over with a little whitewash, which, when dry, will show up a line scribed on it. Machined surfaces may be prepared by brushing them over with copper sulphate solution which leaves a thin film of copper on the surface. This shows up a scribed line very clearly.

The second point we should like to stress is that marking out does not dispense with the use of measuring instruments for the control of machining. However exact a component is marked out, the marking line may be up to 0·20 mm wide. For the simple fractional dimensions on a drawing, which may generally be finished to within $\pm \frac{1}{4}$ mm or 0·25 mm, it is quite in order to work to the scribed line, but for particular dimensions the line must only be used as a guide, the final finishing being done to micrometer, vernier or some other exact measuring method.

Example 1. To mark out the vee'd plate shown at Fig. 193(a).

Here the plate may first be shaped to its finished rectangular dimensions, viz. 90 mm × 50 mm × 10 mm thick.

1. Cover one side with copper sulphate solution, set the odd-legs to 10 mm and mark lines 10 mm from each edge and 10 mm from one end. Open odd-legs and test from each side, adjusting odd-legs until the point marks the centre of plate. Scribe a line down the centre (Fig. 193(b)) and make small dots at (i) and (ii).

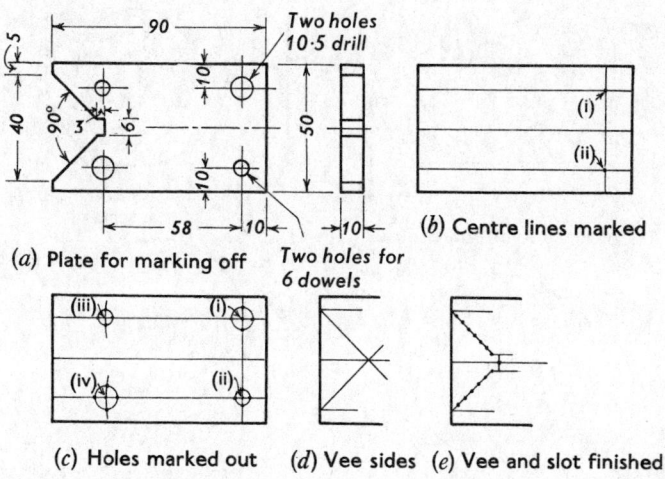

(a) Plate for marking off

(b) Centre lines marked

(c) Holes marked out   (d) Vee sides   (e) Vee and slot finished

Fig. 193 Marking off Vee'd plate

2. Set dividers to an opening of 58 mm, and mark (iii) and (iv) from (i) and (ii). Scribe 6 mm and 10·5 mm circles and dot punch round them (Fig. 193(c)).

3. Set odd-legs to centre line, increase the opening by 20 mm, and scribe a short line each side at the vee end. Set protractor to 45° and scribe lines for sides of vee (d).

4. Open odd-legs to 3 mm more than centre setting and scribe short lines to meet vee lines. With small square from edge of plate, scribe line for bottom of 6 mm slot, 3 mm from where side line cuts vee. Centre dot shape of vee (Fig. 193(c)).

Example 2 (Fig. 194(a)).

This plate will come for marking out already turned to 150 mm diameter, bored 32 mm and faced up to 12 mm thick.

1. Brush one side over with copper sulphate solution and drive a piece of lead, brass or hardwood into the hole. With the odd-legs find the centre and dot punch it lightly. Set the dividers to 50 mm and scribe the 100 mm diameter hole circle (Fig. 194(b)).

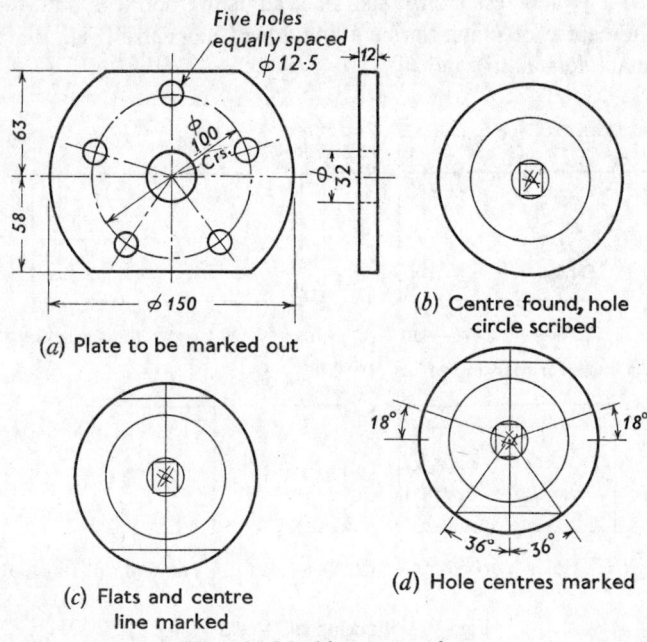

(a) Plate to be marked out

(b) Centre found, hole circle scribed

(c) Flats and centre line marked

(d) Hole centres marked

Fig. 194 Marking out a plate

2. Clamp the disc to an angle plate with fitter's clamps. Set the scribing block point to the centre dot, and from a rule held vertically against the angle plate read off the height of the scriber point. Raise the point 63 mm and scribe a line for the top flat on the disc. Lower the scriber point by 121 mm and scribe a line for the bottom flat. Adjust the edge of a square to intersect the centre dot and scribe a vertical line on the disc, using the square blade as a guide. The plate is now as shown at Fig. 194(c).

3. The angle between any pair of 5 equally spaced holes is $\frac{360}{5} = 72°$. Hence the two upper outside holes will have their radial centre lines inclined at $90-72 = 18°$ with the horizontal centre line, and the line joining the

centre of the plate to each of the lower holes will be inclined at $\frac{72}{2} = 36°$ with the vertical centre line. Set a protractor or bevel gauge so that its blade is inclined at 18° with the base and adjust it up to the work (using a parallel strip for packing if necessary) so that the edge of the blade intersects the centre dot of the disc. In this way mark the radial centre lines for each of the upper outer holes. Set the protractor so that its blade is inclined at 36° with the vertical (54° with horizontal) and in the same way as before, scribe the radial centre lines for the two lower holes (Fig. 194(d)). The work may now be removed from the angle plate.

4. Dot punch the hole centres very lightly, and starting at the top hole set the dividers to the next hole and run round to check that the hole centre distances are all equal. If there is any slight discrepancy it must be corrected, but in doing this the centre position of the top hole must not be moved. The operation is completed by scribing circles for the 12·5 mm holes, and dot punching the surfaces to be machined with small punch marks about 6 mm apart, at the same time making larger punch marks at the centres of the holes.

Example 3. To mark out the bracket shown at Fig. 195(a) which has to be machined on the surfaces marked " " ".

1. Whiten the edges and upper surface of the base, the edges, front and rear faces of the top machined portion. Set the bracket on wedges and packings as shown at (b), and adjust its position until the rough face of the base is approximately square and the other faces of the bracket parallel with the table. If the base and the other faces of the bracket are not quite square, it will be necessary to effect a compromise and split the error when setting it up.

2. Set the scriber point level with the back face of upper portion of the bracket, raise its height by 26 mm and scribe line, (*i*) all round, lower it 13 mm and scribe line, (*ii*) on both sides, lower it a further 15 mm (28 − 13·05) and scribe the machining line for the front edge of the base (line *iii*). Inspect to see that the markings allow a reasonable amount of machining, and that when machined, the base will be approximately 76 mm deep. Set the scriber point 35 mm above the bottom edge of the base and scribe line (*iv*) for the hole centres.

3. Set an angle plate square with the sides of the bracket (Fig. 195(c)), and with a scribing block on the angle plate set the point of the scriber level with the centre of the un-machined sides of the 38 mm slot. Mark this centre line (*v*) on the sides and rear face of this portion of the bracket.

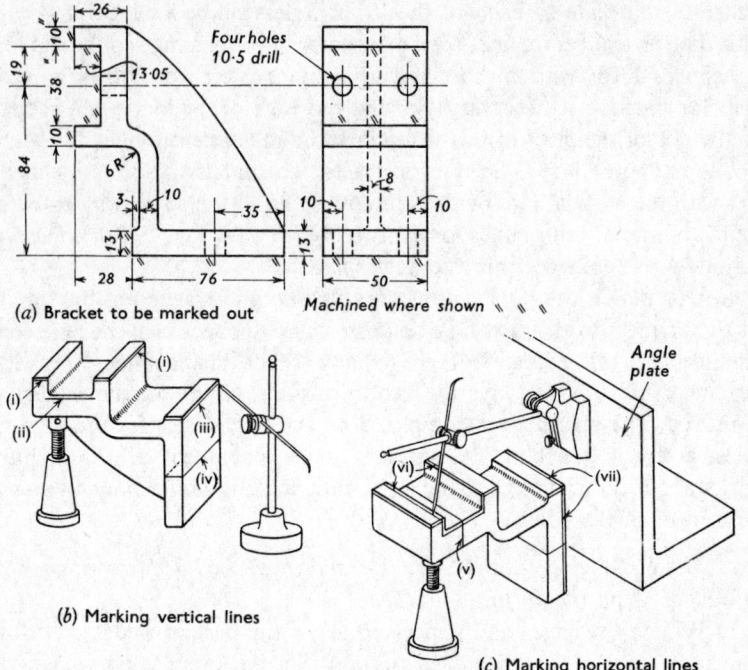

(a) Bracket to be marked out

Four holes 10·5 drill

Machined where shown

(b) Marking vertical lines

(c) Marking horizontal lines

Angle plate

Fig. 195 Marking out a bracket

Increase, and decrease this setting of the scriber point by 19 mm, and scribe lines (*vi*) for the slot sides. Reset point of scriber to slot centre, reduce it by 84 mm and mark the machining line round the edge of the base (line *vii*).

4. From the sides of the bracket, with odd-legs set at 10 mm, mark the other centre line required for each of the four 10·5 mm holes. Dot punch hole centres, scribe hole circles, and dot punch the remainder of the marking out to complete.

Example 4. To mark out the bracket shown at Fig. 196(a) which is to be machined on the base, boss faces and bores.

(The holes in the casting will be cored about 6 mm less than their finished diameters.)

Before we proceed to a consideration of the marking-out methods to be used, it will be well for us to discuss the machining of the bracket, as the method chosen for machining will influence the scheme to be adopted for marking it out. We will summarise the points in connection with the machining:

1. The boss faces may be faced on the planer or shaper, at the same time that the base is being machined or they may be machined at the time of boring.

2. The holes may be bored on a lathe faceplate (the bracket revolving), or they may be done on lathe saddle, milling machine, boring machine, etc. (bracket stationary and tool revolving).

3. It will be easier to get the 58 mm dimension correct by facing the top boss face, and measuring its distance from a plug in the already bored 32 mm hole, than it will be to bore the hole 58 mm from the already faced boss.

4. If the height of the base boss is made $105-68 = 37$ mm from the base, when the top hole is bored 105 mm from the base, the 68 mm dimension will come correct automatically.

5. As we have stated before, the marking out for the holes and for the dimensions between machined surfaces is only intended as a guide for setting up and machining; the actual surface dimensions will have to be measured and finished to gauges and/or measuring instruments.

We will assume, therefore, that the bracket is machined as follows:

1. Set up and machine the base on a planer or shaping machine.

2. Bore the 32 mm hole, and face the 57 mm boss, with the bracket bolted to a lathe face-plate.

3. Bore the 38 mm hole and face the *outer* end of the 70 mm boss with the work secured to an angle plate on the lathe face-plate.

4. Face the inside end of the 70 mm boss on a shaping machine.

Now for the marking-out operations:

1. Plug the holes, whiten the casting round the edges of the base and bosses, and on the faces of the bosses. Pack up on wedges or packing and adjust until top face of base is parallel with marking-out table when tested with the bent end of the scriber (Fig. 196(b)).

2. Using the odd-legs, find the centre of the 70 mm boss, and by measuring from a square blade held in contact with a long edge of the base, check that this is about in the centre of the base. If it is not, re-position it striking a compromise between the centre of the base and the centre of its own boss. Set the point of the scribing block level with this centre, re-set it 105 mm lower and mark the machining line round the base (i). This should show at least 3 mm of machining allowance, and indicate the base as coming about 19 mm thick when finished, otherwise a compromise must be effected,

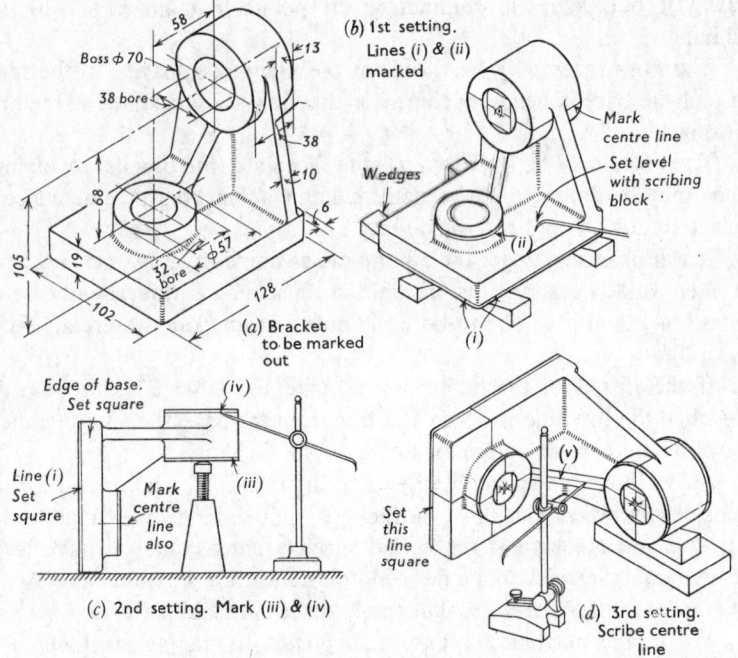

Fig. 196 Marking out a bracket

but the 105 mm must not be varied (i.e. the top hole would have to be thrown slightly off the centre of its boss). After marking the base, raise the scriber 37 mm (105–68) and mark round the edge of the 57 mm boss (ii). At this setting the centre line of the 70 mm boss may be marked round it (see diagram).

3. Set up the bracket with a jack under the boss as shown at Fig. 196(c), and adjust until the smoothest long edge of the base and the machining mark on it are both square with the face of the marking-out table. This involves setting and testing in two planes at right angles. Set the scriber point level with the centre of the 57 mm boss, raise it 58 mm and mark round the edge of the 70 mm boss. Raise the scriber a further 38 mm and mark the upper face of the boss in the same way (lines iii and iv). These should show about 3 mm of machining on each face, and if there is insufficient on one or the other a compromise must be made by throwing the centre of the 57 mm boss to one side or the other.

4. Set the bracket on the smoothest long edge of its base (the one used

for squareness in (3)), pack up the boss and set the base machining mark square with the table. Pick up the centre of the 70 mm boss and mark it all round the bracket (Fig. 196(d), line v).

Finish off by scribing the circles for the bores and dot punching all the lines with punch marks about 6 mm apart.

Referring back to our discussion of the possible machining methods for this bracket the reader will see that if the boss faces are planed or shaped before boring, the marking out for boring must be done *after* the faces have been machined as the facing would remove any previous marking put on them. This procedure must be adopted for any case when a hole is to be bored in a face which has to be previously cleaned up on some other machine.

Drilling operations on the bench

The fitter, during the course of his work, will have much occasion to make use of various drilling operations. We have already discussed the twist drill as a cutting tool (p. 173), so that we may now consider drills themselves and the drilling processes likely to be encountered on the bench.

The drill is a tool for originating a round hole. We say 'originating' to distinguish between cutting a round hole from the solid metal, and enlarging a hole that has already been cut such as is carried out by reamers, counterbores, etc.

The **twist drill** is the most commonly used variety of drill, and twist drills are specified according to the end by which they are held, called

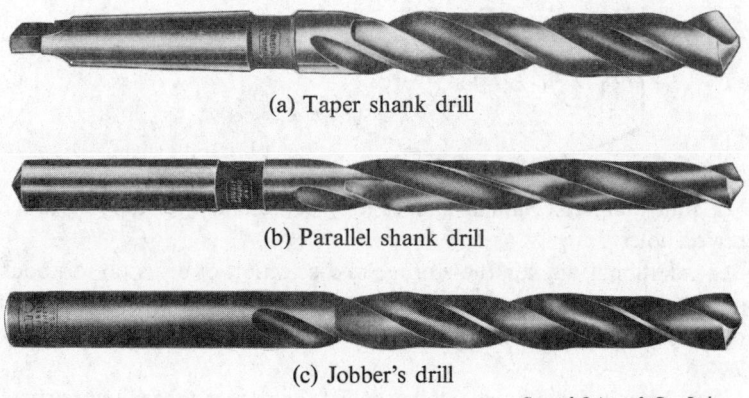

(a) Taper shank drill

(b) Parallel shank drill

(c) Jobber's drill

Samuel Osborn & Co., Ltd

Fig. 197 Twist drills

the *shank*. Drills may be *taper shank, parallel shank* or *jobber's* drills, and are shown at Fig. 197. Jobber's drills are shorter than the other varieties and are usually confined to sizes up to about 12 mm diameter. Above that diameter, the sizes of drills advance by $\frac{1}{2}$ mm. Jobber's drills are held in the machine by special chucks, an example being shown on the electric drill at Fig. 204. Very often, due to slipping in the drill chuck, the shanks of jobber's drills become scored, and the size markings obliterated. The **drill gauge** (Fig. 198) enables any drill to be readily selected by trying in the holes of the gauge. The gauge shown is for drills rising by $\frac{1}{2}$ mm steps

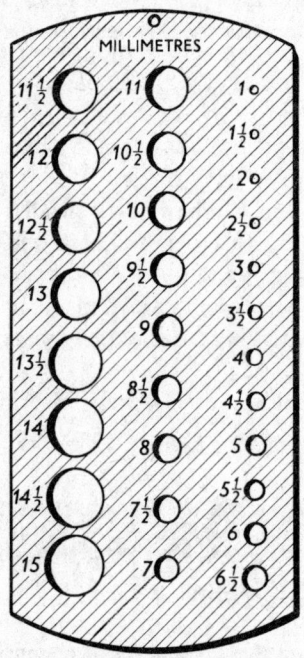

MILLIMETRES

L. S. *Starrett Co.*

Fig. 198 Drill gauge

to 15 mm diameter. Similar gauges may be obtained for tapping drills for screwed holes.

An additional aid for the storage and selection of drills up to about 15 mm is a base having a series of holes the exact size of each drill in the series. The sizes are marked on the base so that any drill can be readily selected.

Taper shank drills have a taper called the Morse taper, and particulars of Morse tapers are also given at the end of the book. The size of taper incorporated on any particular drill will depend on the drill diameter, and the following table shows approximately how these vary.

Table 19. Morse taper shanks on drills

Morse Taper No.	1	2	3	4	5
Range of drill sizes included . . .	Up to 14 mm	14 mm– 24 mm	24 mm– 32 mm	32 mm– 50 mm	Over 50 mm

At the end of the taper shank of a drill is a tongue called the *tang*, and when the taper shank is fitted into the socket or machine spindle, this tang engages in a slot. If the taper itself is in good condition the frictional grip between this and the surface of the taper hole should be almost, if not entirely, sufficient to drive the drill, but if the taper becomes damaged, more load will be thrown on the tang in driving the drill, and if the drill seizes in the hole, the tang may be twisted off. For this reason, taper shanks should be given every consideration in use and always extracted with the proper taper drift (Fig. 199).

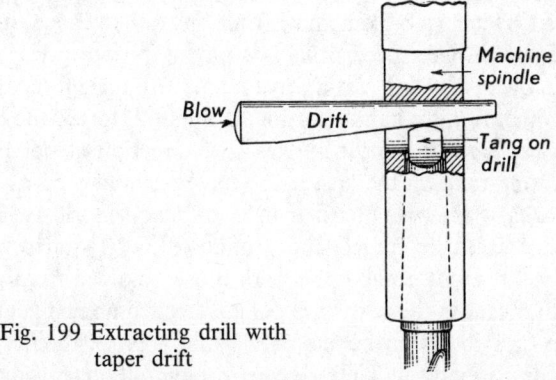

Fig. 199 Extracting drill with taper drift

The number of the Morse taper hole in a machine spindle will depend on the size of the machine, varying from No. 1 in small machines to No. 4 or No. 5 in large ones, and when a drill has to be accommodated in a spindle with a larger taper than its shank, taper sockets must be used (Fig. 200). These should also be cared for, as if they become damaged the drill fitted into them will no longer run true.

Drills may be made of either carbon steel or high-speed steel. The carbon

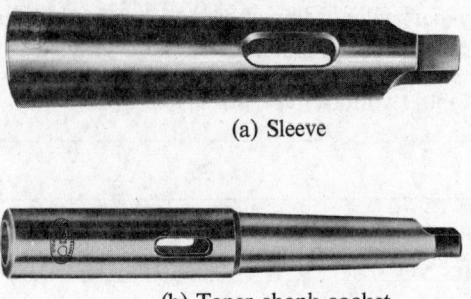

(a) Sleeve

(b) Taper shank socket

Samuel Osborn & Co., Ltd

Fig. 200 Taper sleeve and socket for drills

steel drills are much cheaper, but suffer from the disadvantage that their temper is easily drawn and the edge rendered too soft for further work. When this happens the blued portion should be ground right off and a fresh point made higher up the drill. For high-speed drilling, and cutting hard steel with well-supported drills, the high-speed variety are cheaper in the long run, but for slow speed work, and in cases where the drill is more likely to be spoiled by being broken than from any other cause, it is doubtful whether the use of high-speed drills is a paying proposition.

The flutes of a twist drill are always made with a right-hand helix. This, combined with the right-hand rotation of the drill, forces the chips up the flutes and away from the cutting edges. The direction of slope of this helix also puts cutting rake on the lips, as we have previously discussed.

The **flat drill,** which was the forerunner of the twist drill, is shown at Fig. 201. This drill is easily forged and ground to any required size, but the results it gives are not comparable with those obtained from a twist drill; this is mainly because the twist-drill point is backed up and kept true by the body of the drill following behind, whilst the point of the flat drill has no such influence to guide it. The principal present-day use of the flat drill is for boring very long holes, in which process a cutter in the form of a flat drill is carried at the end of a long bar. The arrangement gives the best

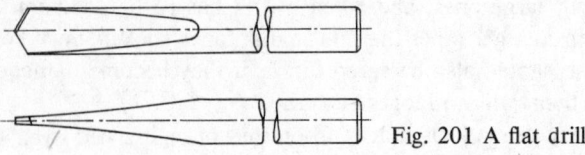

Fig. 201 A flat drill

results when the work revolves, the drill and bar being held stationary and fed into the hole (e.g. on a lathe). Generally the bar which carries the flat cutter has tubes sunk into it for conveying cutting lubricant to flush away the swarf and avoid choking up of the drill.

Drilling machines and appliances

The three principal types of power machine used for drilling are:

(*a*) The *sensitive* machine. A light form of machine used for holes up to about 12 mm diameter.

(*b*) The *upright* or *pillar* machine, in sizes up to that which can drill holes up to 50 mm diameter.

(*c*) The *radial* drill, where the drill head is on an arm and may be swung or moved over the area of the machine table or base.

Fig. 202 Medium upright drilling
machine

Fredk. Pollard Ltd.

Radials may be of the sensitive, or heavy types. A diagram of a medium upright machine is shown at Fig. 202. This machine is provided with a gear box providing nine spindle speeds ranging from 86 to 1920 rev/min and is driven by a 1·5 kW motor at 1405 rev/min. Three rates of automatic feed are provided (0·10, 0·18 and 0·25 mm/rev), but the spindle may also be hand-fed by the lever at the right-hand side, which effects a vertical spindle movement of 130 mm. The drilling capacity of this machine is up to about 30 mm diameter in mild steel.

The fitter will have occasion to use various of the hand-drilling appliances for holes where it is not possible to bring the work up to a machine. The **breast drill** (Fig. 203(a)) may be used for holes up to about 12 mm diameter. Pressure is applied to the drill by pressing the body on the shaped end plate, whilst the drill is rotated by the handle. The hand drill for drills up to 6 mm is similar to the breast drill except that it is smaller, and has a handle instead of a breast-plate. Both breast and hand drills are fitted with a chuck for holding straight-shank jobber's drills.

The **ratchet brace** (Fig. 203(b)) is a hand-drilling appliance, and although slower than the breast drill, is capable of dealing with larger holes. The length of the operating-lever varies from 300 mm to 500 mm, depending on the size of brace, and this rotates the drill during its forward stroke,

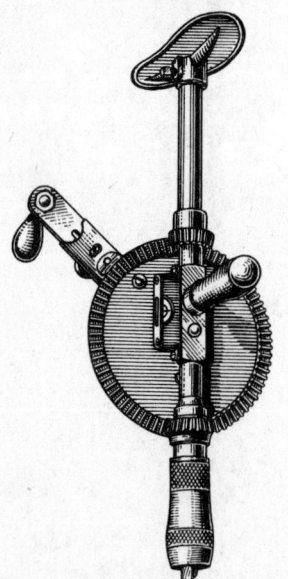

Fig. 203(a) Breast drill

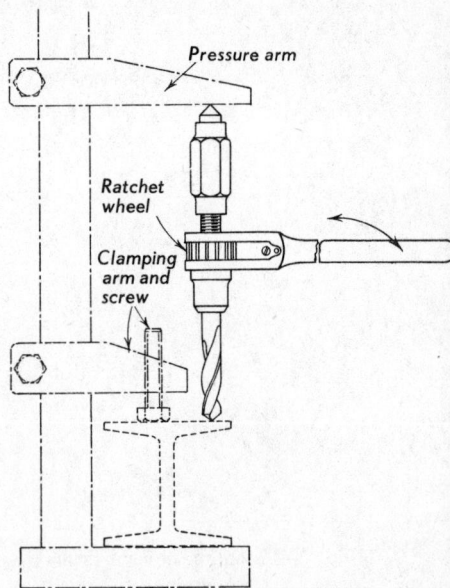

Pressure arm

Ratchet wheel

Clamping arm and screw

Fig. 203(b) Ratchet brace

the pawl slipping over the ratchet as it is taken back for the next stroke. To feed the drill it is necessary to have some form of back plate to take the pressure which is then applied by turning the hexagon feed nut a small amount each time the drill makes a working stroke. The pointed end of the feed nut centres itself on the pressure plate and maintains the alignment of the drill as well as transmitting the feeding pressure necessary. The length of feed that may be given to the drill varies from about 60 mm on the small sizes to 90 mm on the large, and some braces take a taper shank drill whilst others require a square shank. The brace is a slow way of drilling, because only about half a turn may be given to the drill for each stroke, but in awkward situations it is a valuable means of overcoming difficulties.

Electric hand drills. For quickness of operation, the electric drill has a great advantage over the breast drill, for not only does it leave both hands free to guide and feed the drill, but also rotates the drill with more uniformity, and possesses that extra amount of weight necessary to give improved balance and manipulation. A switch is incorporated convenient for the hand so that the drill may be positioned before it is started up. Electric drills of the breast type may be obtained to take drills up to about 16 mm diameter,

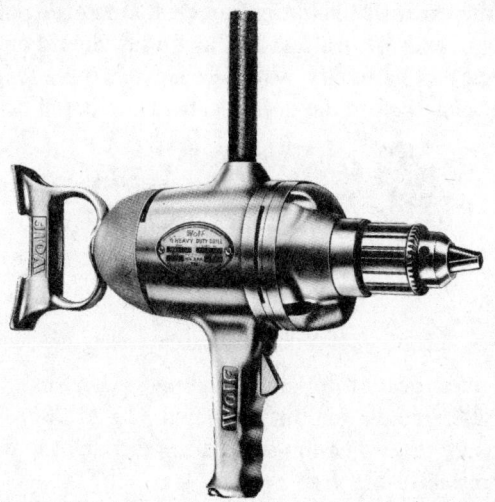

Wolf Electric Tools Ltd
Fig. 204 Electric drill

and an example of one is shown at Fig. 204. For small drills up to 6 mm, a lighter, higher speed type may be obtained rather similar in shape to a pistol, and operated with one hand. Where extremely high speeds are necessary, small hand drills driven by a compressed air motor may be used.

Holding and locating work for drilling

Unless work is so large and heavy that there is no danger of its moving, or being rotated with the drill, it should always be clamped or held by some method, and too much attention cannot be given to clamping because unclamped or insecurely clamped work is not only a danger to the operator, but also a cause of inaccurate work and broken drills. The chief danger in drilling occurs just as the drill point breaks through at the underside of the part being drilled. Whilst the point is being resisted by solid metal, the feeding pressure causes some spring to take place in the machine and the work, putting them into a similar condition to a strong spring which is compressed slightly under a load. As soon as the drill point breaks through, most of the resistance against it suddenly vanishes and the stress in the machine releases itself by imparting a sudden downward push to the drill, just as a sudden relieving of the load from a spring would allow the end of it to jump up. The sudden downward push on the drill generally causes one or both of the lips to dig in, often with disastrous results. When feeding the drill by hand the pressure should be eased off when the point is felt to be breaking through, and for this reason small drills should always be fed by hand. Special care is necessary when drilling thin plate as the drill point often breaks through before the drill is cutting on its full diameter.

J. Parkinson & Son

Fig. 205 Vise for holding work on drilling machine

The vise. A great deal of drilling work may be held in a vise and a good example of a vise suitable for this is shown at Fig. 205. Care should be exercised to ensure that when the drill passes through the work it does not drill into the bottom of the vise.

Clamps and bolts. The tables of most drilling machines are provided either with Tee slots to accommodate bolt heads, or with long slots running through. Whichever be the case the slots enable bolts and clamps to be used as shown at Fig. 206. When using this method of holding, the packing for

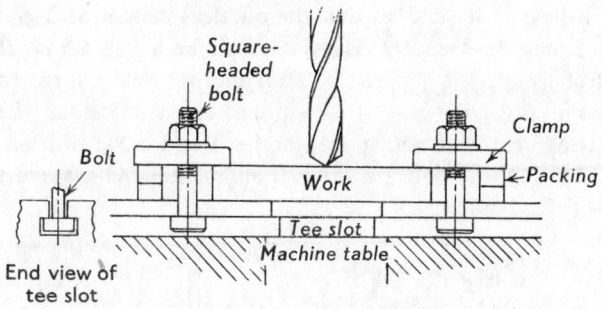

Fig. 206 Clamping work on to drilling machine table

the outer end of the clamp should be as near as possible the same height as the work, and the bolt should be either in the centre or nearer to the work, as shown in the diagram.

Vee blocks. Cylindrical work which has to be drilled perpendicular to its axis is best supported on one or more vee blocks. When the hole has to pass through the centre of the work, the centre line should be scribed across the end of the bar or bush with a scribing block, the line being continued along the curved surface of the cylinder as far as to where the hole must be. A centre dot may then be punched on this line for the hole centre and the circle scribed with dividers. The work is now adjusted so that the centre mark on the end of the cylinder is vertical when tested with a square from the machine table at the same time that the hole centre is under the point of the drill. This is shown at Fig. 207, where is also indicated a U clamp which holds the work and permits the drill to pass through it. When clamping work on to vee blocks or tables always apply the clamp at a point where its pressure is supported directly underneath. (*Note.*—If a bar or

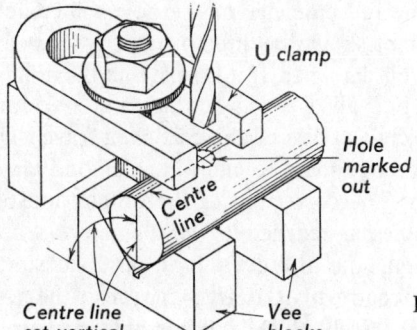

Fig. 207 Setting and clamping for drilling a shaft

bush has a hole in its end so that the odd-legs cannot be used to find its centre, this may be found as follows: 1. Scribe a line across the end as near central as can be judged. 2. Turn the bar half a turn, and try the scribing point along the line. If they are at the same height when the line is level it is on centre; if not, scribe another line parallel with the first one, set the scriber midway between the two and scribe a line which will be the centre line (Fig. 208)).

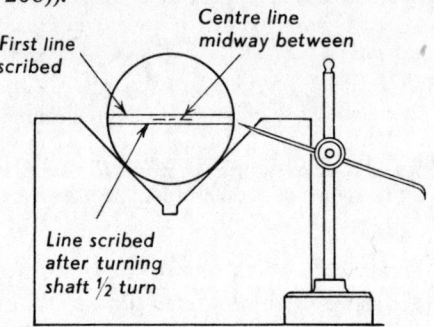

Fig. 208 Finding shaft centre with Vee
blocks and scribing block

Angle plate. When a hole has to be drilled parallel to a face, an angle plate provides a convenient method of support, and such a set up is shown at Fig. 209.

Locating drilled holes. We have already discussed the marking out of drilled holes and might add that a hole should never be started without some form of location even if only a punch mark to indicate the hole centre, and provide assistance for the initial starting of the drill point. When a hole is being drilled to marking out, a careful watch must be maintained at the commencement, to ensure that the full diameter of the hole will coincide with the marking, and the drill point should be lifted once or twice before the point impression reaches the full diameter. If it is seen that the drill has wandered from the centre it must be 'pulled' over by cutting a groove down the side of the impression to which it is desired the drill shall travel. This is shown at Fig. 210, and the groove may be cut with diamond chisel, half-round chisel or centre punch. The correction of the drilling must be accomplished before the drilled hole has reached its full diameter, as after that very little may be done to pull the drill over. It is easier to start a reasonably small drill on a given centre than a large one; also the point and web of a large drill being relatively thick, considerable pressure is

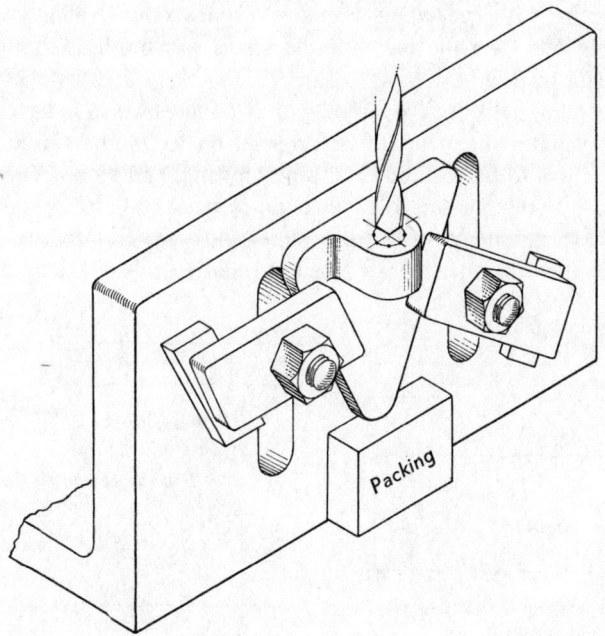

Fig. 209 Drilling with work clamped to angle plate

required to feed a large drill through the work. For these reasons, when a hole to be drilled is greater than 22 mm to 25 mm it is generally advantageous to drill through first with a drill about 8 mm. If this is centred correctly the larger hole will be correct, as the second drill will follow the axis of the smaller one.

Accuracy through method. Most workshop jobs may be completed satisfactorily and with no apparent difficulty if only they are done by the proper method. Tackle them any other way, and the most aggravating

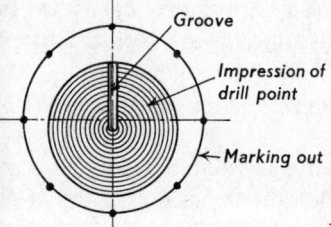

Fig. 210 Correcting for drill starting off centre

hitches occur, with everything seeming to go wrong. Drilling is no exception to this rule, and the most simple job is worth a little thought before commencing to drill.

Many drilling jobs are concerned with two plates which have to be bolted together, a plate which has to be screwed on to another face, a bracket base which has to be secured to another casting, and so on. The fastening is effected by bolts passing right through (Fig. 211(a)), or by some form of screw which passes through one plate and screws into the other. To locate one plate in an accurate position relative to the other, *dowels* are

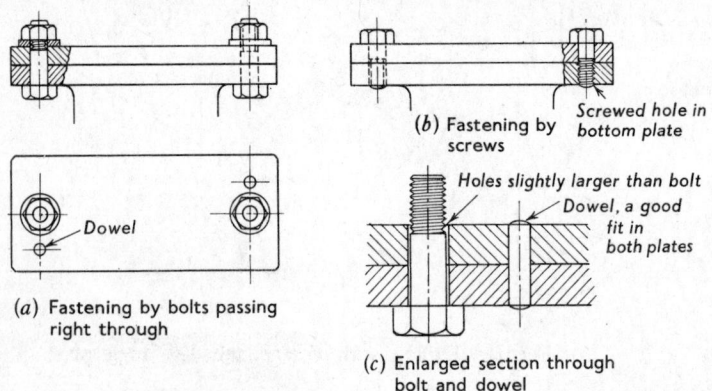

(b) Fastening by screws

Screwed hole in bottom plate

Holes slightly larger than bolt

Dowel, a good fit in both plates

(a) Fastening by bolts passing right through

Dowel

(c) Enlarged section through bolt and dowel

Fig. 211 Fastening by bolts and screws

often used. These are pins made of *silver steel* (about 1% carbon) accurately ground to size, and driven into well-fitting holes in both plates. With such an arrangement, however many times the plates may be dismantled, they can always be put together in the same position. The bolts or screws are not expected to provide this location, because they are clearance in their holes.

Now in cases such as these, the actual positions of the holes are relatively unimportant, but unless the holes in both plates are exactly coincident (in line), the bolts, screws or dowels will not pass through to enable the fastening to be made.

To ensure that the holes shall be in alignment, various methods may be adopted.

1. If bolts are being used, requiring similar holes in both plates, one plate may be marked out, and then clamped in the correct position on top of the other, using fitter's clamps. Both plates are then drilled together. As a

precaution against the plates moving during the process of drilling a number of holes the end ones may be drilled first, and a drill or fitting bolt passed through to lock the plates whilst the remainder of the drilling is completed.

2. When a plate is screwed to another (Fig. 211(b)), the holes in the upper plate must be clearance for the bolts, whilst those in the lower one must be drilled smaller to allowed for tapping the thread. This may be carried out as follows: (a) drill through both plates with the smaller drill as above and then, after taking them apart, open out the holes in the upper plate, or (b) mark out and clamp together, drill upper plate and allow the drill point to penetrate the surface of the lower plate until it just about reaches its full diameter. Take the plates apart and use the drill-point indentations as a guide for drilling the smaller holes in the lower plate. If this second method is employed and there are dowel holes to be put in, these should be drilled in both plates at the same time and not spotted through for subsequent drilling as in the case of the bolt holes.

Collars on shafts. The above principle must be followed when a collar and shaft have to be drilled for the purpose of pinning the collar in position with a pin running right through (Fig. 212). In theory, if a hole is drilled

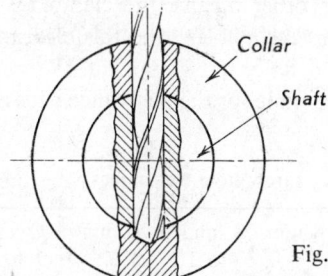

Fig. 212 Drilling assembled collar
and shaft

dead central in the shaft, and a similar one in the collar, they may be assembled, and the pin driven home. In practice, it is extremely difficult to get a hole dead central, and as the pin in the above case must be a good fit in both parts, drilling the parts separately would inevitably lead to failure in entering the pin. If the collar is assembled on the shaft and the two drilled at once, a slight error will not affect the job and there will be no doubt about fitting the pin through.

Location of faces.—Clearance for bolts and screws. It is only in rare cases that the shank of a bolt is made a dead fit in its hole for the purpose of

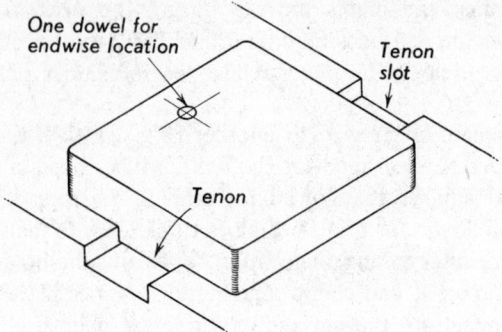

One dowel for
endwise location

Tenon
slot

Tenon

Fig. 213 Locating with tenon and slot

positioning the part, and when this is done the bolt is called a *fitting bolt*.
Generally, it is easier and just as effective to use dowels, or a *tenon* and
tenon slot. A tenon-slot location is shown at Fig. 213, but it should be
noted that if a location parallel to the slot is required an additional dowel
is necessary. When two dowels are used for locating they should be as far
apart as is reasonably possible because in that way they are more
efficient in their function of locating (Fig. 211). Any necessary location
having been provided by dowels or other means, the clamping bolts or
screws are made slightly clearance in their holes, and this clearance provides for slight variations in bolt and hole, mis-alignment of thread, etc.
The following table may be used as a guide for the allowance necessary for
clearance.

Table 20. Bolt and screw hole clearances

Diameter of bolt or screw	Up to 8 mm	8 mm to 20 mm	Over 20 mm
Clearance on diameter of hole	$\frac{1}{2}$ mm	$\frac{1}{2}$ to 1 mm	1 to $1\frac{1}{2}$ mm

Reaming

A drill cannot be relied upon to produce a hole having sufficiently good
qualities of finish and accuracy for many purposes, and when accurate
holes are required a *reamer* must be used for finishing to size. The reamer
does not originate the hole in the same way as the drill, but merely imparts
to the previously drilled hole the necessary smoothness, parallelism, roundness and accuracy in size. Reaming, however, will not correct any errors
which may be in the hole with regard to its position, or direction, because
the reamer merely follows the previously drilled hole, and if this hole were

drilled out of square, or incorrectly positioned, then the reamed hole will have the same defect.

Hand reamers are operated by hand with a tap wrench fitted on the square end of the reamer, with the work held in the vise. Sometimes the square end of the reamer is held in the vise and the work fed on to the reamer by hand. These reamers cut on the flutes which may be straight, or helical, and are tapered at the end for about $\frac{1}{4}$ of their length, the extreme front end being about $\frac{1}{4}$ mm less than the reamer size. The shank is a few hundredths of a millimetre less than the diameter of the flutes. A straight

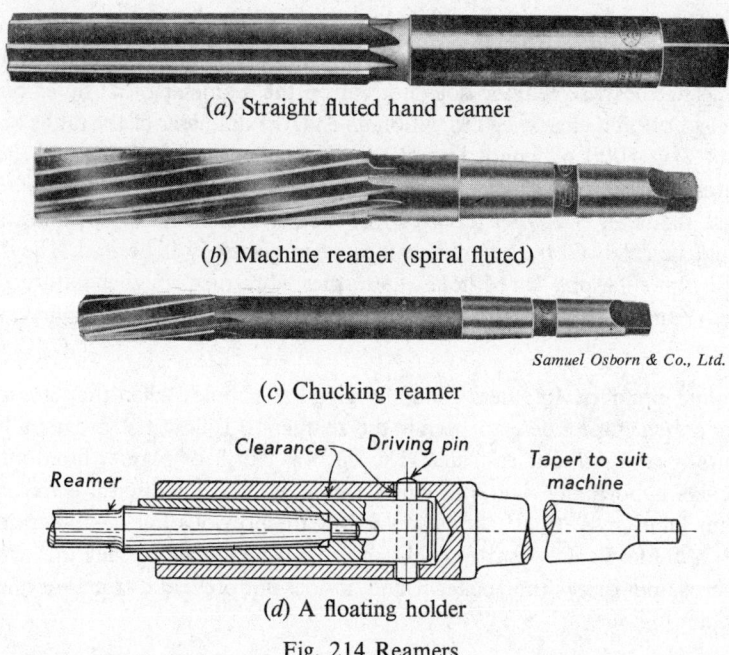

(a) Straight fluted hand reamer

(b) Machine reamer (spiral fluted)

Samuel Osborn & Co., Ltd.

(c) Chucking reamer

(d) A floating holder

Fig. 214 Reamers

fluted hand reamer is shown at Fig. 214(a). The minimum amount of metal should be left for removal by hand reaming, from 0·05 to 0·15 mm being sufficient, as this class of reaming must be regarded essentially as a corrective rather than a metal-cutting operation, and if too much is allowed, not only will the edges of the reamer rapidly wear and lose their size, but also the finished hole may be oversize and inaccurate. Since the problem of high speed in cutting does not enter into hand-reaming operations, these reamers may be made of cast steel, and on the larger sizes case-hardened mild steel gives satisfactory results.

Machine reamers. The solid fluted machine reamer is similar to the hand reamer except that it has a taper shank. These reamers may have straight flutes, but more often they are cut with a left-hand helix as shown at Fig. 214(b), the angle made by the flute, with the length of the reamer, varying from 4° to 8°. The helix is made left-hand to counteract the tendency the reamer would have to screw itself into the hole if the flutes were right-hand, and even straight-fluted reamers have a slight tendency to draw in. Machine reamers may be of carbon or high-speed steel.

Chucking reamers are machine reamers with shorter flutes as shown at Fig. 214(c), and may be either of the type known as *rose* reamers, or *fluted* reamers. The rose reamer does not cut on the diameter of its flutes but is bevelled off and clearanced to cut on its end, the diameter of the flutes being about 1 in 5000 of length less at the back than at the cutting end (body clearance). Reamers of this type will remove greater amounts of metal than fluted reamers, but when doing so do not give such accurate holes, and should be followed by a fluted reamer when accuracy is desired. They are useful for enlarging cored holes, and other work of a rougher nature. The fluted reamer cuts on the flutes in the same way as an ordinary solid reamer.

Floating reamers. Reamers generally give better holes when they are used with a *floating* holder, to permit the reamer to follow the previous hole naturally and without restraint. The end of a drill or reamer fitted into a machine is normally rigid, but when a floating holder is used it is free, and within limits may 'float' about and follow the previous hole. A diagram of such a holder is shown at Fig. 214(d), from which it will be seen that whilst the cross-pin drives the socket round, it does not prevent it from floating or wobbling about.

Expanding reamers. All cutting tools are liable to wear, and to the reduction of their dimensions by the grinding necessary to keep them sharp. This does not matter in the case of many tools, but it is important in a reamer, because we rely on its size to impart accuracy to the hole it produces. When a solid reamer becomes worn, therefore, it must either be ground down to the next size smaller, or the teeth slightly enlarged by softening the teeth, expanding them slightly by hammering their fronts with a punch, re-hardening and grinding up again to size. To avoid this, expanding reamers are made. These are of various types, a common one being shown at Fig. 215. When the reamer has lost its size the diameter of the flutes is increased slightly by springing them open with a tapered pin operated by the screw, after which they are re-ground to the correct size. The amount of expansion possible varies from 12 to 25 hundredths of a millimetre on the diameter.

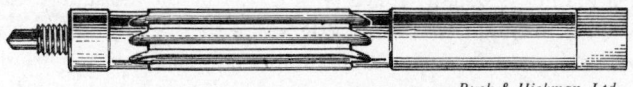

Buck & Hickman, Ltd.

Fig. 215 Expansion reamer

Taper reamers.—Taper pins. Collars, pulleys, etc., are often secured to shafts by a taper pin as shown at Fig. 216(a) and the fitting of these is a bench job. Taper pins may be bought as a standard article, and have a taper of 1 in 48 on the diameter and are specified according to their large end

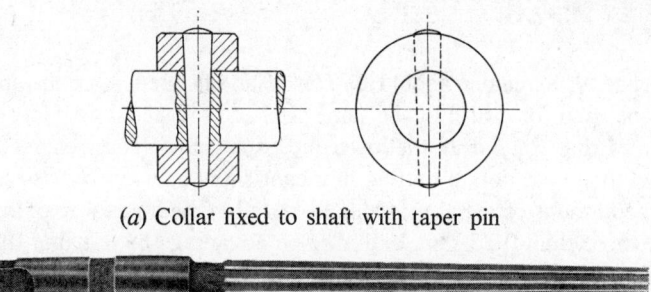

(*a*) Collar fixed to shaft with taper pin

Samuel Osborn & Co., Ltd.

(*b*) Taper pin reamer

Fig. 216

diameter. In fitting such a pin, the collar and shaft to be fitted must be assembled together whilst the taper pin-hole is drilled and reamed to suit the pin. First of all a drill is put through which is slightly smaller than the small diameter of the pin and then the hole is reamed out with the taper reamer (Fig. 216(b)) until the pin fits. The diameter of the drill to put through first may be calculated from the large diameter and length of the pin; for example, on an 8 mm diameter taper pin 50 mm long the taper would be

$$\tfrac{1}{48} \times 50 = \tfrac{50}{48} = 1 \cdot 04 \text{ mm}$$

The small end of the pin would, therefore, be

$$8 \text{ mm} - 1 \cdot 04 \text{ mm} = 6 \cdot 96 \text{ mm}$$

and the nearest drill to put through preliminary to reaming would be 7 mm diameter.

Cutting with reamers. Because it must hold its size as long as possible, the tooth of a reamer is usually made with a reasonable amount of *land* on its top and is not brought to a sharp edge as is the tooth in other tools. This land may be up to 1 mm wide and is sometimes eased off very slightly with an oilstone. The reamer is sharpened by grinding the *front* of the tooth as shown at Fig. 217.

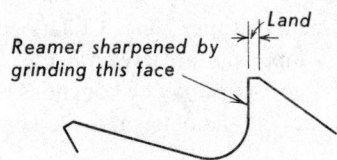

Reamer sharpened by grinding this face

Land

Fig. 217 Reamer tooth

In general, a reamer should be run at about one-half of the speed that would be used for a drill of the same size, and the feed may be from two to five times that for a drill, the lower feed applying to rose reamers. Except when cutting cast-rion, a cutting lubricant should always be used for reaming. The amount of metal that should be left in the hole for reaming should not be more than 0·25 mm, and may be advantage be less, but due to the difficulty in obtaining drills except in steps of $\frac{1}{2}$ mm, it is often necessary to ream out that amount. If trouble is experienced due to this being too much, it may be necessary to rose ream the hole after drilling and before flute reaming.

Counterboring. The preparation of holes for certain purposes involves increasing the diameter of the hole for a certain distance down. This is called counterboring, and is done with a cutter of the type shown at Fig. 218(a) which cuts on its end as shown at (b). The projection at the end of the cutter may be incorporated solid with it, or screwed in as a peg, and by piloting in the hole, serves to steady the cutter, keeping it concentric. The screwed peg type permits a cutter to be used on different holes by having a set of pegs of various sizes. These are often called *peg cutters*, and a common use for them is in cutting the counterbore for the head of a cheese-head screw as at (c). Another use for cutters of this type is in *spot-facing*, an operation often performed around holes drilled in the rough surfaces of castings to provide a flat seating for the surfaces of nuts and washers (Fig. 219).

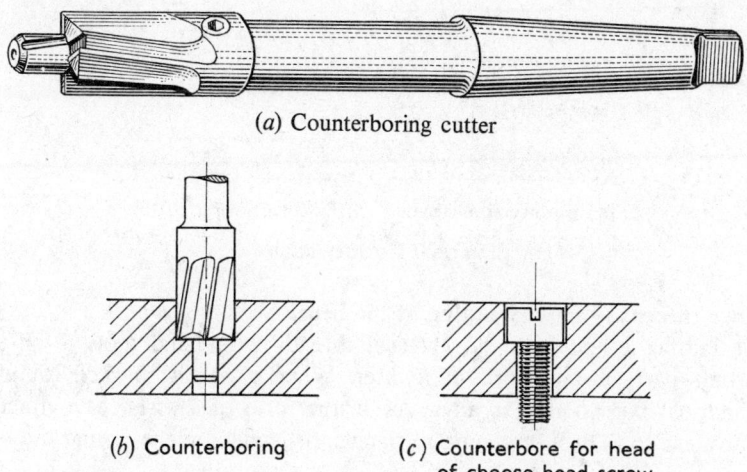

(a) Counterboring cutter

(b) Counterboring

(c) Counterbore for head of cheese head screw

Fig. 218

Countersinking. Countersunk head screws and wood-screws require a 90° chamfer cut round their hole as a seating for the underside of the head. This is cut by means of a countersinking cutter as shown at Fig. 220. More often than not, countersinking is carried out with the point of a drill after it has been ground to an included angle of 90°. This is unavoidable if a countersinking cutter is not available, but involves some drill wastage because the drill has to be ground to 90° to do the countersinking and then reground to 120° for its normal work.

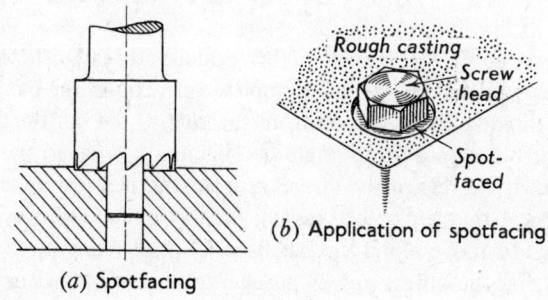

(a) Spotfacing

(b) Application of spotfacing

Fig. 219

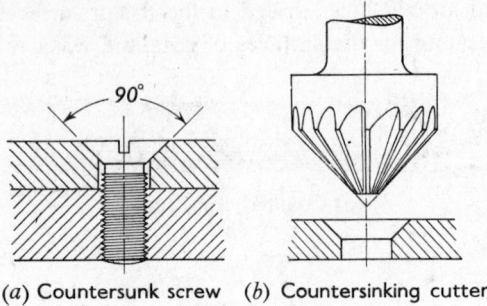

(a) Countersunk screw   (b) Countersinking cutter

Fig. 220 Countersinking

Screw threads.—Thread-cutting at the bench

The cutting of internal and external threads with hand tools is an important part of the work of a fitter. When a line is marked round a cylindrical bar so that it advances in the form of a screw, the effect is equivalent to that of wrapping a triangle, or inclined plane, round the bar.

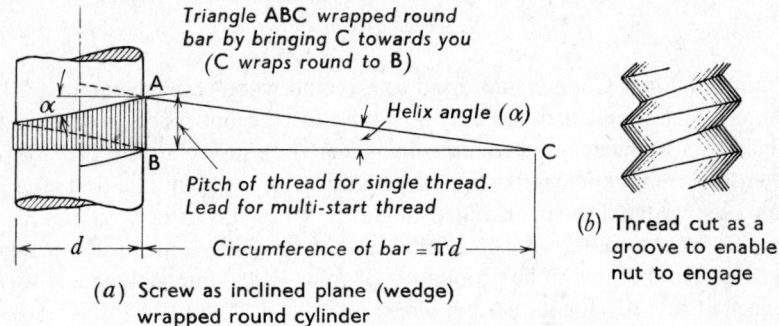

Triangle ABC wrapped round
bar by bringing C towards you
(C wraps round to B)

Helix angle (α)

Pitch of thread for single thread.
Lead for multi-start thread

Circumference of bar = πd

(a) Screw as inclined plane (wedge)
wrapped round cylinder

(b) Thread cut as a
groove to enable
nut to engage

Fig. 221 Development of screw thread

This is shown at Fig. 221(a), and for ordinary threads, the distance the line advances whilst it makes one complete turn round the bar is called the *pitch* of the thread. When viewed from the side of the bar the thread is not perpendicular but slopes at an angle α. This angle is called the *helix angle* of the thread. If we regard the thread as a sloping plane wrapped round the cylinder, we see that an object resting on this plane and moved round it, will be caused to move along the bar because of its rising up the plane, and the distance that the object moves parallel to the bar in one turn will be the pitch. In order to put this into practical effect, our thread line must be made into a groove, Fig. 221(b), so that engagement may be effected with corresponding grooves in the inside bore of the nut, which fits the screw. Thus, when a nut is turned on a screw, for every turn it makes it moves along a distance equal to the pitch.

Locking power of a thread. If we pursue our consideration of a screw as an inclined plane, a wedge is an inclined plane, so that the principle of the screw is identical with that of the wedge. Let us consider the clamping of a plate to a surface as shown at Fig. 222, where the bolt and nut method is shown at the left, and the wedge at the right. (The wedge will require an upper surface to balance the pressure it exerts on the plate, and allow it to

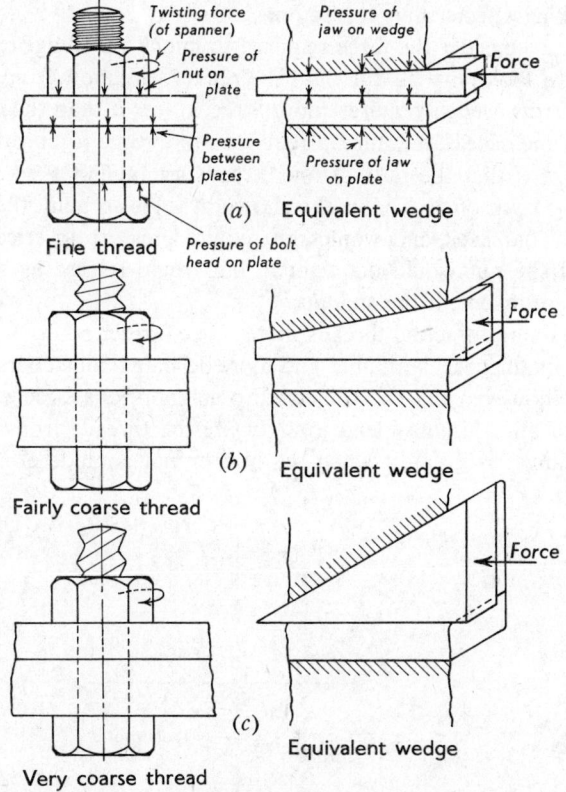

Fig. 222 Diagram to show a screw thread and its equivalent wedge clamping a plate

be 'wedged' in. In the bolt this pressure acts through the threads and its transmitted to the bolt head which presses against the under surface as shown.) A very sharp wedge as at (a) will have a small angle and, if the reader refers to Fig. 222, will be equivalent to a fine pitch thread. At (b) is shown a blunter wedge, and its equivalent thread which is of coarser pitch than at (a). The wedge at (c) is of larger angle still, with a corresponding thread of very coarse-pitch.

Now the reader probably knows from experience, that when a wedge of very slow taper is driven in, much more pressure may be exerted between two surfaces than when the wedge is more blunt; at (a) for example, the pressure of the wedge would be greater than at (b), which in turn would exceed that at (c). We thus see, then, that fine threads are capable of giving greater locking effects than coarse ones.

Now let us consider the force required to knock the wedge out, which is equivalent to loosening off the nut. The reader will know from experience that the narrow wedge requires more force to free it than the coarse one. Such being the case, a fine thread requires more force to unscrew the nut, and this means that the nuts of fine threads are much less likely to work loose than coarse ones. The wedge shown at (c) is so blunt that probably it would fall out itself and would not require pressure to release it! This means that the nut would not tighten, but would release as soon as the pressure were taken off the spanner.

Before passing to actual threads there is one further point. A fine thread, because of its fineness, is thinner and more delicate than a coarse one. This fine thread, however, may be subjected to an enormous force when the nut is tightened, and this may lead to *stripping* the threads from the bolt or nut. The danger is greatest when the bolt or nut is made of a soft, weak

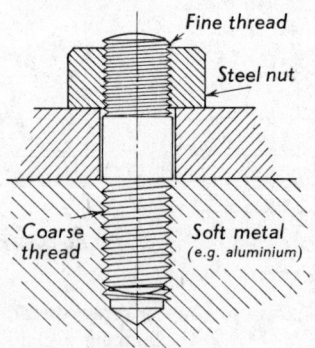

Fig. 223

metal, and very often, if one end of a stud is screwed into a soft metal such as aluminium, whilst the other end takes a steel nut, the end screwed into the aluminium has a coarser thread than the other (Fig. 223).

Standards for screw thread systems. During the past century numerous thread systems have been introduced, but many of them have gone out of use and only a few have persisted. This process of simplification is still going on and we give below the principal survivors of the threads used in this country. As far as we are concerned, our future lies principally with the ISO metric thread, but the others are likely to be in use for some years, until the SI system is fully established.

The ISO metric screw. In 1965 the President of the Board of Trade announced that British industry should progressively adopt metric units until that system could be established as the primary standard of weights and measures in the United Kingdom. This was followed by BSI's major policy statement urging British industry to adopt, for all future designs, either the internationally agreed ISO metric thread or the ISO Unified inch thread. The ISO has reached agreement on recommendations for general purpose screw threads and these are set out in BS 3692, 1967. The thread form and the common sizes are set out in Appendix 2, where it will be seen that the system provides for coarse and fine pitches.

The Unified thread. This was the forerunner of the ISO metric thread and resulted from a series of conferences culminating in the Ottawa Conference of 1945, between Britain, America and Canada. The result of this was that the Unified thread standard was published in 1949. There are two types of this thread; the Unified Coarse (UNC) and the Unified Fine (UNF). (Appendix 3).

The Whitworth thread. Amidst the pressures to modernise and adopt new standards, we should not forget the work of Sir Joseph Whitworth after whom the Whitworth thread was named. In his time he found that the work of engineers was greatly impeded by the lack of standard practice in the use of screw threads and in 1841 he instituted a system. The value of Whitworth's contribution may easily be appreciated if the reader imagines what the position would be like now, as it was then, if every engineering concern adopted their own fancy with regard to the form and pitch they put on screwed parts! No student should fail to read the life of Sir Joseph Whitworth. The Whitworth system is shown in Appendix 4.

The British Standard Fine (BSF) thread. The BSF thread has been in general use in Great Britain for purposes where the Whitworth thread would be too coarse. The form of the thread is similar to the Whitworth and for any given diameter of screw it has a finer pitch. The diameters and pitches are given in Appendix 4.

The British Association (BA) thread. This thread has been in general use in Great Britain for screw diameters under 6 mm, and for instrument work. The thread form and diameter/pitch relationships are given in Appendix 5.

The British Standard Pipe (BSP) thread. (Often called 'Gas' thread). The chief use of this thread is for gas and water pipes but it has been used for other purposes where a larger diameter with a relatively fine thread is required. The thread form is Whitworth and the BSP size is specified according to the *bore* of the tubing upon which it is cut. For example, a $\frac{1}{2}$-inch BSP thread is the thread that would be put on the *outside* of a tube of $\frac{1}{2}$-inch bore and has a top diameter of 0·825 inch (Appendix 6).

The trapezoidal metric thread. This thread, very similar to the British Acme, but with a 30° thread angle, will probably replace the Acme for lead screws when the changeover to the SI metric system gets properly under way.

Cutting internal threads—taps

A tap may be regarded as a bolt with a perfect thread cut on it, which has been provided with cutting edges and hardened, so that when it is screwed into a hole it cuts an internal thread which will fit an external thread of the same size. Taps are made of high-carbon or high-speed steel by turning a screw, and then cutting flutes along it so as to provide cutting edges, as shown at Fig. 224(e). The shank of the tap is left plain and the end is squared to accommodate the *tap wrench* (d), which is used to screw it into the hole. After hardening and tempering, the flute at the front of the threads is ground up to provide the necessary sharpness on the threads for cutting (Fig. 224(e)).

Taps are usually made in sets of three to cut any particular size. These are called taper, intermediate and bottoming; or taper, 2nd and plug. The taper tap has its leading end tapered off for a length of 8 to 10 threads, and this is the first tap to use, the tapered end permitting the tap to enter the hole and cut to the full thread gradually. The second tap has only two or three threads chamfered, and follows the taper tap; if a hole is open at both ends this tap is suitable for finishing the thread in it. The thread on the

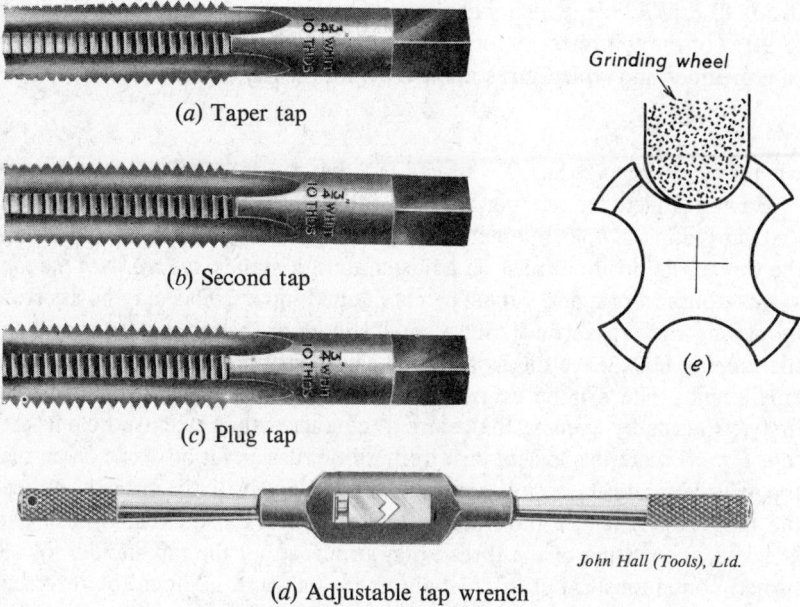

(a) Taper tap

(b) Second tap

(c) Plug tap

Grinding wheel

(e)

(d) Adjustable tap wrench

John Hall (Tools), Ltd.

Fig. 224 Taps and wrench.

plug or bottoming tap runs to its extreme end and this tap should be used as the final tap when a full thread must be cut to the bottom of a blind hole.

Drilling and tapping a hole. For tapping, a hole must first be drilled to a diameter approximately equal to the bottom of the thread, and these *core* diameters are given in the tables of screw threads in the Appendix. The core diameter may be found by subtracting twice the depth of thread from the top diameter, and for the ISO metric shape the depth of thread is $0.5413 \times$ pitch. Thus for, say, a screw of 20 mm diameter, 2·5 mm pitch, the double depth of thread would be $2 \times 0.5413 \times 2.5 = 2.71$ mm, and the core diameter: $20 - 2.71 = 17.29$ mm. The tapping drill must be chosen as the nearest size larger than this.

When the hole to be tapped is 'blind' (i.e. only open at one end) it should be drilled one or two threads deeper than is required for the finished depth of the tapped hole.

[For ordinary purposes, tapping holes are often drilled slightly larger than the theoretical core diameter shown above, as this facilitates the tapping process without prejudicing the thread unduly. Recommended practice for this is to drill a hole, which when tapped will result in a thread about 88% to 90% of the full form. A formula for the tapping diameter which is easy to remember and which gives about 92% of the full thread is

$$T = D - p$$

where T = tapping dia, D = thread top dia, p = pitch, all in mm.

Having drilled the tapping hole, the taper tap is fixed in the tap wrench and started in the hole, but before commencing to screw it round for cutting the thread, its position must be adjusted until it stands *square with the top surface* of the work, and it must be maintained square. This may be assisted by setting the tap vertical with a small square or, better still, after drilling the tapping hole, leave the work in the same position, replace the drill by a small centre like a lathe centre. Enter the tap and keep it straight for the first few turns by keeping the centre in contact with the centre hole in the tap. For all materials except cast-iron a little russian fat or whale oil on the tap will lubricate its action and improve the finish of the threads. When the taper tap is felt to have started its work and its squareness has been checked, the cutting of the thread may proceed, but the tap should not be turned continuously, but after about every half-turn it should be reversed slightly to clear the threads. At the same time, the facility or otherwise with which it is doing its work may be felt from the resistance encountered at the wrench, and if any stiffness develops, no force whatever should be used, but the tap carefully wriggled backwards to clear it. When a blind hole is being tapped the tap should be withdrawn from time to time and the metal cleared from the bottom of the hole. Ultimately, if the hole is straight through, the reduction of resistance will indicate that the taper tap is cutting a full thread and it may be removed from the hole which may be finished with the second tap. When a blind hole is being tapped resistance will be felt when the tap reaches the bottom of the hole, and no force must be used at this point, as the tap may be broken or the thread stripped. Remove the taper tap and take the second one down as far as it will go, finally cutting the threads at the bottom of the hole with the plug tap. Great care should be exercised when using small taps, particularly in blind holes, but in spite of his care the reader will occasionally have the misfortune to break a tap in the hole. When this happens he may be able to get it out with a punch, but if it resists all other means, it must be softened by heating and drilled out, the hole afterwards being re-tapped.

Tapping by machine. When large numbers of holes have to be tapped, hand-tapping is a slow process, and the work is expedited by driving the tap with the drilling machine, running on a slow speed. The main precaution to be observed is to provide some means whereby the tap drive from the machine is not solid, but incorporates a slipping device which will come in to operation when the tap sticks in the hole, or reaches the bottom, and so avoid breaking the tap. One type of machine tapping attachment is

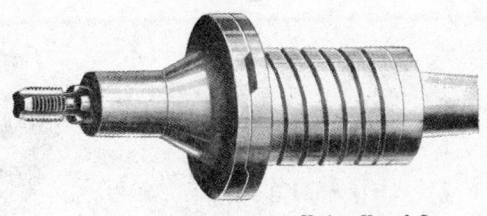

Herbert Hunt & Sons.

Fig. 225 The Pearn tapper

shown at Fig. 225 in which it will be seen that the tap is driven through the dogs (notches) which are kept in engagement by the spring. If the tap sticks, the resistance causes the dogs to ride over one another and compress the spring. With the attachment shown, the machine must be reversed to screw the tap from a blind hole. Other attachments are made with a reversing arrangement incorporated.

Ground thread taps. The ordinary tap suffers from the disadvantage that any hardening, distortion and scaling which may have occurred to the thread is detrimental to its shape, and for accurate work the thread may not fulfil requirements. To overcome this, the threads of ground taps are form ground to their exact shape after hardening, thus allowing a perfect thread to be produced.

Cutting external threads—dies

The tool used for cutting external threads on bars or tubes is called a *die*, and in principle consists of a nut having portions of its thread circumference cut away and shaped to provide cutting edges to the remaining portions of the thread. After hardening and sharpening up on the cutting edges, this is screwed on to the bar upon which the thread is to be cut. In order to hold and manipulate the die it is carried in the centre of a pair of operating handles called *stocks*. There are various forms of dies, and some of the most usual are shown at Fig. 226. At (a) is shown a solid die, whilst

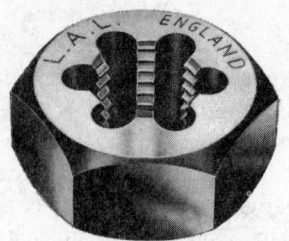

(a) Solid die nut

(b) Split die and stocks

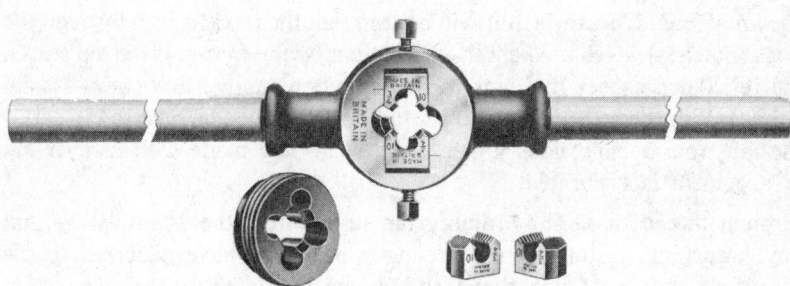

(c) Stocks with loose dies

(d) Stocks and dies for fine threads

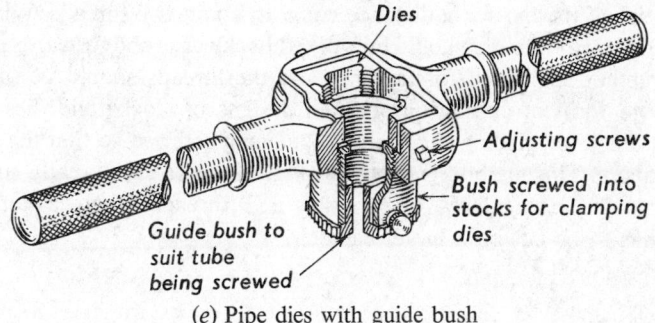

Dies

Adjusting screws

Bush screwed into stocks for clamping dies

Guide bush to suit tube being screwed

(*e*) Pipe dies with guide bush

Lehmann Archer & Lane, Ltd.
John Hall (Tools), Ltd.

Fig. 226 Stocks and dies

at (b) is a split die together with its stocks. The split permits of a certain amount of adjustment in the size the die will cut by springing it a small amount open, or closed, by means of the screws in the stocks. The solid die nut is not usually employed for cutting threads from the solid, but for rectifying damage and knocks to existing screws. The dies shown at (c) consist of a pair of dies or jaws which fit into the stocks and are clamped by a screwed ring. These dies slide and may be adjusted by screws which bear against their outer faces. This permits the dies to be set a small amount open whilst the first cut is taken down a bar and closed in to the correct size for the final finishing cut. A rather similar arrangement is shown at (d), but in this case the dies may be opened sufficiently to admit the un-screwed bar between them. By this means a thread may be cut by threading the dies over the bar, clamping them tightly to the bar and rotating, then passing backwards and forwards over the portion that has to be screwed and gradually closing the dies until the proper diameter is reached. For cutting fine threads on large diameters this method is better than starting from the end, as by clamping the dies on to the bar first, the problem of maintaining the thread square with the axis is facilitated.

The action of dieing a thread is very similar to that of tapping, except that it is more difficult to start the die square. Generally, only one or two of the die threads are chamfered, and this does not give as much assistance as the long taper on the taper tap. In starting the die, great care is necessary, therefore, to maintain its face as near as possible square with the bar, at the same time that the die is pressed on to the end of the bar to help the commencement of cutting. The action is assisted by chamfering off the end of the bar for a distance equal to about two threads. When the cut is under way the die should be worked backwards and forwards similar to the method explained for tapping, and the threads should be supplied with some form of cutting oil. Certain designs of stocks and dies incorporate a bush for guiding the bar or tube being screwed so that the thread is cut square. This method is shown at Fig. 226(e), and is greatly superior to a plain die, particularly for screwing BSP threads, where a fairly fine thread has to be cut on a large diameter.

10   Introduction to the lathe

The lathe is the father of all machine tools and is recorded in the early history of many races, when, equipped with a fixed tool-rest, it was used for wood turning. For its development to the form in which we now know it, we owe much to Henry Maudsley, who developed the sliding carriage, and in 1800 built a screw-cutting lathe on which he turned screws having from 16 to 100 threads per inch, ($1\frac{1}{2}$ to $\frac{1}{4}$ mm pitch) and which were the best screws that had been made up to that time. About 1830 Maudsley constructed a lathe with a 9-ft (3 metre) face-plate, which was used to bore large cylinders and turn flywheels.

In its operation the lathe holds a piece of material between two rigid supports called *centres*, or by some other device such as a *chuck* or *face-plate*, screwed or secured to the nose or end of the *spindle*. The spindle carrying the work is rotated whilst a cutting tool supported in a *tool-post* is caused to travel in a certain direction, depending upon the form of the surface required. If the tool moves parallel to the axis of rotation of the work a cylindrical surface is produced (Fig. 227(a)), whilst if it moves perpendicular to this axis it produces a flat surface as at (b). The tool-post is supported on a *cross-slide* which moves perpendicular to the spindle axis, whilst the cross-slide is integral with a *carriage* which slides along the *bed* of the lathe. The spindle axis is parallel with the slideways of the bed,

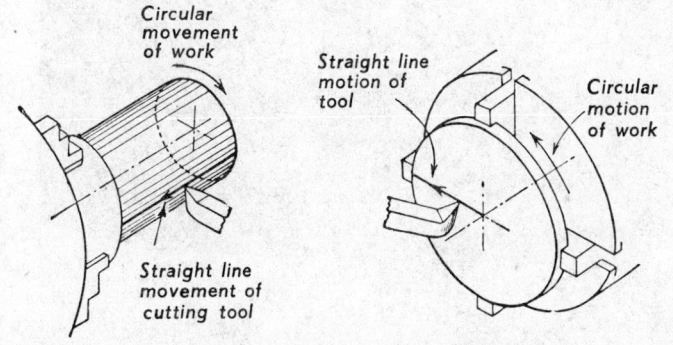

(a) Production of a cylindrical form   (b) Production of a flat surface

Fig. 227 Surface formation on the lathe

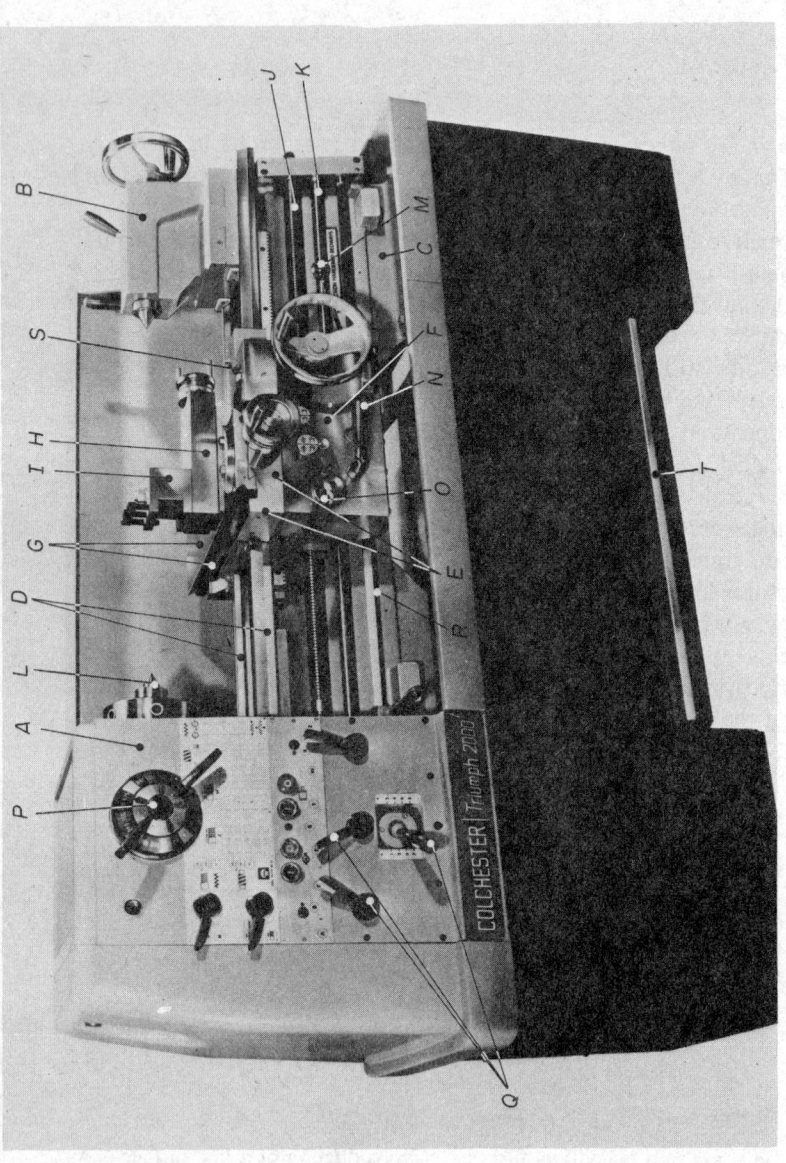

The Colchester Lathe Co. Ltd.

A. Headstock: *B.* Tailstock: *C.* Bed: *D.* Bed slideways: *E.* Carriage: *F.* Apron: *G.* Cross-slide: *H.* Compound slide: *I.* Tool Box: *J.* Lead screw: *K.* Feed shaft: *L.* Spindle nose: *M.* Start and stop: *N.* Feed lever: *O.* Lever for screwcutting nut: *P.* Speed selection: *Q.* Selection levers for feeds and threads:

so that movement of the carriage along the bed causes the tool to move parallel with the spindle axis. The above-mentioned parts of the lathe are all indicated on the diagram shown at Fig. 228.

The size of a lathe. In this country, lathes are specified according to the height of the centres above the bed, and the maximum length that can be accommodated between the centres. Thus a 190 mm × 750 mm lathe would take 750 mm between the centres with the centre of the spindle 190 mm above the bed, and would accommodate work 380 mm diameter *over the bed.* The diameter of work which would clear the carriage would, of course, be less than 380 mm. The length of bed, and the floor space occupied by the machine will be considerably greater than the length between centres because it includes the length of headstock and tailstock. Probably a lathe rated at 750 mm between centres would have an overal length of about 2 metres. In America, the 'swing' of the lathe, and not the height of centres, is specified, and they would call our 190 mm lathe a 380 mm (15 inch) swing.

The bed. The bed of the lathe forms its body structure and is supported at a convenient height on legs. It is cast with a box-like cross-section to give the necessary stiffness for resisting the twisting and other stresses which occur in practice and which, if they strained the bed unduly, would destroy the accuracy of the lathe. The top of the bed is planed to form guides or ways for the carriage and tailstock. The shape and disposition of these ways varies amongst different makers, but two common forms are shown at Fig. 229. The reader should observe, and make note of other lathe beds.

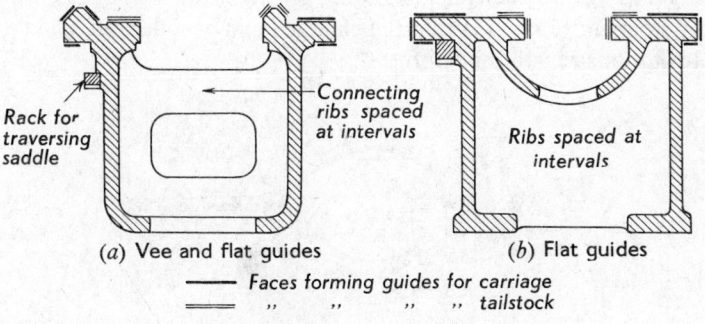

(a) Vee and flat guides   (b) Flat guides

——— Faces forming guides for carriage
═══   ,,   ,,   ,,   ,,   tailstock

Fig. 229. Examples of lathe bed sections.

Gap bed. To accommodate short jobs requiring a larger swing than could be accommodated over the bed, many lathes are made with a gap in the bed, extending for a short distance in front of the headstock. This gap enables a face-plate larger than the normal swing to be used, and widens the range of work for which the lathe may be used. In its gap bed form, the lathe shown at Fig. 228 has a gap allowing a swing of 580 mm diameter for a width of 155 mm in front of the face-plate.

The carriage or saddle. This is a flat-shaped casting, planed on its underside to fit the ways of the bed so that it may slide along. The guides, on which the cross-slide takes its bearing, are planed on the top surface of the carriage, and are perpendicular to the grooves with which the carriage bears on the bed.

To the front of the carriage is attached the *apron* (see p. 304), and the combined carriage and apron is often called the *saddle*.

The headstock is situated at the left-hand end of the bed and serves to support the spindle and driving arrangements. It is constructed of cast-iron, and incorporates bearings for the spindle at each end. The steel spindle is hollow so that bars may be passed through it if necessary, and has a Morse taper hole at the nose end for accommodating the centre. Lathes used to have the spindle-nose threaded to enable chucks and face-plates to be screwed on (Fig. 232), but this practice has given way to other methods. One of these is the flanged spindle (Fig. 233), whilst another, as employed on the machine at Fig. 228, mounts the chuck on to the enlarged end of the spindle (150 mm) and secures it with a Camlock arrangement. The considerations for abandoning the screwed nose are (1) the liability of the thread to wear or become damaged and thus lose its centralising accuracy, (2) to guard against the chuck screwing off when the machine is provided with a stopping brake, (3) that the enlarged end of a flanged spindle contributes to its strength and stiffness.

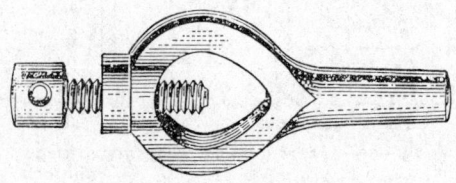

Fig. 230 Lathe carrier

The driving plate is attached to the spindle-nose and carries a projecting pin which drives work held between centres by engaging with a *dog* or *carrier* (Fig. 230) clamped to the work.

Chucks may be of the 4-jaw independent, or 3-jaw self-centring patterns. Each jaw of the 4-jaw chuck is operated by a separate square-threaded screw, whilst in the 3-jaw type all the jaws close in together, actuated by

F. Pratt and Co, Ltd   *Richard Lloyd & Co., Ltd*

(*a*) Independent 4-jaw chuck   (*b*) 3-jaw self-centring chuck (showing scroll)
(Jaws are reversible)

Fig. 231 Lathe chucks

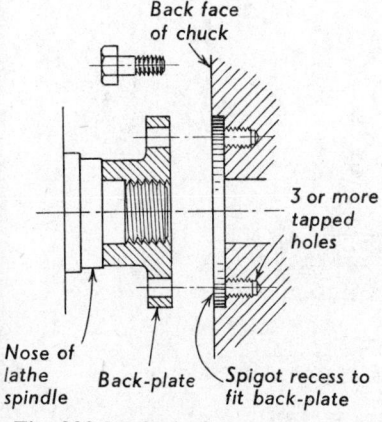

Back face
of chuck

3 or more
tapped
holes

Nose of
lathe
spindle   Back-plate   Spigot recess to
fit back-plate

Fig. 232 Method of carrying chuck
on screwed spindle-nose

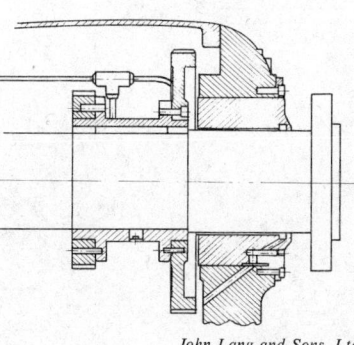

John Lang and Sons, Ltd

Fig. 233 Flanged spindle

the *scroll*, which is a spiral groove cut on the face of a flat disc. The principle is similar to the movement of a nut by a screw except that the 'screw' is cut on the face of a disc. The 3-jaw chuck is easier to operate, but its gripping efficiency is much less than that of the 4-jaw (Fig. 231). The chuck is adapted to a screwed nose spindle by using a *back-plate* as shown at Fig. 232, whilst on the flanged spindle it is bolted direct (Fig. 233). To ensure true running the recess in the chuck is a good fit on the diameter of the plate or spindle flange and is bolted hard on the face (generally the back-

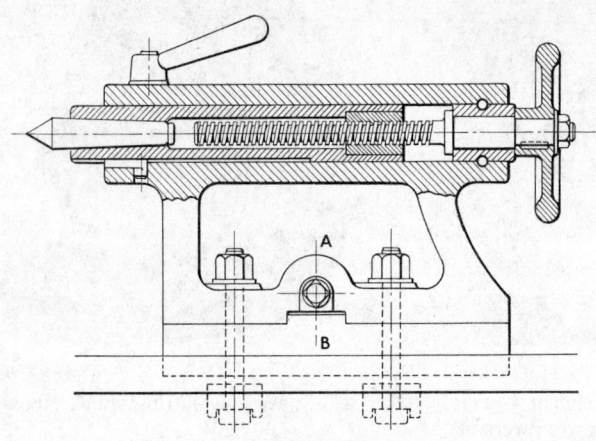

(*a*) Section through barrel

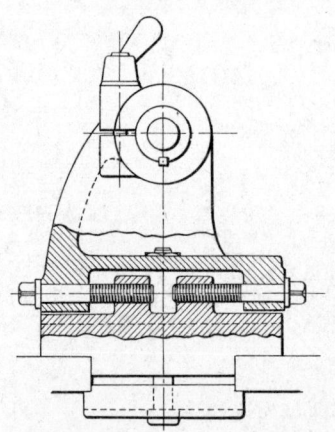

(*b*) Lower half in section on AB to show method of setting over

Fig. 234 Arrangement of tailstock

plate is finish turned in position). This is a common method of locating two large circular plates together so that they register accurately, and is called *spigoting*, the projection on the one plate, which fits into the recess in the other, being called a *spigot*.

The face-plate is used to hold work which can neither be turned on centres nor held in a chuck. It has bolt slots for accommodating the necessary clamping bolts (Figs 257–9).

The tailstock is the counterpart of the headstock, and carries the right-hand centre for supporting work when turning on centres. It is also used for supporting and feeding drills, reamers, etc., when it is necessary to use these for drilling work held in the chuck. The body of the tailstock is a casting, planed on its base to fit the ways of the bed, and when in its normal position, the barrel which carries the centre is in line with the spindle. An adjustment is provided for setting the tailstock centre to one side for taper turning (see later). The barrel of the tailstock is moved by the wheel or handle at its far end and is bored with a Morse taper hole at the end nearest the headstock. A section through a tailstock is shown at Fig. 234(a) and an end view at (b); to move the tailstock over, one screw is loosened and the other tightened.

The compound (top) slide. On most lathes a compound slide is interposed between the tool-post and the cross-slide. This slide may be swivelled to any angle and its use enables the tool to be moved in directions other than those permitted by the carriage and cross-slide. The length of such movement is limited to the travel of the compound slide, but generally the type

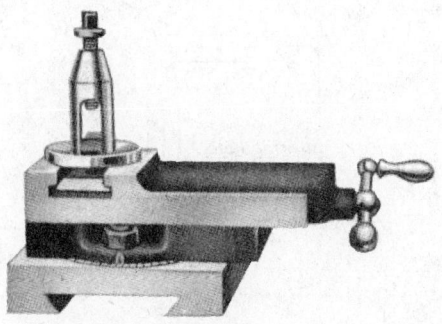

Fig. 235 Compound slide

of work for which it is used does not require a long travel. The compound slide is useful for turning and boring short tapers, chamfers, and other jobs requiring an angular movement of the tool (Fig. 235).

The feed motion. besides holding and rotating the work, and supporting and guiding the tool which operates on it, a lathe must be capable of *feeding* the tool along or across at a predetermined rate. If power feed were not available for the tool, the usefulness of the machine would be greatly curtailed because all the movements would have to be made by operating the wheels or levers by hand. This would lead to irregularity in the turning, and no operator could give his attention to the work at the same time as operating the handles without getting tired very quickly. For screw-cutting, accurate mechanical feed of the tool is necessary to produce threads of correct pitch (see later).

The feed motion is transmitted to the carriage and cross-slide by the *feed-shaft*, a long, plain shaft with a keyway, which runs along the front of the bed. On some lathes, a separate feed-shaft is dispensed with, and the

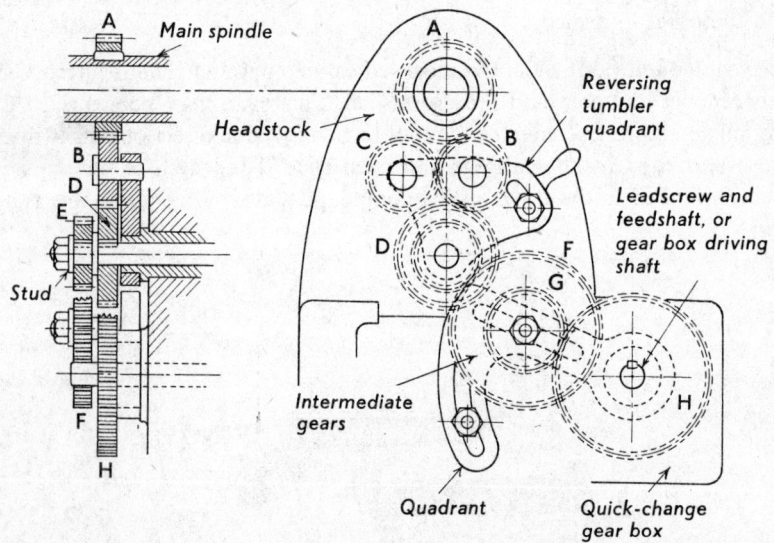

Fig. 236 Diagram showing drive from spindle to headscrew or gear-box
(Upper portion in section in left-hand view)

motion transmitted by a keyway cut along the *lead-screw*, the screwed shaft running along the bed of screw-cutting lathes. The feed-shaft is driven by gearing from the main spindle, and means are provided for varying the ratio of the drive to give different rates of feed. This may take the form of a gear-box (Fig. 272) with one or more levers for moving the gears, or the speed ratio may be altered by fitting different gears. To reverse the direction of rotation of the feed-shaft a tumbler gear arrangement is generally fitted. A diagram showing one arrangement of the drive to the feed-shaft and lead-screw is shown at Fig. 236, which also shows details of the tumbler gear reversing mechanism. On most modern lathes the feed gearing connection is more sophisticated than that shown at Fig. 236, but the principle is unchanged, and the reader should grasp the essentials better from our simple diagram. Whatever may be the form of drive to the feed-shaft, however, the lead-screw must be gear-driven in order that the correct speed ratio may be maintained (see later discussion on screw-cutting).

From the feed-shaft (or the keyway in the lead-screw), the feed motion is transmitted to the carriage and cross-slide by mechanism contained in the *apron*, the front part of the carriage to which is attached the hand wheels, knobs and levers for operating the carriage and cross-slide. The mechanism inside the apron of a lathe is shown at Fig. 237. The pinion A engages with the rack extending along the bed and manual operation of the saddle is effected by pushing in the handwheel to engage its pinion with the large gear on the spindle of A. The feed-shaft and lead-screw both pass longitudinally through the apron and the drive for the feed motions is transmitted to the wormwheel B by a worm, keyed to and sliding along the feed-shaft. Longitudinal feed is obtained when shaft C is pushed in so that its small gear engages with the large gear on the shaft of A. For the cross feed the plunger D is depressed to cause its gear to engage with a gear on the actuating screw of the cross-slide and so to transmit the drive. Lever E actuates a clutch for the feeding drive gears and F actuates the closure of the lead-screw half nut by means of lever G. Reversal of the feed and screw-cutting motions is effected by reversing the rotation of the shafts in the headstock gear-box.

The centres. These are worthy of attention, as the accuracy of centre work is greatly influenced by their condition. The shanks of centres are usually finished with a Morse taper and the point which fits the centre hole in the work is generally made to an angle of 60° (sometimes, for heavy work, the

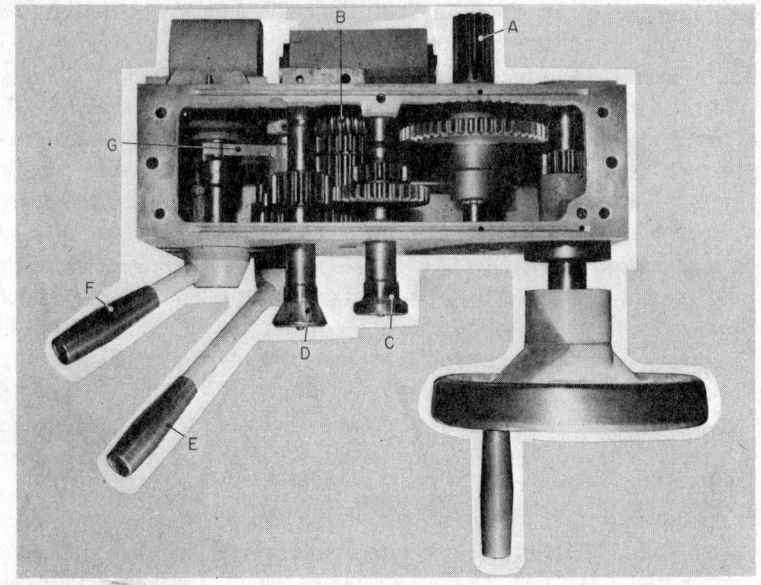

Fig. 237 Lathe apron mechanism

centre angle may be 75° or 90°). The headstock centre is accommodated
to the taper hole in the spindle by means of a taper reduction sleeve (Fig.
238(a)), and as this centre rotates with the spindle it is often called the
'live' centre, as opposed to the tailstock centre which is stationary, and
called the 'dead' centre. The fact that the live centre rotates with the work
imposes the need for special precautions because, if it is not running true,
the turned surface of the work will not be concentric with the centre hole
when the tool approaches the headstock. When the tool is turning near the
tailstock centre, the turned surface there is bound to be true with the centre
hole because the centre is stationary, and the work revolves on it. If, then,
we are turning with the live centre untrue, the surface will be concentric
at one end, gradually becoming out of truth with the centres and increasing
as the tool approaches the headstock. When the work is reversed in the
machine to finish the portion previously covered by the driving dog, we
shall be turning it up to a similar diameter but on a slightly different centre,
and instead of the two turned surfaces matching up, a step will be left, the
effect being as shown exaggerated at Fig. 238(b). Great care should be
taken, therefore, to ensure that the headstock centre is perfectly true, and

the following precautions should bring this about without difficulty (the truth of the live centre may be tested with a dial indicator). The shank of the centre should be perfectly round and straight, and a perfect fit in the bore of the reduction sleeve which should also fit perfectly in the taper hole in the spindle. These surfaces should not be allowed to become knocked about and should always be cleaned before fitting together. A punch mark or scribed line should be put on them so that they may always be fitted in the same relative position.

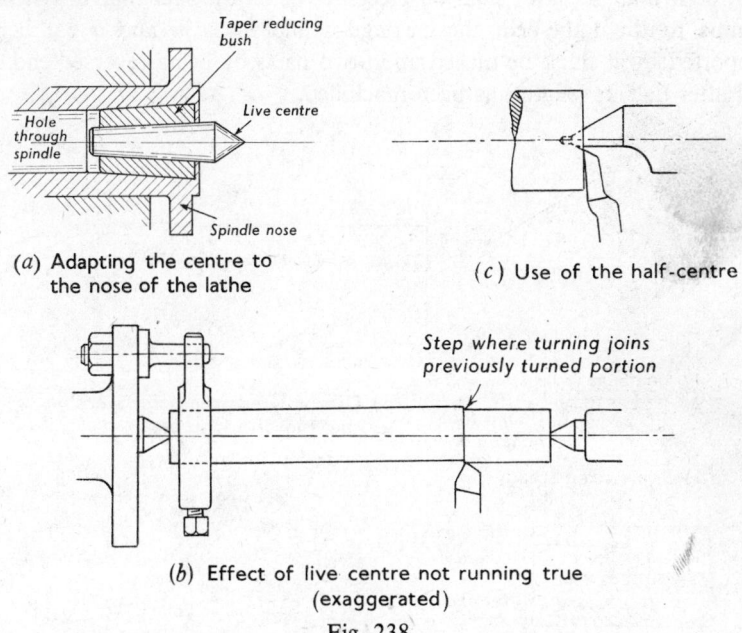

(a) Adapting the centre to the nose of the lathe

(c) Use of the half-centre

(b) Effect of live centre not running true (exaggerated)

Fig. 238

Lathe centres are usually made of carbon tool steel and should be hardened with little or no temper. Extreme hardness is not as necessary for the live centre because it is not subject to wear, but if the tailstock centre is not very hard it will soon seize up. When the centre points need renovating the operation is best performed by putting each one in turn in the spindle, running the lathe, and grinding them with a *tool-post grinder* held in the tool-post and fed with the compound slide set at 30° with the spindle axis. The tool-post grinder is a small electric grinder rather like the electric drill shown at Fig. 204, but with a shank to clamp it in the place of the tool, and a high-speed grinding wheel instead of the drill chuck.

A *half-centre* is a centre which is cut away almost to its point and is used in the tailstock for facing up the ends of work (Fig. 238(c)).

Steadies. For supporting long, slender work against the pressure of the cut, a steady is used. The fixed or 3-jaw steady (Fig. 239(a)) is clamped to the bed of the lathe and supports the bar, being turned by means of three jaws set at 120° with each other. Usually, before this steady is put on, the bar is skimmed up true for a length of about an inch at the point where it is desired that the jaws shall be placed. Due to the fact that this steady clamps to the lathe-bed, the carriage cannot pass it, and a bar being supported by it must be turned up in two parts by being reversed end for end after half its length has been machined.

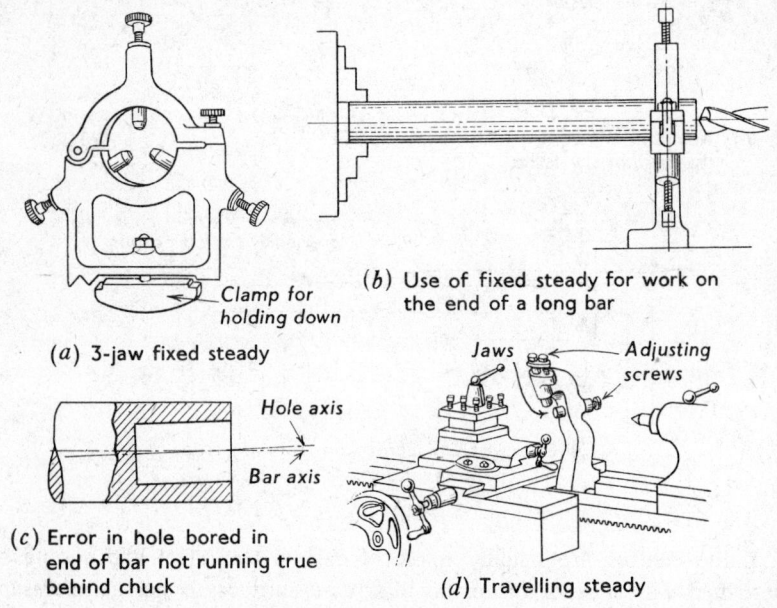

Clamp for holding down

(*b*) Use of fixed steady for work on the end of a long bar

(*a*) 3-jaw fixed steady

Hole axis

Bar axis

Jaws   *Adjusting screws*

(*c*) Error in hole bored in end of bar not running true behind chuck

(*d*) Travelling steady

Fig. 239 Lathe steadies

A very useful application of the 3-jaw steady is for centring, or for turning, drilling or boring operations at the end of bars, and the set up is shown at Fig. 239(b). The advantage of using the steady, even although the bar might have been held in the chuck and pushed up the hollow spindle, is that accuracy is ensured, because if the end held in the chuck is true, then the whole bar is running true and the operation being done is sure to be co-axial with the bar. Imagine, for example, a hole is to be bored in the end of a bar which, for the purpose, is pushed into the hollow spindle and held in the chuck at one end. We can true up the short length protruding from the chuck jaws, but we cannot see whether the whole length *inside* the spindle is true, and if it does not run true, then the hole we bore will have an error which is shown exaggerated at Fig. 239(c).

The *travelling steady* (Fig. 239(d)) is fixed to the carriage and travels along with the tool, supporting the bar against the thrust of the cut by means of jaws which are opposite the tool and slightly behind it, so that the bearing is taken on the round portion which the tool has just finished. This steady is used in cases where it would not be convenient to turn half-way and reverse the bar to complete the turning as is necessary when using a fixed steady. The reader should observe with regard to this steady that when the tool and jaws are set and clamped, the diameter turned is invariable, being that diameter which will just pass through the three points made up of the tool and the steady jaws. The turned bar, then, should be *parallel*, but there is no assurance that it will be *straight*.

Graduated sleeves. The compound and cross-slides are moved by means of a screw which engages with a nut attached to the slide, the screw generally being cut with a square, or acme thread. The hand lever, wheel or knob is pinned or keyed to the end of this screw, and in addition there is a graduated sleeve which has a number of divisions shown on it, with a zero (0)

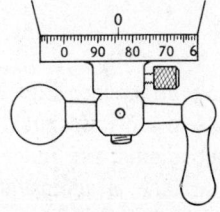

Fig. 240 Indicating sleeve for movement of slide

mark scribed level with its edge on some stationary face (Fig. 240). This is for obtaining an accurate measure of the slide movement and the maximum use should be made of it, but it is first necessary to calculate the value of its divisions. This is easily done by dividing the number of divisions into the movement given to the slide by one turn of the screw. For example, if one turn of the screw moves the slide 5 mm and there are 100 divisions round the sleeve, 1 division will represent $\frac{5}{100} = 0.05$ mm of slide movement. The movement of the slide for one turn of the handle is easily checked with a rule, as it is almost sure to be an even millimetre, corresponding to a whole number of millimetres for the pitch of the screw.

Stops. Many lathes are provided with stops for the carriage and cross-slide, which permit the slide to be brought to the same position every time. They are useful when a number of similar articles have to be machined, as they save the time of constant checking, and free the mind of the anxiety

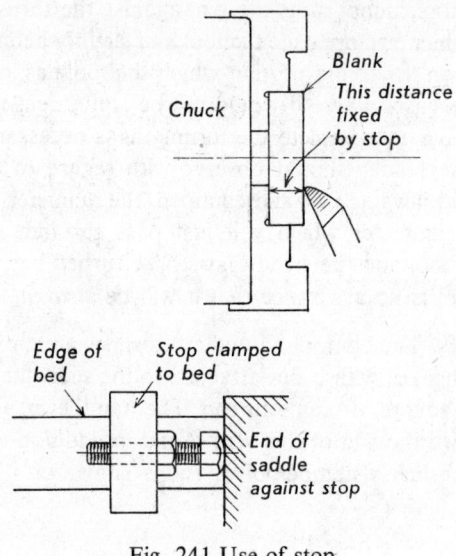

Fig. 241 Use of stop

of machining undersize. Fig. 241 shows a carriage-stop being used for facing blanks in a 3-jaw chuck. The previously machined face of the blank is always set against the inner face of the jaws, and the stop set so that when the carriage is against it, the blank will be the required thickness. When working against a stop, care should be exercised to use the same pressure against it each time or slight variations may occur. We shall refer to cross-slide stops again when discussing screw-cutting.

Turning on centres

The first stage in turning work between centres is that of drilling the ends of the work to accommodate the lathe centres. The correct form of centre hole is shown at Fig. 242(a), and this is best produced by the *combination centre drill* (b). The hole could be produced by first drilling with a small drill

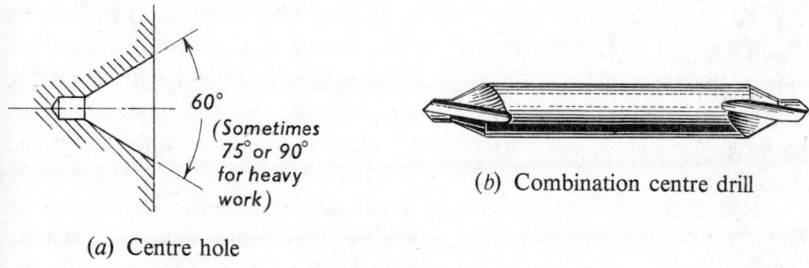

(a) Centre hole

(b) Combination centre drill

Fig. 242

followed by countersinking to the 60° angle, but this is a roundabout method and should only be used if a centre drill is not available. The small hole at the bottom is important as it ensures that the lathe centre bears on the conical part and not at its extreme point. A common fault is to make the centre hole too large, and nothing looks worse than huge unsightly centre holes in the ends of nicely finished work. Centre drills may be obtained in all sizes, and if the lathe centres are looked after (i.e. with good points) there is no reason for having centre holes larger than necessary.

If the centre holes are to be drilled on a drilling machine, the centre of the bar should first be found and centre-punched at each end. To find the centre, odd-legs, vee blocks and scribing block (p. 258), or the centre-finding square of the combination set may be used. The holes may then be drilled by supporting the bar vertically under the spindle of a sensitive drill, care being exercised to hold the bar rigidly and feed the centre drill carefully, as these are rather delicate and soon break. Another method of centring is to use the 3-jaw chuck of the lathe to hold the bar, whilst the drill is held in a Jacob chuck in the tailstock. If the bar is too large to go up the hollow spindle, the 3-jaw steady may be used as shown at Fig. 239(b). The lathe method is convenient, because the end of the bar can be faced to length at the same time. When a large amount of centring has to be done a special centring machine is a more convenient method than any other. This generally consists of a short bed with a self-centring chuck mounted at one end. A drill spindle is mounted on the bed so that its axis is central with the chuck.

After the bar has been centred, a driving dog must be clamped to it, and the lathe prepared by attaching the driving plate, fitting the centres and setting the machine to a suitable speed for the size of the bar to be turned. If the ends of the work were not faced to length with the centring operation, the half centre should be fitted to the tailstock first, and the end facing carried out before any other turning is done (Fig. 238(c)). If this facing is left until later, and the ends of the bar are not square with the axis, one side of the centre hole will have less bearing surface than the other, and the centre, in bedding itself in, will wear more off that side than the other, causing the centre hole to become slightly eccentric with the turned diameter.

Before fitting the bar between the lathe centres a little russian fat or other suitable lubricant should be put in the tailstock centre hole, and then the tailstock centre adjusted, until when the bar is held and turned, it can just be felt that the centres are gripping it. The feel must be neither too easy nor too tight, and the adjustment should be felt from time to time as it may get tight due to the bar expanding with the heat of the cutting or loose due to wear of the tailstock centre hole. When adjusting the tailstock do not loosen the clamping screw right off, but have it so that it grips the barrel very slightly. It should be clamped up tightly after the proper adjustment has been obtained. When the ends of the bar have been faced, the facing tool should be taken out and a straight rougher, with cutting angles suitable for the material being turned, set in the tool-post. The nose of the tool should be adjusted to be level with the centre, and its overhang from the tool-post support as small as possible. At the same time it should be possible for the tool to be brought to the end of the bar without the necessity for a large projection of the tailstock barrel to avoid the tool or carriage fouling the tailstock body.

Everything is now ready for taking a cut along the bar, and after calipering the rough bar, and determining from the drawing, or pattern, how much material there is to be removed, a roughing cut may be put on and the carriage feed engaged. The depth of the roughing cut will depend on how much metal there is to remove, on the size and strength of the machine and on the rigidity of the work. If there is not much metal to come off, a small cut should be taken, and as soon as possible a trial made of the diameter being turned. On the other hand, with plenty of material to be removed, the cut should be as large as the machine will take, and if all the precautions regarding the clamping of the tool, tailstock and driving dog, lubricating the tailstock centre, etc., have been taken, there is nothing to be feared about taking a reasonable cut. The length for which the cut may be allowed to proceed along the bar will depend on the job. If it has to be turned parallel for its whole length, the cut may proceed until the tool, or some part of the carriage, approaches close to the driving dog, when the traverse must be stopped. If work is being turned up to a shoulder a rule must be held on the job, and the cut stopped when within 1 mm to 2 mm of the finished dimension. This process may be repeated until the portions that may be turned with the dog on the far end are within about 1 mm to 2 mm of their finished sizes, when the work must be reversed, and the other end roughed out.

Finishing. When the roughing out has been completed it will be advisable to touch up the tool for finishing, or change it for another. Also, it may be necessary to look out other tools for such purposes as facing and cornering shoulders, etc. At this point also, if the bar has been turned a fair amount smaller than when starting, the speed may be increased to the next step, and to obtain a good finish, the feed reduced. The same general procedure as for roughing may now be followed for finishing, except that more attention must be paid to finish, and to the measurement of the diameters and lengths. Important sizes should be turned using a micrometer for measurement and less important ones measured carefully with calipers. If lengths are required to closer limits than may be obtained with a rule, the micrometer, vernier, depth gauge or some other suitable instrument must be used.

The reader should cultivate the habit of finishing his work with the *tool* and not with a file or emery cloth. A limited use of emery cloth is permissible to obtain a good finish, but a file should *never* be used to finish a cylindrical surface. First-class turners regard the use of a file with the contempt it deserves, and the reader should not consider himself in the higher grades of the skill until he is able to share the same opinion. An old file may be kept handy for removing the very sharp corners from the edges of shoulders and ends of diameters, but this should be its only application. Plenty of practice in using the calibrated sleeve on the cross-slide, and the development of his confidence, will eventually train the reader to finish work to dead size with the tool.

Correcting for non-parallel turning

We have not mentioned the possibility of the lathe turning out of parallel. If the head and tailstock centres are not exactly on a line parallel with the travel of the carriage, when a cut is taken along a bar held between them the bar will not be parallel. The error may be rectified by moving the body of the tailstock a small amount to one side or the other, by means of the adjustment shown on Fig. 234(b), and we leave it as an exercise to the reader to ascertain from the error on his work whether the tailstock must be moved towards or away from him. What we do advise him is, to find out, and remember, which way the tailstock moves when the adjusting screws are turned in a certain direction, so that when he wishes to move it away from him, he does not do the reverse. This is because the movements generally required for correction are so small that no check (e.g. sight or feel) is possible, and the process of trial and error necessary to bring the job parallel is often a trial of one's patience. If the work is not parallel, and there is not much material on it with which to make the necessary experimental cuts, adjust the tailstock a little and then feed in the tool until it just skims the work at the smaller end. Take the reading on the graduated dial, move the tool to the other end and feed it in until it cleans up the diameter to the same as that at the smaller end. Observe the reading on the graduated dial which, if the setting is correct, should be the same as before. If, to bring the two ends the same, the tool has been moved *beyond* the previous reading, then *more* movement of the tailstock is needed, whilst if the previous reading is not reached, the tailstock movement has been too much. Do not try to remember these rules but use the method, and reason out the correct solution.

We will now discuss one or two typical centre turning jobs, but must impress on the reader that for every job there is generally more than one method of approach. We will endeavour to give what we consider to be good standard practice, leaving any improved modifications for adoption by the more expert of our readers.

Example 1. To turn up the mandril shown at Fig. 243(a) from mild steel 35 mm diameter, 240 mm long.

Mandrils are used on the lathe, as will be explained later, for finishing work on centres which have been partly finished in the chuck. They should be hardened and ground on the top diameter (as is called for in this case), and the centre holes are recessed as shown to prevent any damage to them when pressing, or driving the mandril into the work. We must leave a grinding allowance of 0·5 mm on the top diameter. The lengths on the mandril are unimportant, and the reduced diameters at the ends, which are for accommodating the driving dog, need not be worrried about as to accurate measurement.

(i*a*) Hold the bar in the 3-jaw chuck, face one end, centre drill and recess centre hole. Reverse in chuck and treat the other end the same way (Fig. 243(b)).
Or
(i*b*) Mark out centres with odd-legs, centre punch and drill centre holes on sensitive drilling machine.

If (ib) is adopted, before proceeding to the turning, face the ends and recess the centre holes, using the half centre to support the work at the tailstock end (Fig. 238(c)). As metal is faced away from the centre hole, it will probably be necessary to go back to the drilling machine and deepen it somewhat.

(ii) To allow 0·5 mm grinding, the top diameter of the mandril must be turned to $32·16 + 0·5 = 32·56$ mm approximately. If it is brought to between 32·5 and 32·6 and reasonably parallel, it will be satisfactory.

Fit a driving dog at one end, place between centres and using a straight roughing tool, turn the bar to 33 mm diameter as far as the dog will allow. Two cuts along should be sufficient and at this stage a check may be made for turning reasonably parallel (Fig. 243(c)).

(iii) Change the dog to the other end of the bar, clean off the un-

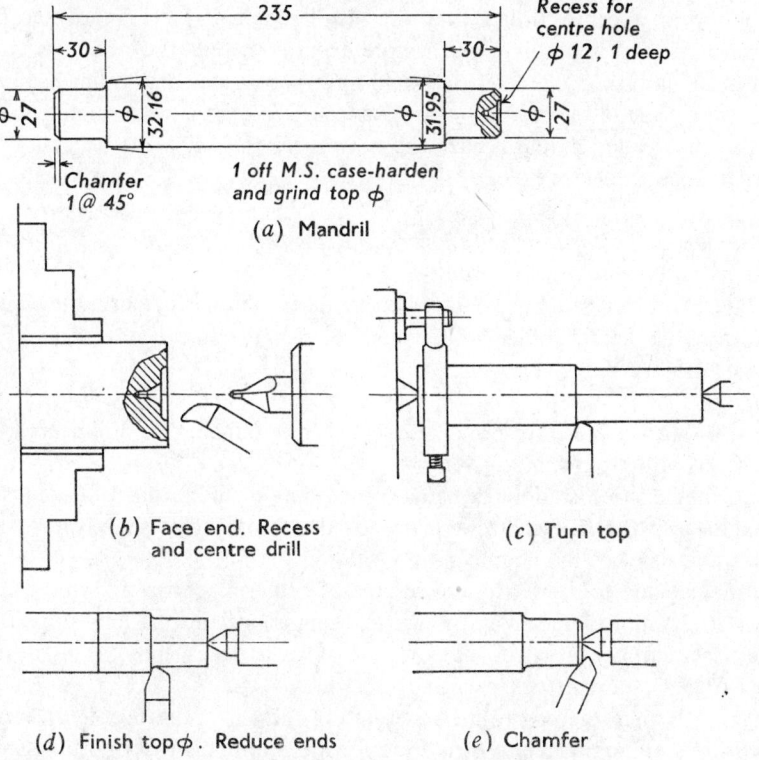

(a) Mandril

(b) Face end. Recess and centre drill

(c) Turn top

(d) Finish top φ. Reduce ends

(e) Chamfer

Fig. 243 Operations in turning a mandril

turned portion, then turn down to 27 mm diameter for a length of 30 mm (d).

(iv) Swing the tool round in the post, or fit another tool with a straight face, set at 45° and chamfer the end (e).

(v) Clamp the driving dog on the 27 mm diameter portion, using a piece of sheet copper or brass under it to avoid damage to the turned surface. Place between centres, reduce and chamfer the other end as at (iii) and (iv).

(vi) Finish off by turning the top of the mandril to 32·5/32·6 mm diameter. Do not worry unduly about the taper as the grinder will set his machine to give it correctly.

(vii) Remove sharp edges from all corners with an old file. Inspect centre holes, and if any burrs have been thrown up round them fit the half centre and take a small cut with the side tool to remove them. Clean out the centre holes with a centre drill if they are at all rough. If the mandril had not been finished by grinding, much more accuracy would have been required in finishing its top diameter. It would have been necessary to finish this to drawing sizes after making sure that the live centre was running dead true (the reader should realize the reason for this, also the necessity for good, smooth centre holes).

Example 2. To turn the pin shown at Fig. 244(a), from a piece of mild steel 32 mm diameter, 180 mm long.

(ia) Hold the bar in a 3-jaw chuck, face one end and drill the centre. Reverse in chuck, face other end to length and centre drill (Fig. 244(b)).

Or

(ib) Mark out centres with odd-legs, centre punch, and drill centres on sensitive drilling machine.

If method (ib) is adopted, before proceeding to the turning, face ends to length with the half centre and deepen the centre holes slightly if facing removes most of the countersunk portion.

(ii) Fit dog or carrier to one end, place between centres and turn top of bar to 30 mm diameter up to carrier, using a straight roughing tool. Two cuts should be sufficient, and at this stage a check may be made for parallelism, the tailstock being adjusted if necessary.

(iii) Change the dog to the other end of the bar, clean off the unturned portion and commence to turn down for the 19 mm diameter portion, stopping the tool when the turned length is 104 mm. Continue this until the

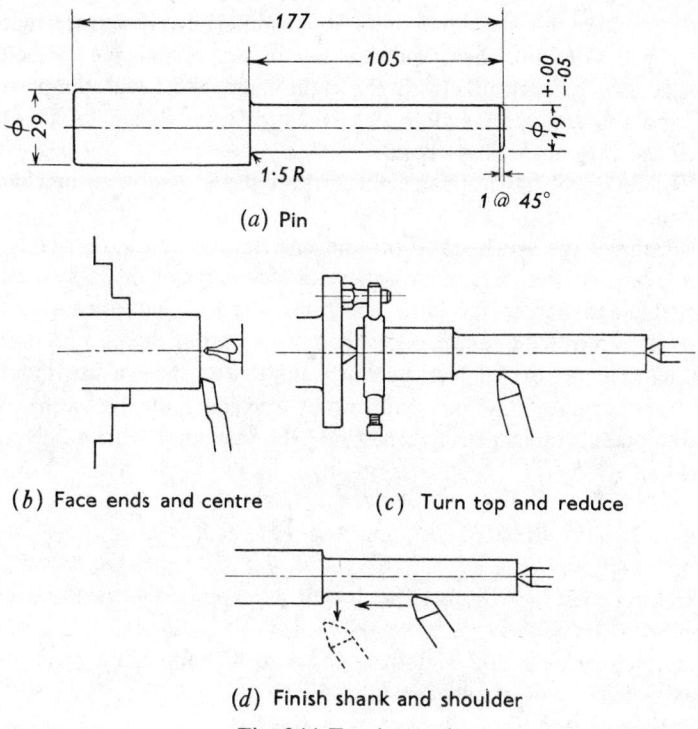

(a) Pin

(b) Face ends and centre   (c) Turn top and reduce

(d) Finish shank and shoulder

Fig. 244 Turning a pin

reduced diameter is about 20 mm (19 + 1), and level up the shoulder as near as possible with the same tool (Fig. 244(c)).

(iv) Grind a tool to a 1·5 mm nose radius, and set it so that the shoulder may be faced, and the shank turned at the same setting (Fig. 244(d)). Square up the shoulder as far in as the turned diameter, and to barely 105 mm from the end. Use the micrometer to turn the shank to about 19·25 mm diameter, disengage the feed when the tool is about 1 mm to 2 mm from the shoulder and continue by hand (do not let the carriage pause) until it just touches the shoulder. Return the tool and set it to cut the shank 19 mm diameter (not more), by turning a length of about 4 mm, and 'mic'ing', until the correct size is reached. Engage the feed, and when the tool approaches to within about 1 millimetre of the shoulder, disengage the feed with the right hand, allowing the left hand to continue the rotation of the traverse wheel so that the carriage movement is at the same rate as before. Transfer the right hand to the cross-slide handle, and as soon as the tool touches the shoulder, press it in sufficiently to ensure that the shoulder will skim up, then, without any pause, commence to feed the cross-slide slowly outwards with the right hand. This will complete the shank and the shoulder, and by working this way a clean-looking joint between the two surfaces is obtained.

(v) Set any tool having a straight portion at 45° and turn the chamfer on the end.

(vi) Remove the work, take off the carrier and put a carrier on the smaller diameter, first wrapping a piece of sheet copper or brass round to prevent damage. Put in the lathe and finish the top diameter to 29 mm. With a file remove the sharp edges from each end of the 29 mm portion. If, during the process of turning, burrs have been thrown up round the centre holes, these may be removed by replacing the half centre and using the ending tool again, or they may be taken off with a half-round scraper.

Example 3. To turn the pin shown at Fig. 245(a) from steel 50 mm diameter, 140 mm long.

(i) Centre, and face the ends to length as explained in (i*a*) or (i*b*) for Example 1.

(ii) Attach driving dog and turn down to 46 mm diameter as far as the dog (3 cuts along should be sufficient for this).

(iii) Reverse end for end, and turn the top down to the previously turned portion. Now rough down the 22 mm diameter portion until it is about 23 mm diameter, with the shoulder roughly squared up to 54 mm from the end of the bar (Fig. 245(b)).

(iv) Reverse, and put a dog on the reduced portion. Rough down to 33 mm (32 + 1) diameter, leaving the collar roughly squared up, and 20 mm (19 + 1) wide. Rough the end to 20 mm diameter, leaving 38 mm between the two shoulders (Fig. 245(c)).

(v) Reverse with dog on the small end, and finish the 22 mm diameter as explained in Example 2 (iv). This diameter may be finished to calipers as the absence of size limits means that it may be finished to within 0·25 mm of size. The collar will now be 1 mm over its finished size. With file, remove sharp edges from end corner of bar.

(vi) Reverse, put dog on 22 mm portion with brass to prevent damage. Finish the large diameter to $32 {}^{+ \cdot 00}_{- \cdot 02}$ mm (micrometer) and face collar (if necessary) to 1 mm over 19 mm. Finish the small diameter to 19 mm (calipers) and face collar to leave a *full* 38 mm between them. Finish the top to 45 mm diameter (calipers).

(vii) Check that the work is nice and firm between the centres, grind up a side tool to 1 mm nose radius, and face the collar to 19 mm $+0.02\atop-0.00$ wide. For this operation lock the carriage so that it cannot move

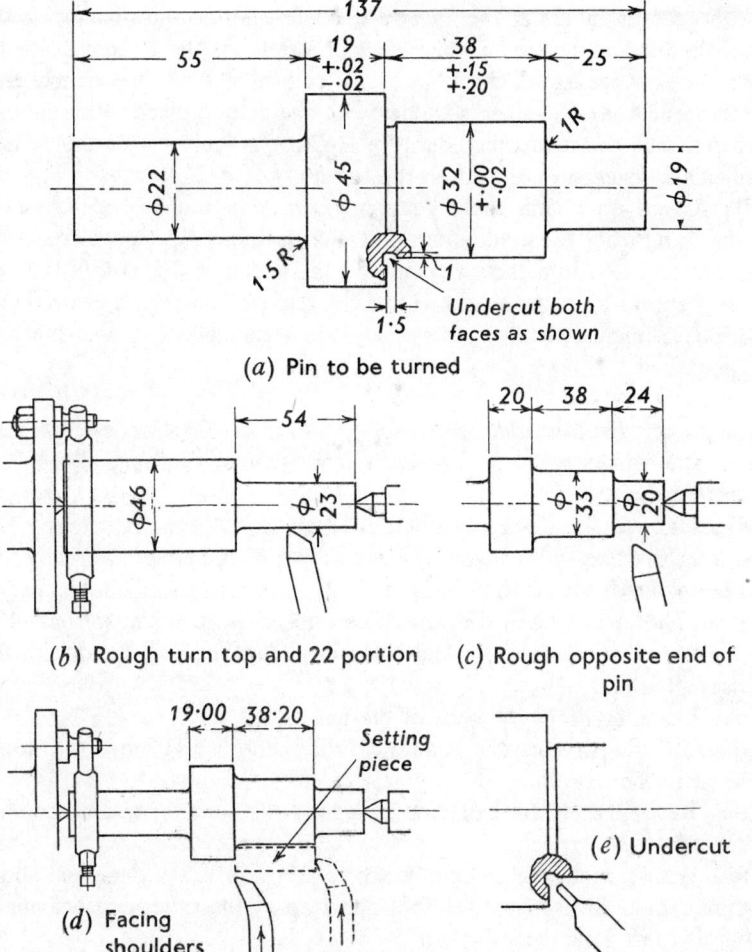

(*a*) Pin to be turned

(*b*) Rough turn top and 22 portion

(*c*) Rough opposite end of pin

(*d*) Facing shoulders

(*e*) Undercut

Fig. 245 Operations in turning a pin

along the bed; on most lathes a screw or handle is provided for this purpose. Feed the tool towards the shoulder with the compound slide set parallel to the bed, taking small cuts down the shoulder and testing with a micrometer until the size is reached. Leave the corner with a 1 mm radius blending in to the 32 mm diameter (Fig. 245(d)).

(viii) File a piece of material to 38·20 mm (micrometer or vernier) for setting the tool. Loosen the carriage and with the same tool as in (vii) clean out the corner radius at the 19 mm diameter, noting the reading on the cross-slide indicating sleeve when the tool just skims the 19 mm diameter. Lock the carriage again, check the work for firmness between centres, and set the tool from the other shoulder with the setting piece. With the tool thus set, face down the shoulder (Fig. 245(d)), stopping when the 19 mm diameter is reached (as noted on the sleeve).

(ix) Grind up a thin tool $1\frac{1}{2}$ mm wide, rounded on its end. Lock the carriage and using the graduations on the cross and compound slide sleeves feed the tool in 1 mm each way, to give the undercut (Fig. 245(e)).

(x) Remove all sharp edges with a file, remove burrs from centre holes and polish diameters with fine emery cloth and oil, if a nice finish is required.

Example 4. To turn the limit gauge shown at Fig. 246(a) from cast steel. The gauge is to be hardened, and ground on its end, gauging diameters.

We shall require a piece of material 32 mm to 34 mm diameter by 133 mm long to make the gauge. To finish the knurled handle a knurling tool will be required. This is shown at Fig. 246(c) and consists of a holder carrying two hardened wheels with their faces milled at an angle, so that when they are pressed against a round surface the surface is marked with the crossed knurling.

(i) Centre, and face the ends of the bar.

(ii) Fit a dog to one end, clamp between centres and rough turn to 31 mm, up to the dog.

(iii) Reverse end for end, and complete the turning of the top diameter.

(iv) With a tool sink in (and rough to $22\frac{1}{2}$ mm) leaving the end about $32\frac{1}{2}$ mm long. Swing over the tool and turn to the other end, leaving it about $18\frac{1}{2}$ mm long (Fig. 246(b)).

(v) Touch up the tool, grinding about $1\frac{1}{2}$ mm radius on the nose, finish turning the centre to 22 mm diameter and face the end shoulders to leave

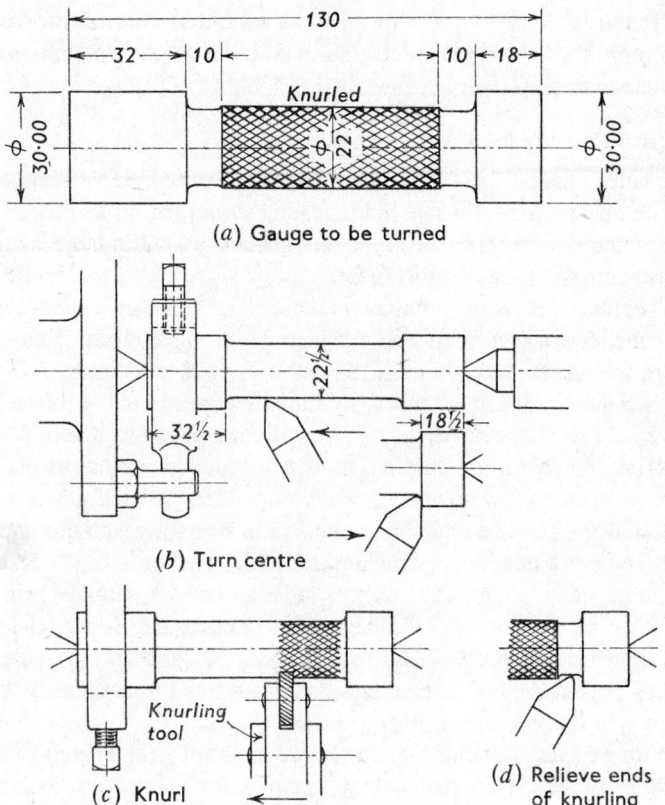

Fig. 246 Turning a limit gauge

the gauging portions 18 mm and 32 mm long. Be careful to match up the cut where meeting in the centre.

(vi) Set the knurling tool in the tool-post and adjust so that it is square with the work. Press the rollers *hard* on to the handle portion of the gauge until the markings are seen to form; engage the feed and allow the tool to traverse to the other end. If, after the first attempt, the markings are not deep enough, repeat the process (Fig. 246(c)).

(vii) Remove the knurling tool and with the tool used in (iv) and (v) relieve the ends of the knurling (Fig. 246(d)).

(viii) Finish the top of the gauging diameters to 30·5 mm diameter (i.e. 30 mm + $\frac{1}{2}$ mm grinding allowance).

(ix) Remove sharp edges with an old file, and any burrs thrown up at the centre holes. Inspect the centre holes and remove any roughness on the countersunk portion.

Chuck work on the lathe

If the reader studies the full possibilities of the chuck, he will find that its technique opens up a wide field of interesting work, and offers much greater scope for the development of the higher grades of skill in lathe work than does turning on centres. Unfortunately, many turners are too prone to the use of centres and many jobs are machined in this way which could be turned much more accurately and efficiently by using a chuck. If the reader, therefore, wishes to make himself master of the full technique of the lathe, he should give sufficient attention to chuck work, and allow himself to be guided by those of his turner advisors whom he knows to make full use of their 4-jaw chuck, rather than by those who always incline to the use of centres, or to the 3-jaw chuck. A chuck may often be used in conjunction with the tailstock centre and the combination forms a very rigid and satisfactory method of holding, particularly for heavy cuts (see Fig. 255).

To be properly equipped, a centre lathe should be supplied with both 3-jaw self-centring and 4-jaw independent chucks, the 3-jaw having two sets of jaws for the purposes shown at Figs. 241 and 243(b). This is not necessary on the 4-jaw because the same jaws may be reversed. The 3-jaw chuck is quick, and easy to use, because the work is automatically self-centred (or at least it should be, but never is for the greater part of the life of such a chuck). For this reason many turners, and particularly the impatient young ones, use a 3-jaw chuck, when in the interests of accuracy and efficiency they should be using the 4-jaw pattern, and, consequently, never master the art of quickly setting up a job in the 4-jaw, and never become sufficiently skilled in chuck work to realise the superiority of the one over the other.

The self-centring chuck is very useful for light turning jobs, but the chucks of this pattern usually supplied with centre lathes cannot be relied upon to grip sufficiently well for heavy cuts, or when a job requires holding on a short length. Their 'self-centring' characteristic is not to be relied upon after the newness has worn off, and if this is counted as an important property, the gripping portions of the jaws should be ground from time to time with a tool-post grinder. This grinding also corrects the distortion which occurs to the jaws due to holding work near their fronts, causing their gripping faces to depart from squareness with the chuck face and only gripping

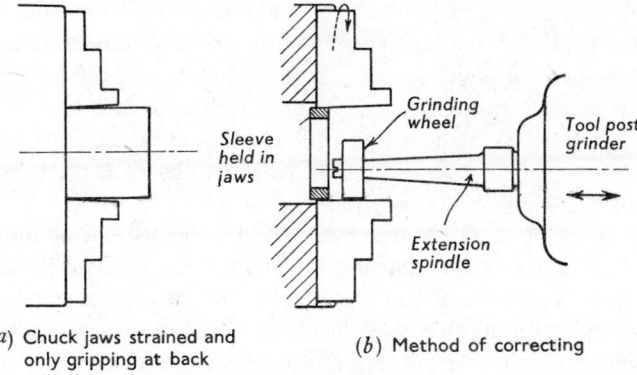

(a) Chuck jaws strained and
    only gripping at back

(b) Method of correcting

Fig. 247 Holding in the chuck

with their back portions on work extending for their whole length (Fig. 247(a)). To grind the jaws they should first be gripped on to a piece of material of a size representing the average opening of the jaws when in use. This should be pushed as far to the back of the jaws as possible, so that all but the short length gripping it may be ground. An extension spindle will be required for the tool-post grinder which should be clamped with its spindle parallel to the lathe bed. With both the lathe and the grinder running, the grinding wheel is traversed by hand backwards and forwards, and the cut applied by the cross-slide in amounts to suit the capacity of the wheel to take it. When the whole length of the jaws has been cleaned up they should be taken out and the small unground portion finished off by hand (Fig. 247(b)).

Setting work in the 4-jaw chuck

The 4-jaw chuck is not always fully appreciated because apparently, the setting up of a round bar is considered to be laborious. Actually, with a little practice, this becomes the work of a few minutes, and the superior gripping power means added safety, greater peace of mind and less possibility of the work moving during an operation.

Marked on the front of the chuck will be found a number of rings about 10 mm apart, which should be used as a guide for the preliminary setting of the jaws, and which will enable the work to be set to within about 1 mm of centre. From that point, take a piece of chalk in the right hand, rest the hand on a block so that the chalk is about at centre height, and grip the right wrist with the left hand for additional steadiness. Run the lathe and hold the chalk until the high side of the work is marked by it. Loosen slightly the jaw (or jaws) opposite the chalk mark, and tighten the side of the chalk mark. Repeat this process until the work runs true and a continuous chalk line is made round it. With diligent practice the reader will soon master this, and when proficient, he should be able to set up to a few hundredths of a millimetre very quickly.

When work has to be set true both on the diameter and on the face, set the diameter true without tightening too much, and then set the face true by tapping with a spanner or hammer; finally tighten all jaws equally. Often, when one side of a disc has been faced up, it must be reversed in the chuck for facing the other side parallel. A good method of setting this is to use inside calipers from the face of the chuck to the machined face of the disc, tapping the disc until its inner machined face is parallel with the chuck face.

His approach to chuck work will bring the reader into contact with two aspects of lathe operation which we have not yet discussed. These are: boring and the use of mandrils.

Boring

To form the surface of a hole truly, it must be bored with a single-point tool. A hole which has been drilled and reamed may be cylindrical, parallel and accurate to size, but there is no guarantee that its centre line will be true with the rotational axis of the work because the drill may have run out a little, and the reamer will have followed the drilled hole. Such a hole, therefore, may not be true with other surfaces which have been turned on the work.

A hole may be finished to size by reaming, after boring, with confidence, because if only one or two hundredths are left for the reamer to remove it will follow the bored hole, and the accuracy will be preserved. This method, in fact, is often a quicker and better way of finishing a hole to size than by boring it right out, because when boring, the fear of overshooting the mark and making the hole oversize, tends to make progress very slow on the last one or two cuts.

The set up for boring is as shown at Fig. 248(a) and the tool used should be as large as will conveniently pass clear through the hole. A good and easily made tool is one forged up from round material, and supported on a small vee block as shown. For large holes a *boring bar* should be used, as this gives greater rigidity, and the tool may be renewed cheaply and easily.

Boring bars may be obtained in numerous forms, an example being shown at Fig. 248(b).

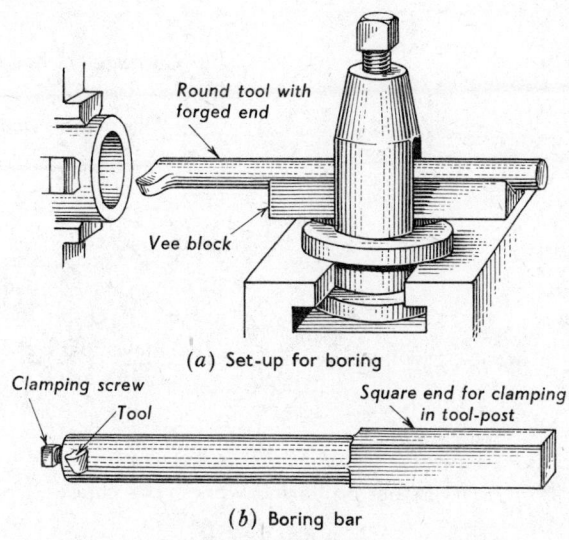

(a) Set-up for boring

(b) Boring bar

Fig. 248 Boring

Lathe 'boring taper'. If the headstock has been accurately located on the bed, so that the axis of rotation of the spindle is exactly parallel with the bed ways, when a job is held in the chuck, and bored with a single-point tool, it will be parallel. At some time in his career, however, the reader may encounter a lathe which does not bore parallel. The conditions causing this are shown exaggerated at Fig. 249, where SS is the axis of rotation of the spindle and work, and BB the axis of the bed, parallel to which the tool moves. In the diagram, the spindle axis is inclined to the front of the lathe and the hole bored will be large at the back end; if the error is the other way, the hole will be large at the front.

The fault may only be corrected by adjusting the headstock until the spindle axis is set parallel with the bed, and on flat bed lathes there is often a screw adjustment underneath the head for the purpose. When the headstock fits on the inverted vees of a vee bed, however, the correction can only be made by scraping up the vee grooves in the headstock, a difficult and tedious operation. If, therefore, a vee bed lathe is boring taper, and it is verified that the fault is not caused by tool wear, or tool off centre, the user will have to make the best of it. In any case the error will be very small and may generally be corrected by removing the excess metal with a half-round scraper. Small holes may be corrected by finishing them with a reamer.

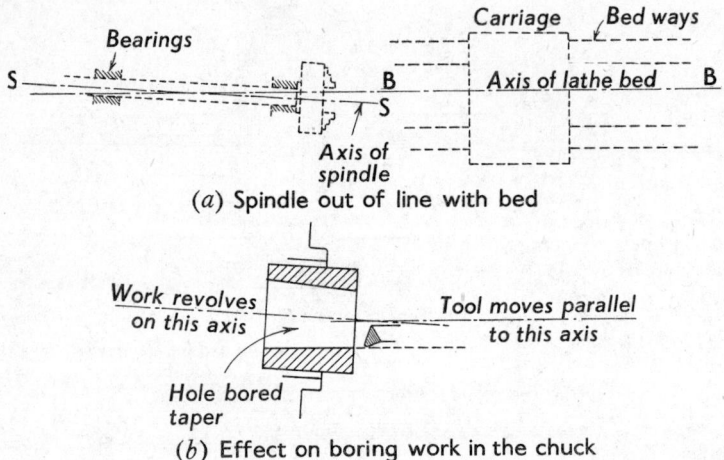

(a) Spindle out of line with bed

(b) Effect on boring work in the chuck

Fig. 249 Showing cause of lathe boring taper

The mandril

For completing work which has been bored and partly turned in the chuck a mandril is used. This is a straight bar with its ends reduced to accomodate the driving dog, its top diameter being ground to a good finish, and accurately true with the centre holes. The centre holes are recessed so that they may not be damaged when the mandril is being driven in the work. (see Fig. 243(a)) A taper of about 1 in 1000 of its length is put on the top diameter, the small end being small enough to enter the size of hole for which the mandril is made, and the other end large enough so that the mandril will not pass right through the hole. To cause sufficient frictional contact for the work to be driven against the cut the mandril is forced into the hole on a *mandril press* (Fig. 183). Mandrils are usually made of mild steel, and case-hardened.

A mandrill should not be used if, by using a little extra forethought in setting up in the chuck, a surface may be finished that way. Although mandrils are true when new, they tend to become damaged, bent or worn on their centre holes, and in using a worn mandril the reader may be putting his confidence in something which is worthless.

Examples of chuck work. We will now discuss the production methods for a few chuck jobs, adding the proviso, as before, that alternatives are possible, but the methods we give represent good standard practice. As guiding rules in chuck work, as well as in many other machining operations, the reader should observe the following:

(*a*) Always do as many operations as possible *at the same setting*. This ensures accurate results automatically. As a simple example, consider the turning and boring of a bush. If turning and boring are both carried out at the same setting in the chuck, the bore is sure to be true with the outside diameter. Any other method, using separate settings, always opens up the possibility of error.

(*b*) Rough out all surfaces to within $\frac{1}{2}$ to 1 mm of size before commencing to finish. Taking the bush again; if the top diameter is finished before the hole is started, the heavy drilling and boring to get the hole out may distort the bush with heat, or move it in the chuck. Both faults would be fatal.

Example 5. To turn the bush shown at Fig. 250(a) from mild steel 56 mm diameter, 66 mm long.

First Setting

(i) Set up the material in the 4-jaw chuck, holding on a length of about 10 mm. Set the lathe speed to turn 50 mm diameter (i.e. about 150 rev/min).

(ii) Face the end until it cleans up. Put a drill, about 12 mm diameter, in the tailstock, and drill through the work. If the drill tends to wobble about when it is pressed against the work, press the tool against it to hold it still. This may be taken away when the drill has started.

When the 12 mm drill has been taken through, replace it by a 23 mm drill and enlarge the hole (Fig. 250(b)).

(iii) Turn the top of the bush to 44·5 diameter, for a length of 51 mm (Fig. 250(c)).

(iv) Set a boring tool so that its point is on centre, and its body clears the hole when passed through. Commence to bore the hole by putting on a cut about $\frac{1}{2}$ mm deep. Obtain inside calipers, micrometer and 26 mm plug gauge. After taking one cut through, measure the bore, then proceed with the boring until the finished size is reached (Fig. 250(d)). At this stage the reader may finish bore to the plug gauge, or he may bore to within about 0·05 mm of size, and finish with a reamer. If he feels confident enough to finish the hole by boring, that may be done now; but if he decides

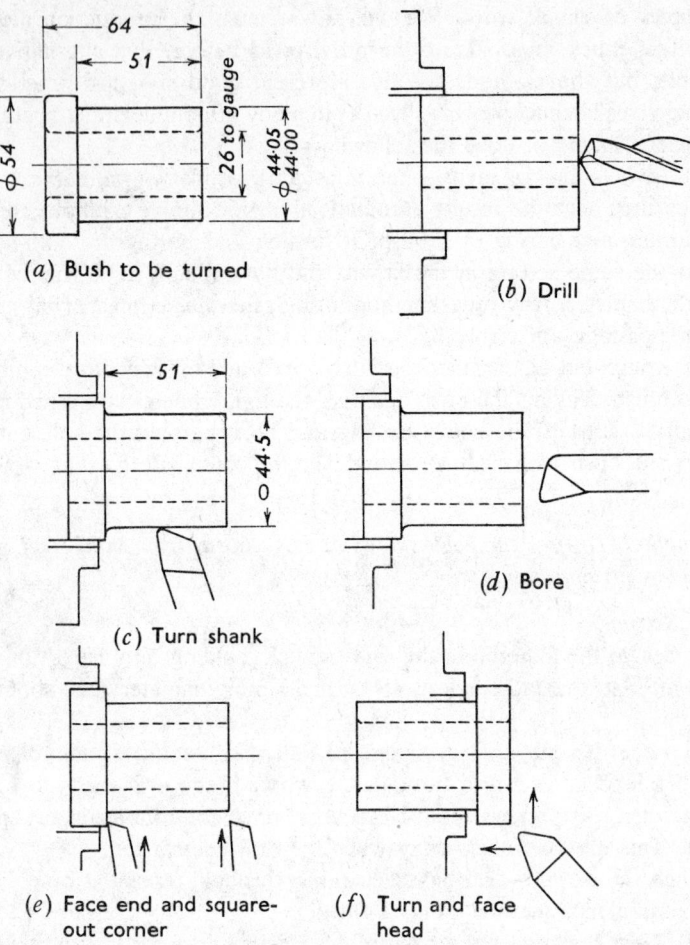

(a) Bush to be turned

(b) Drill

(c) Turn shank

(d) Bore

(e) Face end and square-out corner

(f) Turn and face head

Fig. 250 Turning a bush

to use the reamer, that operation should be postponed until the work is otherwise finished, and ready to be taken from the chuck. If *boring* to the plug gauge, bore to the inside calipers and micrometer, until only a few hundredths remain in the hole. Put on the cut estimated to bring the hole to size, and traverse it in for about 3 mm, run back the carriage and try the plug gauge. If it enters easily, take off some of the cut, run the remainder of the cut slightly further in, and try again with the plug, which should, this time, either fit or be too large. If the plug is too large at the first trial put on a little more cut and try again. Eventually, the end of the plug may just be entered, and then the cut should be fed through. When this has been taken through, the plug may still not enter the hole beyond the first trial length, because the tool has sprung slightly, and if it is fed through again, *without putting on any more cut,* the gauge may be entered.

(v) Finish turn the top of the bush to 44·05/44·00 diameter, using a finishing tool with about a $1\frac{1}{2}$ mm nose radius.

(vi) With a knife tool face the end, square out the corner of the shoulder, and face up to the 51 mm dimension (Fig. 250(e)).

(vii) Finish ream the hole if this has been left for reaming.

(viii) With an old file remove sharp edge from bottom corner of bush, with a half-round scraper do the same on the inside bottom corner. (A half-round scraper is an essential item in the kit of turner's tools.)

Second Setting

Either 1 (*a*). Set up in chuck, holding fairly gently. Turn top to 54 mm diameter and face to length (Fig. 250(f)). Remove sharp edges with file outside, and half-round scraper from bore.

Or 1 (*b*). Press on to a mandril. Turn top to 54 mm diameter and face end to length. Remove sharp edges with file. Hold lightly in chuck. Remove sharp edge left from not being able to face right up to the mandril.

Parting-off

The reader will find that his work will be facilitated by being able to cut off a job after he has turned it up. This is called parting-off, and is carried out with the parting-off tool as shown at Fig. 251.

In theory, the process of parting-off in the lathe is simple. The tool is clamped in position and carefully fed to the work which it proceeds to part off without difficulty. In practice, particularly on a lathe which is beginning to show signs of wear, the process is far from easy, and when the reader has experienced the tool digging in and breaking, he will realise that theory and practice require considerable manipulation to make them agree. However, in spite of any difficulties, let him not descend to the humiliation of cutting off in the lathe with a hacksaw, but persevere until he has overcome the difficulties which confront him.

The parting tool should have less rake than a turning tool for the same material, and should be narrower at the back than at the cutting edge, with clearance all round. The edge may be slightly tapered so that it cuts deeper at the side away from the chuck. The width of the tool will vary according to the size of work, but should be between 3 mm and 8 mm. It should be set slightly *below* centre and clamped tightly, with the minimum of over-hang. Before parting-off, the carriage should be locked against lengthwise movement, and the compound slide set in line with the bed. Reduce the speed to about one-half that for ordinary turning and feed in the tool

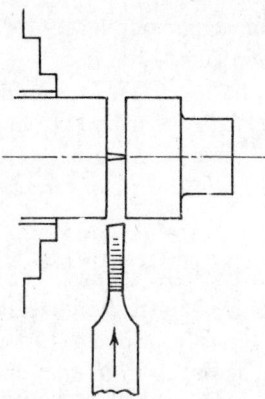

Fig. 251 Parting-off

slowly by hand. When it has penetrated to about the depth of its width, withdraw it, and move it a few thousandths sideways with the compound slide; feed in again, and after a further similar penetration into the solid metal, bring it out and move it sideways a slight amount to the other side of its original position, and so on. This minimises the tendency of the tool to dig in and cause trouble. With these precautions and care in feeding, the tool should do its work without trouble, but if there is end play (length-wise slackness) in the lathe spindle the tool may still dig in, and the thrust arrangements on the spindle should be adjusted. Often, difficulties in the operation of parting-off may be overcome by working with the tool upside down, and either locating the tool at the back of the work or running the lathe reversed. The tool should always be as near as possible to the chuck, and the cranked parting tool (as shown at Fig. 126(d)) helps to achieve this

Example 6. To turn the pin shown at Fig. 252(a) from 25 mm diameter bar.

First Setting
   (i) Obtain a piece of 25 mm diameter bar and grip it tightly in the 3- or 4-jaw chuck with about 42 mm protruding.

(ii) Turn down the top to 22·5 mm diameter for a length of about 40 mm. Face end.

(iii) Reduce to 16·03/16·00 mm diameter for a length of 26 mm (Fig. 252(b)). Finish top to 22 mm diameter.

(iv) Set straight tool edge to 45° and chamfer end of pin.

(v) With tool 1½ mm wide, clean out corner, turn undercut and skim underface of head.

(vi) Part off, leaving head 7 mm thick.

Second Setting

(i) Hold on shank in chuck, push underside of head against chuck jaws and set true.

(ii) Face top of head and chamfer (Fig. 252(c)).

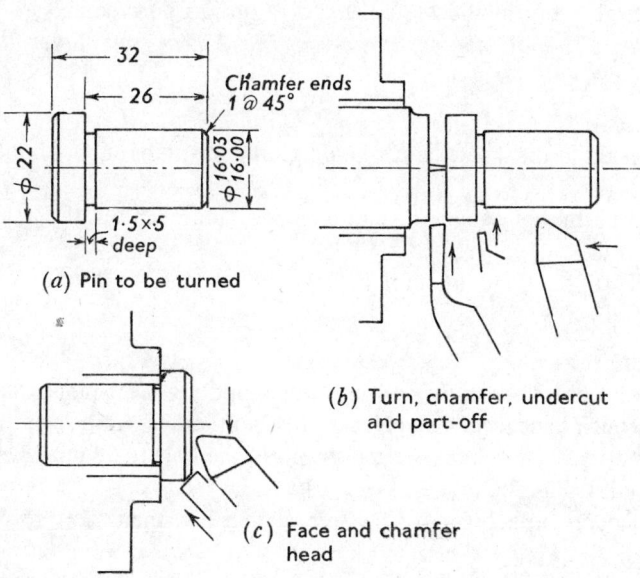

(a) Pin to be turned

(b) Turn, chamfer, undercut and part-off

(c) Face and chamfer head

Fig. 252 Turning a pin

11 The chuck (*cont.*)—the face-plate—taper turning —screw-cutting

The production of more complicated parts often requires several settings to obtain the desired accuracy and, in addition, resort must be made to various other methods and aids. The following example will serve to illustrate this, and it will be observed that although the bush does not greatly differ from the one we have already discussed, the shoulder at the bottom of the larger bore modifies the method of turning considerably.

Example 1. To make the bush shown at Fig. 253(a) from material 60 mm diameter, 70 mm long.

1st Setting

    (i) Set up in the 4-jaw chuck holding on a length of about 10 mm. Face end.

    (ii) Drill through, about 12 mm diameter, and open out the hole with a 21 mm drill.

    (iii) Turn shank to 46 mm diameter $(45+1)$ for a length of 58 mm (Fig. 253(b)).

2nd Setting

    (i) Hold on shank in 3- or 4-jaw chuck and set reasonably true.

    (ii) Rough face end, leaving about 1 mm on the total length; rough turn head to 58 mm diameter and open out the hole to 58 mm deep with a 30 mm drill (Fig. 253(c)).

    (iii) Rough bore hole to 31·5 mm ($\frac{1}{2}$ mm less than size), to depth of drilling.

    (iv) With specially ground boring tool, square out the bottom of the hole to a depth of 58 mm $+\frac{1}{2}$ mm. If a bed stop for the carriage is available set this to the carriage when the tool face is the correct distance from the end of the bush as shown at Fig. 253(d). Feed the tool by hand, controlling the point at which to stop winding the cross-slide, by the graduated sleeve. Take a preliminary note of the reading when the tool just skims the surface of the bored hole, and when winding in to cut away the metal, stop at the observed mark.

(v) With the same tool as (iv), finish boring the hole to the plug gauge. The bed stop will look after the depth, and use the power feed until the carriage is about 3 mm from it, when the travel should be completed by hand. If no stop is available, feed the last 3 mm by hand, stopping when the tool is felt to contact the shoulder.

(vi) Finish face bush end to a distance of 58 mm from bottom of hole; finish turn head to 57 mm diameter. Remove sharp edges from end of hole and head.

3rd Setting

(i) Obtain a piece of scrap material about 35 mm diameter, 90 mm long, and set it in the chuck with about 60 mm protruding. Turn this down

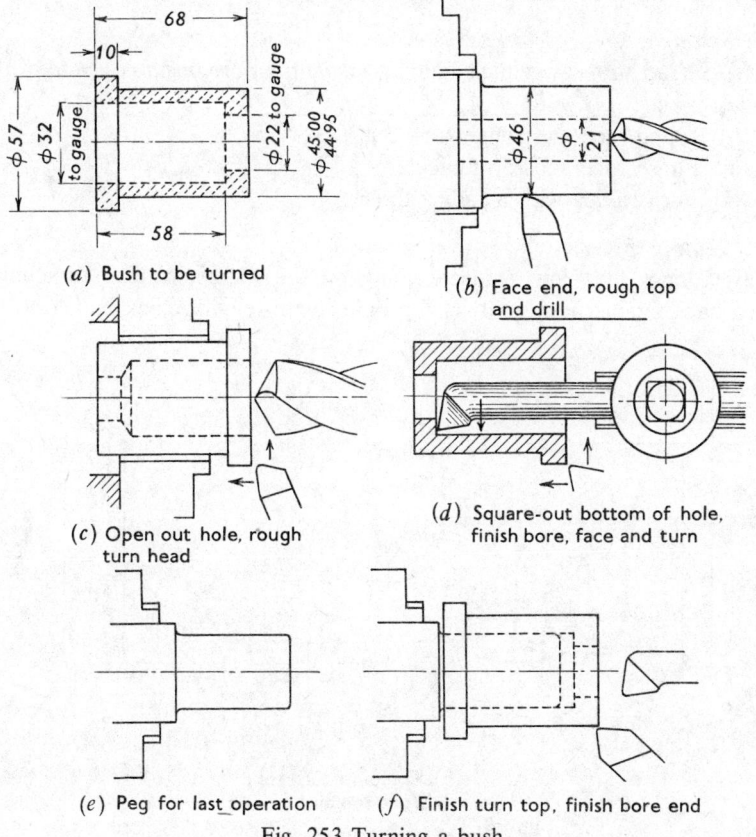

(a) Bush to be turned

(b) Face end, rough top and drill

(c) Open out hole, rough turn head

(d) Square-out bottom of hole, finish bore, face and turn

(e) Peg for last operation

(f) Finish turn top, finish bore end

Fig. 253 Turning a bush

up to the chuck until its diameter is such that the 32 mm bore of the bush will be a tight press fit. This is to be used as a peg upon which the bush is to be supported for its final turning (Fig. 253(e)).

(ii) Press the bush on to the peg by winding the tailstock barrel against it (take care to keep the axes of the bush and peg in line).

(iii) Turn the shank of the bush to 45·00/44·95 mm diameter, face the head to thickness and face the end to length (Fig. 253(f)).

(iv) Bore the end of the bush to 22 mm plug gauge.

(v) With old file and half round scraper, remove sharp edges from all corners. Remove from chuck, and press peg out of bush.

Example 2. To turn the ring shown at Fig. 254(a) from a casting having 3 mm of machining all over.

1st Setting

(i) Set up in 4-jaw chuck holding *inside* the bore on the *outside* of the chuck jaws.

(ii) Rough face and rough turn top⎫
(iii) Finish face and top diameter ⎬ Fig. 254(b).
(iv) Turn radius with a radius tool.⎭

2nd Setting

(i) Clamp in 4-jaw chuck holding on top diameter. Press machined face back against jaws and set top diameter true with clock indicator.

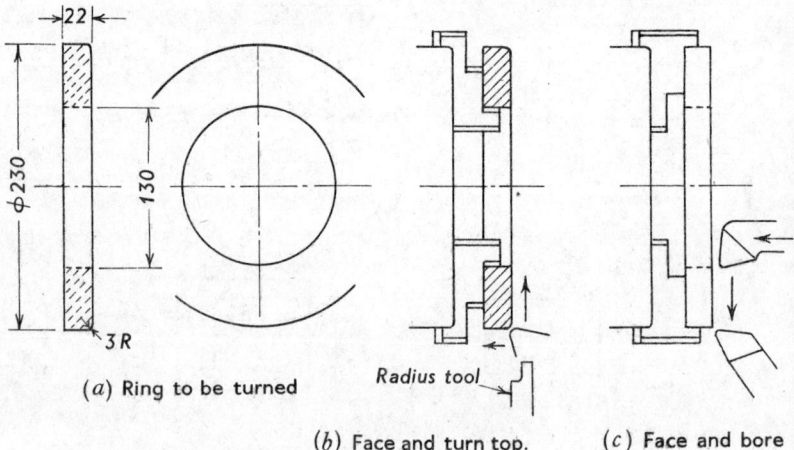

(a) Ring to be turned

Radius tool

(b) Face and turn top. Turn radius

(c) Face and bore

Fig. 254 Turning a ring

(ii) Rough face and bore ⎫
(iii) Finish face and bore ⎬ Fig. 254(c).
                          ⎭

(iv) Remove all sharp edges from corners.

Combined use of chuck and tailstock centre

When a bar is held in the lathe and subjected to a heavy roughing cut, a considerable deflection is likely to take place under the pressure of the tool. It should be our object to support the work in such a manner that it is as stiff as possible in order that the deflection may be reduced to a minimum. When a bar is held between centres, the centres merely provide

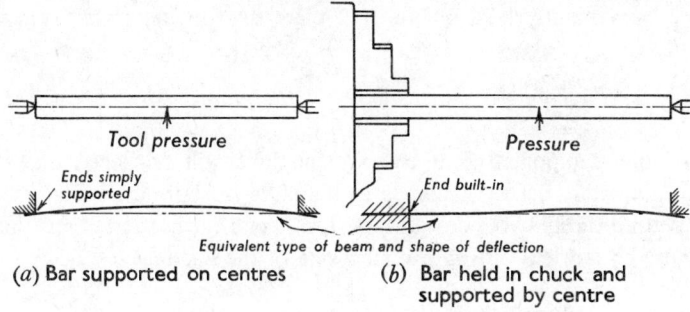

(a) Bar supported on centres   (b) Bar held in chuck and
                                        supported by centre

Fig. 255 Deflection of work supported in the lathe

supports but do not contribute to the stiffness of the shaft, and its ability to remain straight under pressure. Such a shaft, when loaded, will be bent to the circular shape shown exaggerated at Fig. 255(a). If, now, one end of the shaft is held in a chuck, whilst the other is supported by a centre, the chuck end of the shaft is kept horizontal by the restraining effect of the jaws, and is unable to take up an angular direction. Such a shaft will bend to the shape shown exaggerated at Fig. 255(b), and whilst there is still some deflection, its amount will be less than that for the case where the ends are simply supported by centres. For heavy roughing cuts, therefore, we can add to the stiffness of the work by holding one end in the chuck whilst the other end is supported by the back centre.

Thus, for a job of the type shown at Fig. 256, the most efficient way of working would be as follows:

(i) Centre one end, hold the other end in chuck with tailstock centre to support. Rough to within 1 mm of finished sizes.

(ii) Reverse in chuck, centre the other end and again rough out.

(iii) Finish turning to size on both centres.

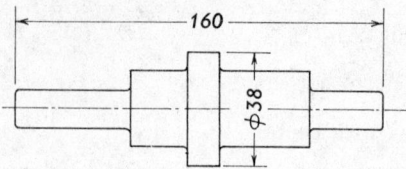

Fig. 256 Pin to be turned

The face-plate

The face-plate is used for holding work which cannot be held conveniently in a chuck, for doing operations not suited to the chuck and for finishing work which has been partly turned in the chuck. An example of a job which is more conveniently held on the face-plate than in the chuck is given in Example 3.

Example 3. To bore the hole shown in the plate casting shown at Fig. 257(a).

Since the component is a casting, the hole will be cored out about 40 mm diameter. Before it comes to the lathe for boring the plate should be shaped up on its two sides, and on the step at the bottom. After this the hole must be marked out on the step side of the casting.

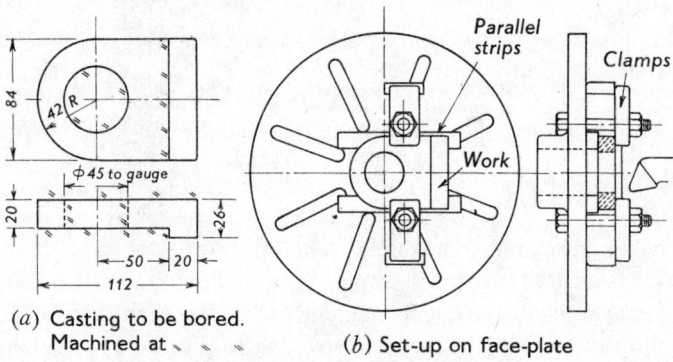

(*a*) Casting to be bored.
Machined at

(*b*) Set-up on face-plate

Fig. 257 Boring on Face-plate

(i) Clamp the work to the face-plate, packing it up on parallel strips if the face-plate hole is not large enough to clear the boring tool when it passes through. Set the bore approximately central, and attach the face-plate to the nose of the lathe. Clamp a piece of steel or lead to the face-plate opposite to the work, so that the face-plate is approximately in balance.

(ii) Pull the lathe round and tap the work until the marking out is true.

Set the scribing block on the bed or carriage and use the scriber point to test the truth of the marking out. Pull up the clamping bolts tight.

(iii) Clean up the hole with a boring tool (or preferably a boring bar). Check the 50 mm dimension by measuring from the step to the edge of the hole, and adding half the hole diameter to the measurement observed. If not correct, adjust, take another trial cut and check again. Check also that hole is central, by calipering between each edge, and the edge of the plate.

(iv) When the hole is correctly positioned, finish bore it to a 45 mm plug gauge (Fig. 257(b)).

The next example illustrates the use of the face-plate to complete the machining of a component which, for its first setting, has been held in the chuck.

Example 4. To turn the casting shown at Fig. 258(a) which has to be machined on the faces shown "—"—"

In work of this type it is important that the faces of the machined casting shall be parallel, and that the spigots on each side of the plate shall be exactly true with one another, and with the 38 mm central hole (i.e. all of them concentric on the same axis).

1st Setting

(i) Hold in 4-jaw chuck on the 185 mm boss. Set true on the diameters, and on the un-machined top face of the base. Set lathe at about 25 to 30 rev/min.

(ii) Rough face, leaving 1 mm to $1\frac{1}{2}$ mm on base thickness. Rough turn spigot recess to 179 mm diameter, $7\frac{1}{2}$ mm deep. Rough bore central hole to 37 mm diameter.

(iii) Square out recess to 179·5 mm diameter (Fig. 258(b)).

(iv) Finish face base to thickness.

(v) With same tool as in (iii) finish bore recess. Traverse the carriage by hand. Measure its diameter with a vernier, allowing for the double jaw thickness (Fig. 163(b)). Face the bottom of the recess with the same tool.

(vi) Finish bore the 38 mm central hole to the standard plug gauge (lathe speed may be increased for this boring). Remove all sharp edges.

2nd Setting

(i) Fit the face-plate, screw a ring of cast-iron or steel to it and set approximately true. The ring should be 200 mm to 250 mm diameter and 14 mm to 20 mm thick with tapped holes for screws. If a number of such rings is available, a suitable one may be selected for any job such as this.

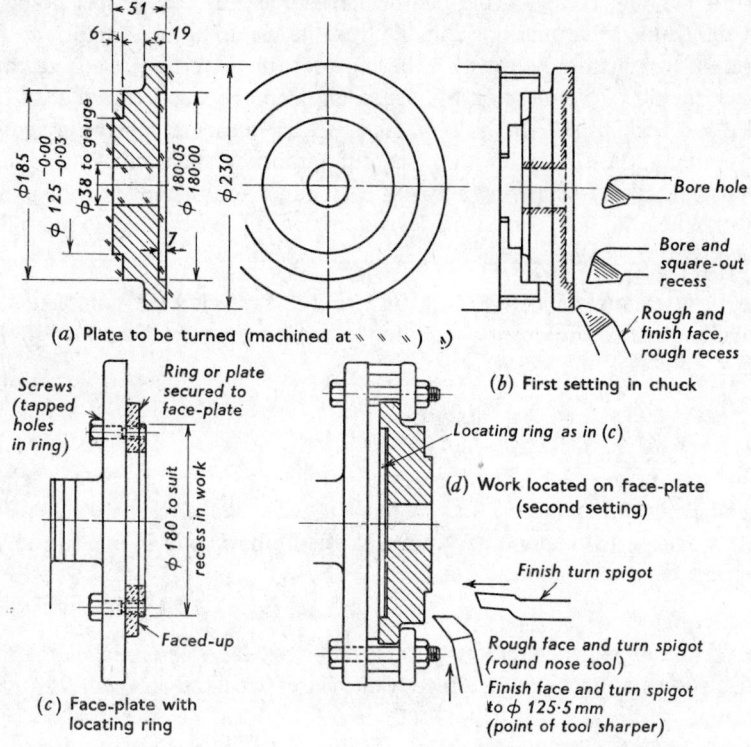

(a) Plate to be turned (machined at ＼ ＼ ＼) ＼)

Bore hole

Bore and square-out recess

Rough and finish face, rough recess

(b) First setting in chuck

Screws (tapped holes in ring)

Ring or plate secured to face-plate

Locating ring as in (c)

(d) Work located on face-plate (second setting)

Finish turn spigot

Rough face and turn spigot (round nose tool)

Finish face and turn spigot to φ 125·5 mm (point of tool sharper)

Faced-up

(c) Face-plate with locating ring

Fig. 258 Operations for turning round plate

(ii) Turn a spigot on the ring to fit the recess just turned in the job, and face up true. The spigot should be about 5 mm high so that the work seats on its bottom face (Fig. 258(c)).

(iii) Clamp the job to the face-plate, locating it on the ring.

(iv) Rough face front, and rough turn 125 mm spigot to 126 mm.

(v) With side tool, finish face to 51 mm thick, turn spigot to 125·5 mm diameter and square out corner, finish lower face to make spigot 6 mm high.

(vi) Finish turn spigot to $125^{-\,\cdot00}_{-\,\cdot03}$ diameter with a knife tool. Measure with a vernier (Fig. 258(d)).

(vii) Remove all sharp edges with file and scraper.

The component shown at Fig. 259(a) is essentially a face-plate job, as it could not be set up in the chuck with sufficient accuracy to bore the hole parallel with the base.

Example 5. To bore the hole and face the bosses on the component shown at Fig. 259(a).

The bracket is a casting, and the hole will be cored about 36 mm diameter. The base should be machined before the bracket comes to the lathe for boring.

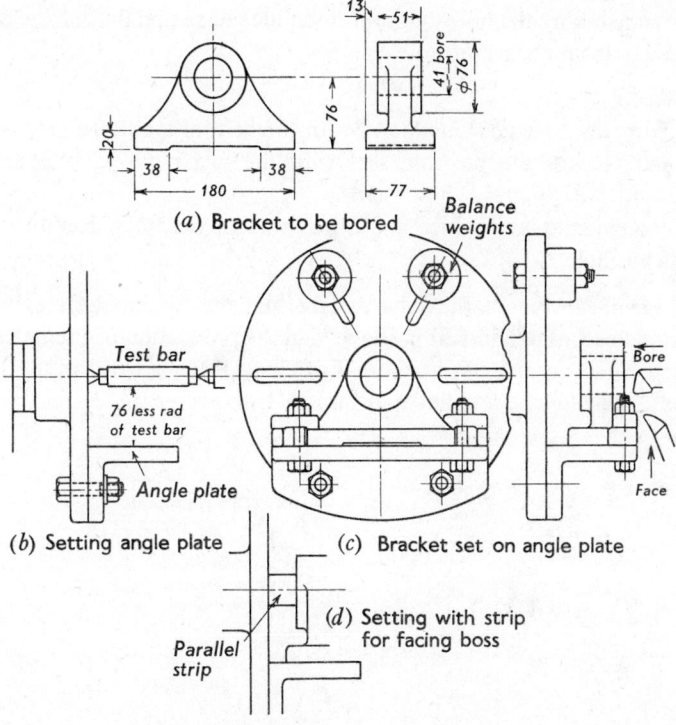

(a) Bracket to be bored

(b) Setting angle plate

(c) Bracket set on angle plate

(d) Setting with strip for facing boss

Fig. 259 Boring and facing boss of bracket

1st Setting

(1) Screw the face-plate to the lathe, and bolt an angle plate to it. Set the angle plate 76 mm from the centre of the lathe, by clamping a mandril between the centres, and adjusting the angle plate until the distance between

the mandril and the angle plate is 76 mm less half the mandril diameter (Fig. 259(b)).

(ii) Clamp the bracket to the angle plate and set the edge of its base flush with the edge of the angle plate. This will bring the bored hole square with the length of the bracket. Sideways adjustment may be effected by threading a mandril through the cored hole, holding it between the centres, and setting the bracket until the clearance between the mandril and the hole is the same at each side. Tighten up all clamps, and balance the face-plate with a lump of lead bolted opposite to the angle plate.

(iii) Rough face the outer boss and rough bore the hole (Fig. 259(c)).

(iv) Finish bore the hole to the 41 mm plug gauge and finish face. Scrape sharp edge from corner of hole.

2nd Setting

(i) Turn the bracket round, and with the help of a parallel strip set the machined face of the boss parallel with the face-plate. Clamp up tight (Fig. 259(d)).

(ii) Rough and finish face the other side of the boss. Remove sharp edges from hole.

Taper turning. Many lathe jobs require that the turned surface shall be conical instead of cylindrical in shape, and the production of such a surface is called taper turning. We saw in Fig. 227(a) that when the tool is fed parallel to the spindle a cylinder is turned. If, however, the tool is still kept

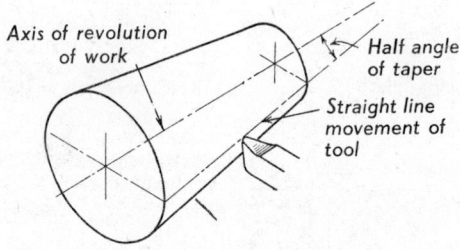

Fig. 260 Taper turning

in a horizontal plane through the work axis, and moved at an angle with this axis, a conical surface will be produced (Fig. 260). The same effect will be obtained if the tool moves in its orginal direction and the *work axis* is altered to be at an angle with the line of the tool motion.

There are various methods of turning taper surfaces, the chief of which being as follows:

(i) *Compound slide method.* Guiding the *tool* at a suitable angle by feeding it with the compound slide set at the correct angle. This method may be used for external turning, or boring, but the length of surface that may be turned is limited to the travel of the compound slide. Another factor which

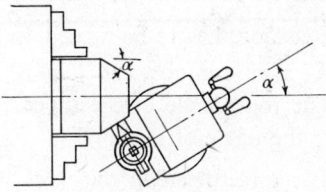

Fig. 261 Taper turning with compound slide

limits this method to short jobs is that the slide may only be fed by hand, which tends to cause fatigue to the operator, and irregularity in the surface being turned. A diagram showing this method of taper turning is shown at Fig. 261.

(ii) *Forming tool.* A short external taper may be turned with a flat tool set at the correct angle as shown at Fig. 262. This method is only practicable for short work, as if a long surface is being turned there is a tendency for the work to vibrate and chatter, resulting in a rough finish. The edge of the tool must be exactly straight if the work is to be accurate. We have

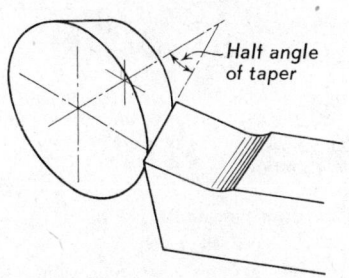

—Half angle of taper

Fig. 262 Turning a taper with a straight tool

already used this method for turning small chamfers on the corners of examples under discussion.

(iii) *Setting over the tailstock.* This method is limited to the production of slow external tapers on work held between centres. The tool is fed in its normal direction parallel to the bed, whilst the *work* is caused to rotate on an axis inclined at an angle. This is achieved by setting over the tailstock centre, so that instead of the line of the centres being parallel to the bed, it is inclined at a small angle. Then, when the tool moves along the bed, it will cut a taper on the work, the angle of which will be *twice* the inclination of the centre lines (Fig. 263). We have already discussed, and illustrated, the method by which the tailstock is set over (p. 301 and Fig. 234(b)); the amount of set-over is determined as follows:

When a taper has to be turned, its proportions will be given either:

(*a*) A: 1 in so much, e.g. 1 in 10 on the diameter.

(*b*) The total angle of the taper will be given in the same way as the the angle of a wedge.

(*a*) The total taper on the work will be found by dividing its length by the length in which unit taper occurs, and the tailstock set-over is *one-half* this amount.

For example: To find the tailstock set-over to turn a taper of 1 in 20 on a job 235 mm long.

$$\text{Taper on diameter} \qquad = 1 \text{ in } 20$$
$$\text{Taper on a length of 235 mm} = 235 \div 20 = 11 \cdot 75 \text{ mm}$$
$$\text{Tailstock set-over} \qquad = \frac{11 \cdot 75}{2} = 5 \cdot 88 \text{ mm}$$

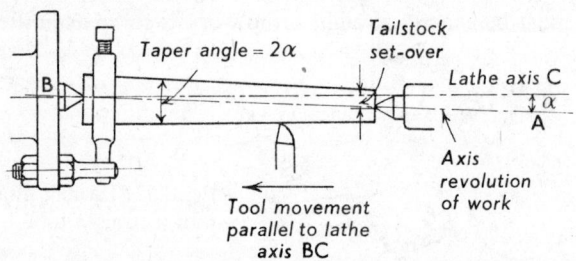

(*a*) Taper turning by setting-over tailstock

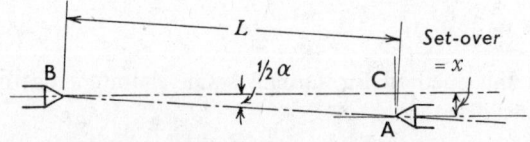

(*b*) Diagram for example (*b*) p. 341

Fig. 263 Taper turning with tailstock set-over

(*b*) When the taper is given as an included angle, the tailstock centre must be set so that the line joining it to the headstock is inclined at one half this angle.

In Fig. 263(b) let L = length of work; x = tailstock set-over and α = angle of taper.

Then the angle ABC will be $\frac{1}{2}\alpha$, and $\dfrac{AC}{AB} = \sin \frac{1}{2}\alpha$, i.e. AC = AB $\sin \frac{1}{2}\alpha$.

For example: To find the tailstock set-over to turn a taper of 6° on a job 235 mm long.

$$\text{In this case AB} = 235 \text{ mm, and } \tfrac{1}{2}\alpha = 3°$$
$$\text{AC} = \text{set-over} = 235 \times \sin 3° = 235 \times 0\cdot0523 = 12\cdot3 \text{ mm}$$

When the tailstock has been set over as near as possible to the calculated amount, the turning is performed in the usual way, until the small end of the taper will enter the gauge (see later), after which the tailstock must be adjusted to correct any error present, and a second trial cut taken.

Although the line joining the centres is inclined at an angle, their directions are not along this line but still parallel with the bed. This causes their disposition relative to the centre hole in the work to be as shown at Fig. 264, and leads to some distortion in the holes, particularly when the taper is large. Another point in connection with this method is that the set-over depends on the length of the job, so that if the same taper has to be turned on a number of pieces whose lengths vary slightly, then there will be some variation in the tapers unless the tailstock is adjusted to suit. The safest way, if possible, is to ensure that both length of work and depth of centre hole are kept the same on all bars. Still bearing this point in mind, the reader should visualise the effect on the turned taper, when the tailstock end of the work is badly out of square, and has not been faced up.

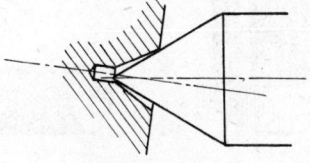

Fig. 264 Effect of setting over
tailstock centre

(iv) *Long taper holes.* When a taper hole is relatively small and long, it is not practicable to bore it with a tool, and it is formed with a taper reamer. We have already discussed the drilling of small taper pin-holes with a taper reamer, and the reamer for larger holes is very similar.

Generally, for larger work, the hole is first roughed out with a roughing reamer, the teeth of which are designed to remove metal quickly, as at its best, taper reaming is a slow job. Diagrams of roughing and finishing reamers are shown at Fig. 265, where it will be seen that the teeth of the rougher are notched to help cutting and break up the chips. When reaming a taper hole it should first be drilled as true as possible to a diameter slightly smaller than the small end of the taper.

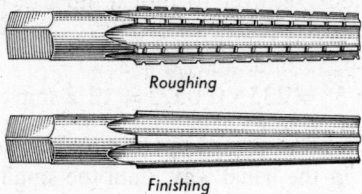

Roughing

Finishing

Fig. 265 Taper reamers

The testing and measurement of tapers

The most satisfactory method of testing a taper which has to fit a similar taper, is either to turn the one to fit the other, or to work to taper plug and ring gauges of proved accuracy. When a taper has to be produced without the mating taper or gauge for reference, it can be measured either by checking its angle with a protractor or sine bar, or by using precision balls or rollers. These methods are given in the author's *Senior Workshop Calculations.*

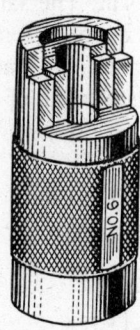

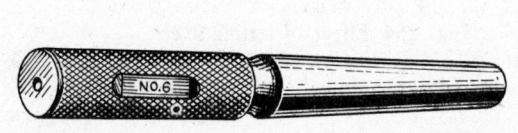

Brown & Sharpe

Fig. 266 Taper gauges

Diagrams showing taper plug and ring gauges are shown at Fig. 266, and when they are used they must gauge not only the taper, but also the diameter of it at some point. On the plug gauge this is often arranged for

·y marking a line round the gauge at the large diameter of the taper, and he plug must be let into the hole being gauged until this line is flush with its arge end face. The ring gauges are often made with their large end diameter qual to a stated dimension, and when using this gauge, its end must stand t a certain distance down the taper, from its largest dimension. Instruc- ions for this are often given on the working drawings, but if they are not, t is a fairly simple matter to work it out; for example, Fig. 267(a) shows he drawing of a taper, together with particulars of the gauge which is to be ·sed for it. The difference between the drawing dimension and the large end ·f the gauge is $24 - 22 = 2$ mm.

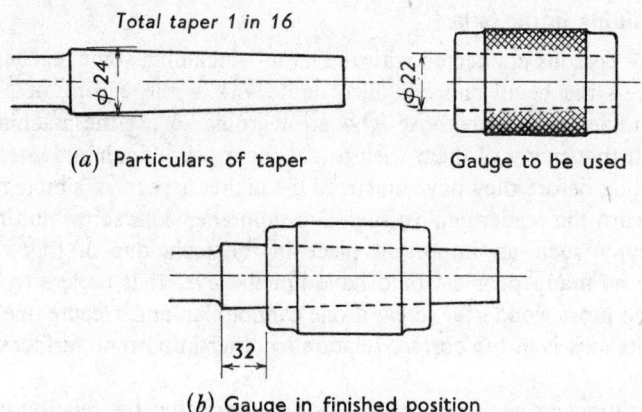

(a) Particulars of taper   Gauge to be used

(b) Gauge in finished position

Fig. 267 Gauging a taper

Since the taper is 1 in 16 on the diameter, a difference of 2 mm ·n the diameter will correspond to a length difference along the taper of $6 \times 2 = 32$ mm.

Hence when the taper is finished the end of the gauge must stand 32 mm ·om the shoulder on the work (Fig. 267(b)).

Jse of taper gauges

Vhen a taper is being turned or bored, the gauge should be tried as soon as .bout an inch length of engagement can be obtained. Put a longitudinal halk mark on the male portion (i.e. plug gauge, or bar to be tried), fit the vork and gauge together and rotate them once or twice. When they are aken apart the chalk will be seen to have been rubbed away more at one nd if the fit is not correct, and the setting must be adjusted until the chalk mark is rubbed equally all along its length. After re-adjustment of the setting, be careful not to remove too much when taking trial cuts because as we have seen from the above example, 2 mm taken off the diameter of a 1 in 16 taper, allows the gauge to move a further 32 mm along the work and at this rate the taper may be down to size before it has been corrected

Morse tapers are commonly used in the workshop and particulars of these are given in the Appendix.

Tapers are turned on the more expensively equipped lathes with the aid of a *taper turning attachment,* and when we come to a further discussion of the lathe we shall give particulars of this method.

Screw cutting in the lathe

To many persons connected with the lathe—including some teachers of its technique—the be-all and end-all of lathework is the cutting of a screw. This attitude spreads to those who are learning to use the machine, with the result that many of them wish to cut the most difficult and uncommon screws long before they have mastered the higher aspects of simple turning. Let us warn the reader against such a misapprehension; screw-cutting does not occupy such an important place in the technique of this versatile machine as many people would have him believe. It is useless to be able to cut the most wonderful screw if one cannot plan and execute one's work so that its axis is in the correct relation to other important surfaces on the same job.

For cutting an accurate screw, it is necessary that the relation between the movement of the saddle, and the turns of the work, should be carefully controlled. This is brought about by means of the *lead-screw,* the long screwed shaft which runs along the front of the bed. This screw is driven by a train of gears from the spindle as shown at Fig. 236. Usually the drive is first carried to a spindle called the *stud,* which for all purposes may be assumed to be the spindle itself, as it rotates at the same speed, and in the same direction, unless caused to reverse by the *tumbler* mechanism. From the stud, the drive is conveyed to the lead-screw by a train of gears, and to vary the relationship between the turns of the lead-screw and those of the stud, these gears may be varied. In Fig. 236, gear A on the main spindle is shown driving the stud gear D through B, and since A and D are the same size, the stud revolves at the same speed and in the same direction as A. To reverse the stud for the purpose of reversing the feed or lead-screw, the nut holding the top quadrant is loosened and the quadrant swung so that B comes out of engagement with A and C goes in. (B is in permanent mesh with D.) A now drives D through C and B which, by introducing an additional gear, causes a reversal in the direction of D. To drive the lead-screw or gear-box shaft another gear E is put on the stud and, as shown in the diagram, E then drives F, and G, which is keyed to F, drives H, giving a compound train. In the arrangement shown a simple train from E to H may be obtained by packing H out level with E and putting a single connecting gear on the lower quadrant. To provide adjustment for correct meshing, the pin carrying F and G may be moved along the quadrant and the whole assembly swung about the centre of H.

The connection between the lead-screw and the saddle is effected by a nut, fixed to the inside of the apron, and screwed to suit the lead-screw. This nut is made in one or two halves and arranged in such a way that, by operating a lever at the front of the apron, the halves may be engaged with the lead-screw (Fig. 237). When the nut is engaged, the saddle moves along the bed a distance equal to the pitch of the lead-screw for each turn of the screw, and since the tool which is used to cut the thread on the work is for all purposes solid with the carriage, this moves the same distance. We thus have a rotation of the work, combined with a fixed longitudinal movement of the tool for each turn the work makes and the result is a screw formed on the work as shown diagrammatically at Fig. 268.

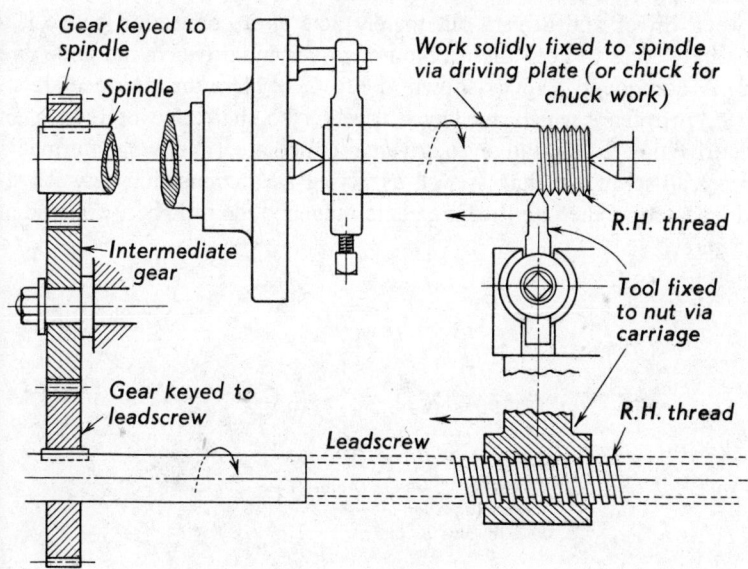

Gear keyed to spindle

Spindle

Intermediate gear

Gear keyed to leadscrew

Leadscrew

Work solidly fixed to spindle via driving plate (or chuck for chuck work)

R.H. thread

Tool fixed to nut via carriage

R.H. thread

Fig. 268 Diagrammatic representation of screw-cutting

Gears for screw-cutting. We may now proceed with the determination o screw-cutting ratios. The speed ratio between the lead-screw and stud i controlled by the gear drive connecting them, and as we have previousl seen, this relation depends only on the number of teeth in the gears, witl small gears turning more times than large ones. The types of gear connec tions on a lathe are *simple* and *compound*. In the simple train shown at Fig

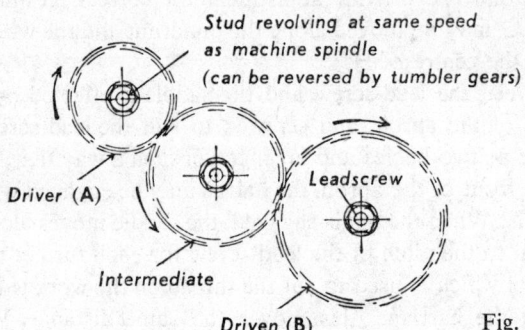

Stud *revolving at same speed as machine spindle (can be reversed by tumbler gears)*

Driver (A)

Intermediate

Leadscrew

Driven (B)

Fig. 269 Simple gear train

269 the gear on the stud drives direct through the *intermediate* gear to th gear on the lead-screw. The intermediate gear has no effect on the rati between driver and driven, but merely acts as a connection between th two, and serves to keep the rotation of driver and driven in the same direc tion. A compound train is shown at Fig. 270. Here the intermediate stu carries two gears which are keyed together so that they rotate as a uni The drive now is, (*a*) stud on to driven intermediate, (*b*) driving intermediat on to lead-screw, so that as well as acting as a connection between stu and lead-screw, the intermediate gears influence the ratio between stud an lead-screw.

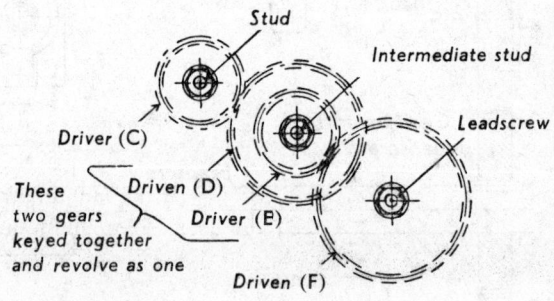

Stud

Intermediate stud

Driver (C)

Leadscrew

These two gears keyed together and revolve as one

Driven (D)

Driver (E)

Driven (F)

Fig. 270 Compound gear train

We shall appreciate most readily the determination of the ratio to cut any given screw if we first consider what happens when we cut a screw with a $\frac{1}{1}$ ratio between stud and lead-screw. When the spindle turns once, the lead-screw turns once, the tool moves along one pitch of the lead-screw, and therefore cuts a thread of identical pitch on the work. If the ratio is $\frac{\text{spindle turns}}{\text{lead-screw turns}} = \frac{2}{1}$ the carriage will move 1 pitch of the lead-screw whilst the work turns *twice*, and the thread will have a pitch of *one-half* that of the lead-screw.

But $\frac{\text{spindle turns}}{\text{lead-screw turns}} = \frac{2}{1}$ means that we must have $\frac{\text{driver teeth}}{\text{driven teeth}} = \frac{1}{2}$, since a small gear rotates faster than a large one with which it is engaged.

Hence as a general condition we may say:

$$\frac{\text{driver teeth}}{\text{driven teeth}} = \frac{\text{lead-screw turns}}{\text{spindle turns}} = \frac{\text{pitch to be cut}}{\text{pitch of lead-screw}}$$

or just:

$$\frac{\text{drivers}}{\text{driven}} = \frac{\text{pitch to be cut}}{\text{pitch of lead-screw}}$$

When the fraction representing $\frac{\text{drivers}}{\text{driven}}$ has been found, it must be thrown into one containing numbers equal to the numbers of teeth in whatever gears are available to make up the drive. Often lathes are equipped with a set of gears ranging from 20T to 120T in steps of 5 teeth.

Example 6. Calculate the gears for cutting the following screws on a lathe with a lead-screw of 6 mm pitch.

(a) 3 mm pitch, (b) 1·75 mm pitch, (c) 0·8 mm pitch, (d) 6·25 mm pitch.

(a) 3 mm pitch

$$\frac{\text{drivers}}{\text{driven}} = \frac{\text{pitch to be cut}}{\text{pitch of lead-screw}} = \frac{3}{6} = \frac{30}{60}$$

A simple train with 30T on the stud driving 60T on the lead-screw.

(b) 1·75 mm pitch

$$\frac{\text{drivers}}{\text{driven}} = \frac{1 \cdot 75}{6} = \frac{175}{600} = \frac{35}{120}$$

A simple train with 35T on the stud driving 120T on the lead-screw.

or

$$\frac{35}{120} = \frac{7 \times 5}{12 \times 10} = \frac{35}{60} \times \frac{25}{50}$$

A compound train with 35T and 25T as the drivers, and 60T and 50T as the driven.

(c) 0·8 mm pitch

$$\frac{\text{drivers}}{\text{driven}} = \frac{0 \cdot 8}{6} = \frac{8}{60} = \frac{4 \times 2}{10 \times 6}$$
$$= \frac{40}{100} \times \frac{20}{60}$$

A compound train with 40T and 20T as the drivers, and 100T and 60T as the driven.

(d) 6·25 mm pitch

$$\frac{\text{drivers}}{\text{driven}} = \frac{6 \cdot 25}{6} = \frac{625}{600} = \frac{25}{24}$$
$$= \frac{5 \times 5}{6 \times 4} = \frac{75}{90} \times \frac{50}{40}$$

A compound train with 75T and 50T as the drivers, and 90T and 40T as the driven. (Note that in this case the lead-screw turns faster than the work.)

Note.—When the reader has made a fraction of the respective thread pitches, he need not worry in remembering which are drivers and which driven if he realises that to cut a *finer* thread than that on the lead-screw, the lead-screw must turn *slower* than the spindle, and vice versa.

Metric-English (inch) conversions

When we, in this country, were using the Imperial system of measurement, it was sometimes necessary to cut a metric screw on an English lathe since the metric system was being used by many of our neighbouring states, and to some extent here. During the period that we are changing over it is fairly certain that we shall have to cut metric screws with lead-screws threaded in inch units and vice-versa.

The conversion from inch to millimetre is that

$$1 \text{ inch} = 25 \cdot 4 \text{ millimetres}$$

so that the ratio between them is:

$$\frac{1}{25 \cdot 4} = \frac{10}{254} = \frac{5}{127}$$

This means that to convert from one to the other by means of a gear train we must introduce a wheel of 127T. (The ratio $\frac{2}{51}$ only differs from $\frac{5}{127}$ by about $0 \cdot 4\%$. No doubt in many cases such a small variation could be tolerated and if the ratio $\frac{2}{51}$ were used a 51T gear is much less unwieldy than one of 127T, particularly when it is a driver.)

We will consider the two variations of the conversion we may have to make.

Metric threads from inch lead-screws. The ratio

$$\frac{\text{drivers}}{\text{driven}} = \frac{\text{pitch to be cut}}{\text{pitch of lead-screw}}$$

will be the same as before.

Let p (mm) = pitch of thread

P (inch) = pitch of lead-screw

$$\left(P = \frac{1}{\text{threads/inch}} \right)$$

then

$$\frac{\text{drivers}}{\text{driven}} = \frac{p \text{ (mm)}}{P \text{ (inch)}} = \frac{p \text{ (mm)}}{25 \cdot 4P \text{ (mm)}}$$

$$= \frac{10p}{254P} = \frac{5p}{127P}$$

showing that one of the *driven* gears will be 127T.

Example 7. Determine suitable gears for cutting the following screws on a lathe lead-screw of $\frac{1}{4}$ inch pitch (4 thread/inch).

(a) 3 mm pitch, (b) 1·75 mm pitch, (c) 0·8 pitch, (d) 7·5 mm pitch.

(a) 3 mm pitch $\quad \dfrac{\text{drivers}}{\text{driven}} = \dfrac{5p}{127P} = \dfrac{5 \times 3}{127 \times \frac{1}{4}} = \dfrac{5 \times 3 \times 4}{127}$

$$= \frac{30 \times 2}{127 \times 1} = \frac{30 \times 40}{127 \times 20} \text{ compound train}$$

(b) 1·75 mm pitch $\quad \dfrac{\text{drivers}}{\text{driven}} = \dfrac{5p}{127P} = \dfrac{5 \times 1\frac{3}{4}}{127 \times \frac{1}{4}}$

$$= \frac{5 \times 7}{127} = \frac{35}{127} \text{ simple train}$$

(c) 0·8 mm pitch

$$\frac{\text{drivers}}{\text{driven}} = \frac{5p}{127P} = \frac{5 \times 0 \cdot 8}{127 \times \frac{1}{4}}$$

$$\frac{4}{127 \times \frac{1}{4}} = \frac{16}{127} = \frac{40 \times 2}{127 \times 5}$$

$$\frac{40 \times 30}{127 \times 75} \text{ compound train}$$

(d) 7·5 mm pitch

$$\frac{\text{drivers}}{\text{driven}} = \frac{5 \times 7 \cdot 5}{127 \times \frac{1}{4}} = \frac{150}{127}$$

$$\frac{75 \times 2}{127 \times 1} = \frac{75 \times 60}{127 \times 30} \text{ compound train}$$

Let us work out (a) above on the basis of a $\frac{2}{51}$ ratio instead of $\frac{5}{127}$

3 mm pitch

$$\frac{\text{drivers}}{\text{driven}} = \frac{2p}{51P} = \frac{2 \times 3}{51 \times \frac{1}{4}} = \frac{24}{51}$$

$$\frac{60 \times 2}{51 \times 5} = \frac{60 \times 20}{51 \times 50} \text{ compound train}$$

The actual pitch of the thread cut would be

$$\frac{24}{51} \times \frac{1}{4} \text{ inch} = \frac{24}{51} \times 6 \cdot 35 \text{ mm}$$
$$= 2 \cdot 988 \text{ mm}$$

An error of 0·012 mm

Inch threads from metric lead-screws. This will be the reverse of our previous consideration.

Let p (inch) = pitch of thread

P (mm) = pitch of lead-screw

$$\left(p = \frac{1}{\text{thd/in}} \right)$$

Then

$$\frac{\text{drivers}}{\text{driven}} = \frac{p \text{ (inch)}}{P \text{ (mm)}} = \frac{p \times 25 \cdot 4 \text{ (mm)}}{P \text{ (mm)}}$$
$$= \frac{127p}{5P}$$

and one of the *driving* gears will be 127T.

Example 8. Determine suitable gear trains to cut the following screws on a lead-screw of 6 mm pitch.

(a) 8 thread/inch, (b) 19 thread/inch, (c) $2\frac{1}{4}$ thread/inch.

(a) 8 thread/inch: pitch $= \frac{1}{8}$ in

$$\frac{\text{drivers}}{\text{driven}} = \frac{127p}{5P} = \frac{127 \times \frac{1}{8}}{5 \times 6} = \frac{127}{240}$$

$$\frac{127 \times 1}{80 \times 3} = \frac{127 \times 20}{80 \times 60}$$

(b) 19 thread/inch: pitch $= \frac{1}{19}$ in

$$\frac{\text{drivers}}{\text{driven}} = \frac{127 \times \frac{1}{19}}{5 \times 6} = \frac{127 \times 1}{95 \times 6} = \frac{127 \times 20}{95 \times 120}$$

(c) $2\frac{1}{4}$ thread/inch: pitch $= 1/2\frac{1}{4} = \frac{4}{9}$ in

$$\frac{\text{drivers}}{\text{driven}} = \frac{127 \times \frac{4}{9}}{5 \times 6} = \frac{127 \times 4}{45 \times 6} = \frac{127 \times 40}{45 \times 60}$$

It may generally be assumed that to have a gear of 127T as a driver creates a very difficult, if not impossible, situation. This leaves the alternative method of cutting a very near thread.

If we work out the wheels for 8 thd/inch on the basis of a $\frac{2}{51}$ ratio instead of $\frac{5}{127}$ we get:

$$\frac{\text{drivers}}{\text{driven}} = \frac{51p}{2P} = \frac{51 \times \frac{1}{8}}{2 \times 6} = \frac{51}{96}$$

$$= \frac{51 \times 10}{120 \times 8} = \frac{51 \times 50}{120 \times 40}$$

This gives an actual pitch of $\frac{51 \times 6}{96} = \frac{306}{96}$ mm

$$= \frac{306}{96} \div 25 \cdot 4 \text{ in} = 0 \cdot 1255 \text{ in}$$

An error of $0 \cdot 0005$ in

Another method used to provide an approximate conversion from English to metric pitches is to use a 63T wheel. This involves a ratio of $\frac{2 \cdot 5}{63}$ as compared with the exact $\frac{5}{127}$ or the $\frac{2}{51}$ we have used above. This ratio, although slightly less accurate than the $\frac{2}{51}$ ratio gives results quite accurate enough for many purposes. Comparing the three possibilities we have:

$$\frac{50}{127} = 0 \cdot 39370 \text{ exact conversion}$$

$$\frac{20}{51} = 0 \cdot 39216 \text{ error about } 0 \cdot 4\%$$

$$\frac{25}{63} = 0 \cdot 39683 \text{ error about } 0 \cdot 8\%$$

The lead of a thread—Multi-start threads

So far we have only considered the case where a threaded shaft has a single thread running along it as shown at Fig. 271(a). It is possible, however, for a piece of work to have several separate and independent threads run-

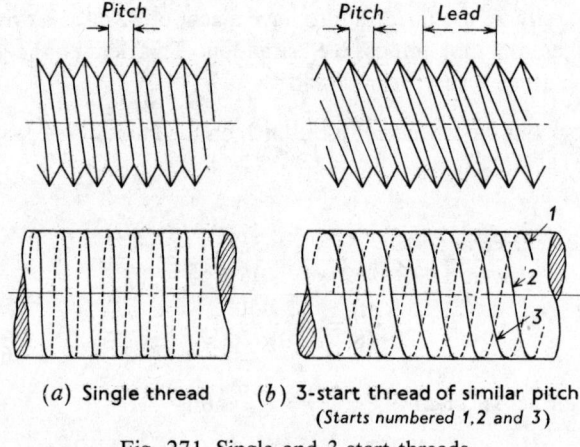

(a) Single thread   (b) 3-start thread of similar pitch
(Starts numbered 1, 2 and 3)

Fig. 271 Single and 3-start threads

ning along it, and Fig. 271(b) shows a cylinder with three threads. Such a screw would not look much different from one having only a single thread, but if one thread is followed round, it will be seen that there are two more fitted in between, and in one complete turn round the bar the thread advances *three times* as far as if it were a single thread. This distance that any one thread advances along the bar whilst it makes one complete turn is called the *lead*. The different threads are called *starts*, and we may have single-start, two-start, three-start, etc., threads. Multiple-start threads are commonly used on fountain-pen caps so that the cap may be screwed up quickly. In any thread the *pitch* is the distance between two adjoining threads, so that in a three-start thread, since there are three separate threads, the lead will be three times the pitch. In general, we may say that if there are 'n' starts, then lead $= n \times$ pitch. A single-start thread has pitch equal to lead.

Example 9. Determine the change wheels for the following threads, on a lathe with a 6 mm lead-screw:

(*a*) 1·25 mm pitch, 3 start, (*b*) 4·25 mm pitch, 2 start (*c*) $\frac{3}{8}$ inch pitch (English), 2 start.

(*a*) 1·25 mm pitch, 3 start has a lead of 3·75 mm

$$\frac{\text{drivers}}{\text{driven}} = \frac{\text{lead of thread}}{\text{pitch of lead-screw}} = \frac{3\cdot75}{6}$$
$$= \frac{375}{600} = \frac{15}{24} = \frac{5 \times 3}{6 \times 4}$$
$$= \frac{50}{60} \times \frac{30}{40} \text{ compound train}$$

(*b*) 4·25 mm pitch, 2 start has a lead of 8·5 mm

$$\frac{\text{drivers}}{\text{driven}} = \frac{8\cdot5}{6} = \frac{85}{60} \text{ simple train}$$

(*c*) $\frac{3}{8}$ inch pitch, 2 start has a lead of $\frac{3}{4}$ inch. Using the approximate solution with a 63T driver we get:

$$\frac{\text{drivers}}{\text{driven}} = \frac{63 \times \frac{3}{4}}{2\cdot5 \times 6} = \frac{63 \times 3}{60} = \frac{63}{20}$$

i.e. a simple train with 63T driving 20T,

or,
$$\frac{63 \times 3}{60} = \frac{63 \times 3}{30 \times 2} = \frac{63 \times 60}{30 \times 40}$$

A compound train with 63T and 60T as the drivers and 30T and 40T as the driven.

Lathes with screw-cutting gear-box

On the more expensive lathes it is not necessary to fit up the drive to the lead-screw every time a thread has to be cut, because the ratio required can be selected with a gear-box. Full instructions are generally given on such

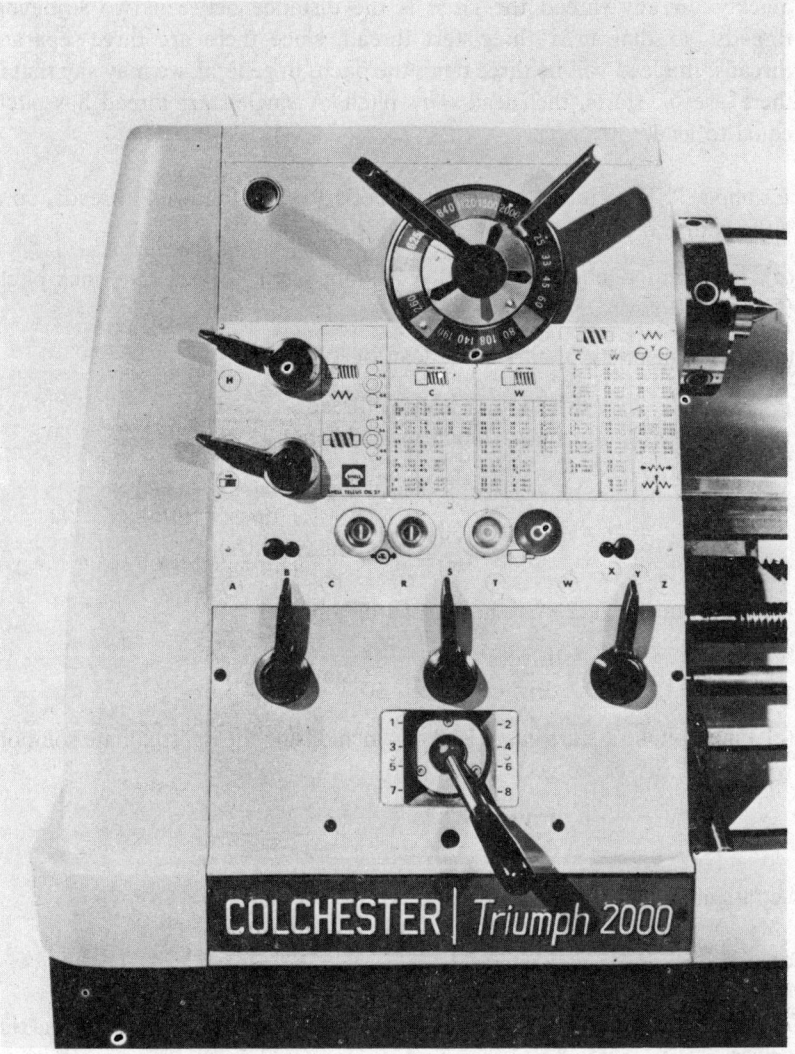

Fig. 272 Outside view of gear-box for selecting feeds and threads. (The two upper LH levers vary the drive ratio from the main spindle. The four lower levers change the gears inside the box.) (See also Fig. 228)

lathes and a diagram of such a box is shown at Fig. 272. The reader will notice that this box enables different feeds as well as threads to be obtained.

Cutting a thread

The form of thread which we shall usually require to cut is the ISO metric, and particulars of this thread are given in the Appendix, p. 398. Screwing is usually the last operation to be performed on an otherwise finished piece of work, and when the diameter of the screwed portion has been turned down to the top size of the thread and the change wheels fitted, a screw-cutting tool must be obtained. This is similar to a parting tool, with its end ground to the vee form of the thread, and rounded off at the point to

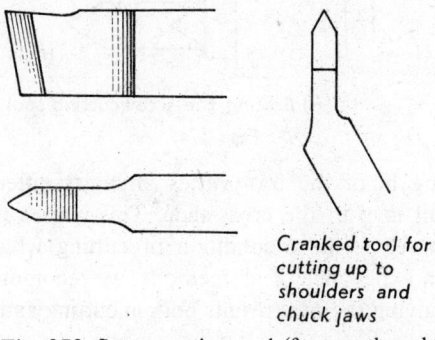

Cranked tool for cutting up to shoulders and chuck jaws

Fig. 273   Screw-cutting tool (for vee threads)

suit the radius at the bottom of the particular thread to be cut (Fig. 273). For obtaining the correct angle on the tool a *screw-cutting gauge* (Fig. 274(a)) is necessary, whilst to get the radius at the end, a *pitch gauge* (Fig. 275) is useful. This latter gauge, which incorporates templates of the usual range of threads, is also useful for inspecting the correctness of pitch at the commencement of cutting, and for determining the pitch of unknown threads. For screwing cast-iron and brass, the screw-cutting tool should have little or no top rake, whilst for steel, the rake may be about one-half that for an ordinary turning tool.

When the tool has been ground satisfactorily to form it must be set on centre, and with its axis square with the axis of the work. This second setting is performed with the aid of the screw-cutting gauge, as shown at Fig. 274(b). The tool must now be traversed a number of times along the work under the control of the lead-screw and fed in a little deeper each time until the thread has been cut to the correct depth (or slightly less, to leave for finishing with a *chaser*). The procedure adopted for this

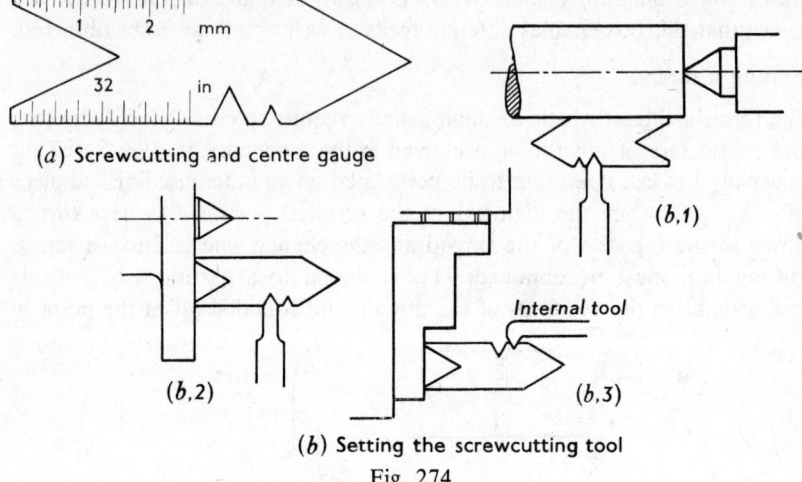

(a) Screwcutting and centre gauge

(b,1)

Internal tool

(b,2)   (b,3)

(b) Setting the screwcutting tool
Fig. 274

progressive feeding in of the tool varies amongst different turners, and many feed straight in with the cross-slide. This causes the tool to cut on both faces of its vee form, a condition of cutting which is not always satisfactory. From experience of all methods, we recommend the following as being the one giving the best results both in cutting and in manipulation.

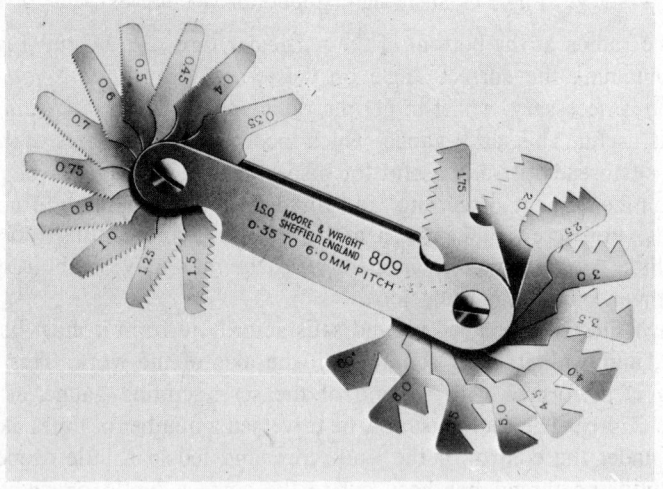

Moore & Wright (Sheffield) Ltd
Fig. 275 Screw pitch gauge

Set the compound slide so that it makes half the thread angle with the axis of the cross-slide (i.e. 30° for ISO metric and Unified, 27½° for Whitworth). This slide should be well adjusted so that its movement has a firm feel about it. Set the cross-slide so that the tool just scrapes the diameter to be screwed, and set the cross-slide stop to prevent the tool moving any further in. If there is no stop, chalk the graduated sleeve so that the cross-slide may be wound to the same place each time. After putting the machine on a slow-speed, engage the nut, and the tool will move along, scraping

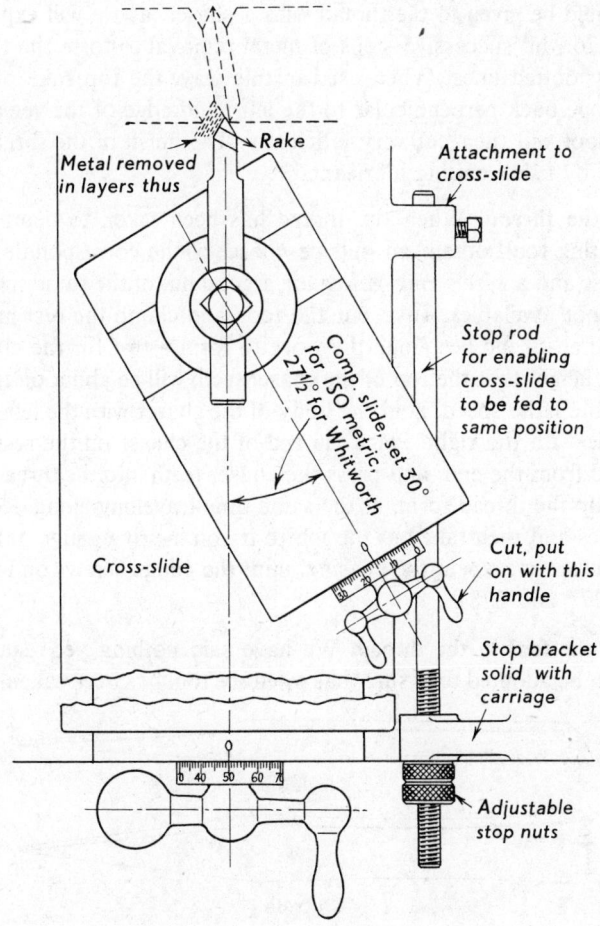

Fig. 276 Set-up for cutting vee threads

a thin line for the thread. When it reaches the limit of the thread, withdraw the cross-slide quickly and disengage the nut (one hand should be kept on the cross-slide handle and the other hand on the lever for the nut). Return the carriage to the starting position, wind in the cross-slide to the stop or chalk mark, and put on a small amount of cut with the compound slide. Engage the nut, and allow this cut to travel along, and withdraw the cross-slide and nut at the end as before. Repeat the process until the thread is almost to its full depth when, to clean it out to form, the last cut or two may be taken by feeding straight in with the cross-slide. The final finish should be given to the thread with a chaser, as we will explain later. In Fig. 276, the successive steps of metal removal to form the thread are shown by dotted lines. When used in this way, the top rake of the tool should slope back perpendicular to the left-hand edge of the vee as shown, and the tool will then cut very efficiently. The finish of the thread will be improved by using cutting lubricant.

Chasing the thread. When the thread has been taken to depth with the screw-cutting tool, obtain an outside chaser of the corresponding pitch, a chaser rest and a *screw ring gauge* (or a good nut of the same size if a ring gauge is not available). Take out the tool, and clamp the rest in the tool-post, so that its end lies along the work to form a rest for the chaser, and at such a height that the top of the chaser teeth will be about on the centre. Increase the lathe speed, hold the body of the chaser with the left hand and the handle with the right. Place the end of the chaser on the rest, pick up the thread from the end, and press the chaser teeth into the thread, when it will trim up the thread form, at the same time travelling along. Repeat this a few times and then take out the job to try on the ring gauge. If the gauge will not screw on, repeat the chasing, until the gauge screws on with a nice fit (Figs 277 and 278).

Locating the tool in the thread. We have said nothing yet regarding the method to be adopted to ensure that when the tool has been taken along the

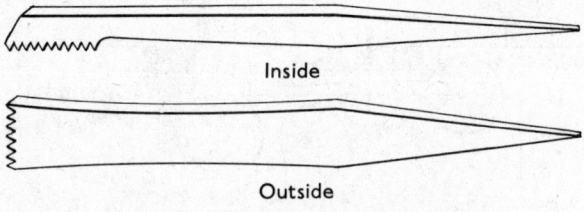

Inside

Outside

Fig. 277 Internal and external chasers

thread and brought back to its starting position, the nut may be engaged to bring the carriage into such a position that the tool follows the same thread again. If the pitch of the lead-screw is a multiple of the pitch being cut (e.g. 1, $1\frac{1}{2}$, 2, 3, 6 mm for a 6 mm lead-screw) the nut may be engaged anywhere and the tool will follow the original thread. If, however, this does not

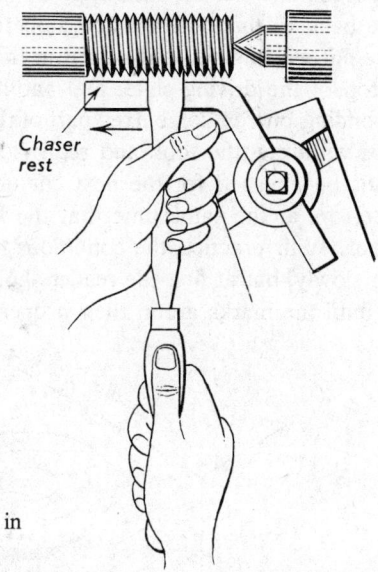

Chaser
rest

Fig. 278 Chasing a thread in
the lathe

apply, the reader will find that if he engages the nut indiscriminately, the tool will probably commence to cut a separate thread alongside the first one and the job will be spoiled. There are several methods for ensuring that the tool shall follow the same thread each time.

(a) *Reversing the machine.* If the lathe is equipped with a means for reversing its spindle, the lead-screw nut may be kept in permanent engagement and the tool returned to its starting position by reversing the machine. The main precaution to be observed if this method is used is to ensure that all the lost motion (called 'backlash') has been taken up before the tool is fed in for each successive cut. This backlash results in a pause in the movement of the saddle at each end, and if the tool is fed on to the work whilst the carriage is stationary, the thread will be spoiled. Always allow the tool to return beyond the end of the screw before reversing and feeding in. Some lathes have a lever on the apron which reverses the lead-screw without reversing the machine, and this method is about the most convenient of all for bringing the tool back to its starting position.

(*b*) *Marking the lathe.* By means of a bed stop for the carriage, or a bar placed in front of the tailstock, make some provision to allow the carriage to be returned to the same position each time, with the tool beyond its starting position. Before engaging the nut for the first time, wind the carriage back to this position and stop the lathe. Pull the machine round until the nut will engage, and with it in this position, make one chalk mark at the top of the driving-plate, and another on the lead-screw opposite a corresponding one on some fixed part of the machine. Start up and take the first cut, withdraw the tool, and return the carriage to the stop. The nut must not be engaged for the next cut until the driving-plate chalk-mark is at the top, at the same time that the lead-screw mark is level with its neighbour. With practice, the conditions may be caught whilst the lathe is running slowly, but at first the reader should inch it round with the starting handle until the marks are in their proper positions.

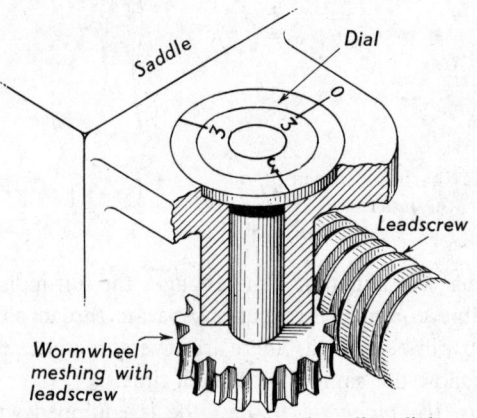

Fig. 279 Diagram showing threading dial
(*See* also Fig. 228 S)

(*c*) *The threading dial.* The threading dial is a fitment to the saddle which performs a duty similar to that of marking the lathe as we have just discussed. The dial is shown at Fig. 279, from which it will be seen that the visible indicating face is connected to the lead-screw by means of a wormwheel and the nut should be engaged when the requisite numbered division on the dial reaches the zero line.

The function of the wormwheel is to meter off, along the lead-screw, a length which is an exact multiple of the pitch of the thread to be cut. Thus, as we have stated above, a 2 mm pitch may be cut using a 6 mm lead-screw by engaging anywhere, but for, say, a 2·5 mm pitch thread it would be necessary to work on intervals of 5 pitches (30 mm) along the 6 mm lead-screw to achieve a satisfactory result. This could be arranged by using a dial with a 15T wormwheel and engaging at any ⅓ revolution, or a 20T wormwheel at the ¼ positions. For a 6 mm lead-screw, a good compromise using a single wormwheel is to employ a 15T wheel with the dial divided into ⅓rds. With this combination all pitches which divide into 90 can be engaged at the zero position (every full rev) and all pitches which divide into 30 can be engaged at any of the ⅓ divisions (Fig. 279). A rather more comprehensive coverage is possible by using a set of interchangeable wormwheels with a dial divided into four parts. With this method, of course, it is necessary to allow for the variable centre distance between the lead-screw and the dial spindle to accommodate various sizes of wormwheels.

Table 20 shows the scope of this latter method.

Table 20. Operation of threading dial
(6 mm pitch lead-screw. Interchangeable wormwheels)

Pitch (or lead) of thread to be cut (mm)	Wormwheel on dial spindle	Where half-nut can be engaged.
0·7, 1·75, 3·5, 7, 14.	14T	1 and 3.
4·5, 9.	18T	or
5·5, 11.	22T	2 and 4.
0·25, 0·5, 0·75. 0·8, 1, 1·5, 2. 3, 4, 6, 8, 12.	16T	1, 2, 3, 4.
1·25, 2·5, 5, 10.	20T	

Cutting left-hand threads

Left-hand threads must be cut by reversing the lead-screw so that it rotates opposite to the work, and hence cutting with the tool moving from left to right. This will often involve turning a groove in the work at the end of the thread to provide a space for starting the tool. The groove should be to the depth of the thread, and of a width equal to 1 to 2 threads.

Cutting internal threads

For cutting internal threads with a single point tool, the notes we have given for external work regarding the manipulation of the lathe apply. In general, internal work is more difficult, because the tool is often less rigid, and its progress cannot always be observed. The hole for the thread should first be bored to the core diameter of the thread, this being determined by subtracting twice the depth of the thread from its top diameter (see The Appendix, p. 398). When the hole is 'blind', a recess should be turned at the bottom as shown at Fig. 280, the depth of this being equal to the thread, and

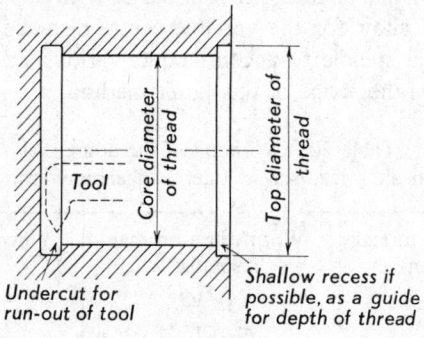

Fig. 280 Preparation of hole for blind internal thread

its width about 1 to 2 threads. Care will have to be exercised to avoid running the tool into the metal at the bottom of the hole in such cases, and a very slow speed, together with guiding chalk, marks or stops, must be used as much as possible. For internal work it is helpful if a short recess can be bored at the front end, having a top diameter equal to the outside diameter of the screw (Fig. 280). This serves as an indicator, and when the tool point just scrapes its diameter we know that it has been taken in to the full depth of thread. Often, if there is any length to spare on the work, the material used for the length of this recess can be faced off afterwards.

It is not usual to feed with the compound slide when cutting internal threads, but to feed straight in with the cross-slide, and use the graduated sleeve for putting on a suitable amount of cut each time. The compound slide may be set parallel with the bed, and occasionally a few thousandths of cut put on with it, the object being to clear the sides of the tool.

Finishing the thread

Internal threads may be chased in the same way as external ones, an internal chaser being shown at Fig. 277. Owing to the greater difficulty of manipulating an inside chaser, some turners prefer to clamp it in the tool-post and feed it with the lead-screw after matching up the threads on the chaser and work. If this is done the chaser should be set so that it is in line, and this may be done by adjusting the tops of its threads to the blade of a small square from the face of the work. When the inside chaser is used by hand the thread should first be tooled to its full depth so that the work of the chaser only consists of putting the radius on the tops of the threads.

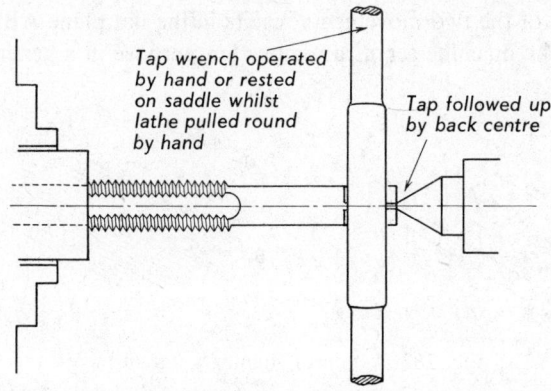

Fig. 281 Tapping in the lathe

If a tap is available of the size being cut, this forms a good way of finishing the thread. Cut the thread to within a few thousandths of its full depth and after fitting a wrench to the tap, start it in the hole. Now bring up the tailstock centre to the centre hole in the end of the tap, and as the tap advances into the work, keep the centre in contact with the tap centre hole by winding the tailstock (Fig. 281). The tapping operation is best performed by pulling the lathe round by the belt with the left hand, whilst the right hand attends to the feel of the wrench, and the screwing up of the tailstock centre. This method of using the back centre to keep a tap in line should always be used when tapping holes in the lathe, and when a hole is tapped from a drilled or bored hole without any previous screwing by a tool it is of great help in obtaining a true thread.

12   The shaping machine

The main function of the shaping machine is the production of flat surfaces, which are obtained by combining a line tool cut with a perpendicular feed. The arrangement of this is shown at Fig. 282(a) and the result at (b), where the tool repeatedly travels along the line AB whilst the work is fed a small distance each time, in the same plane as the line of tool motion, and perpendicular to it. The tool line eventually reaches a position A_1B_1 and the combination of the two movements results in the flat plane ABB_1A_1 being machined. The machine serves an important purpose in a general machine

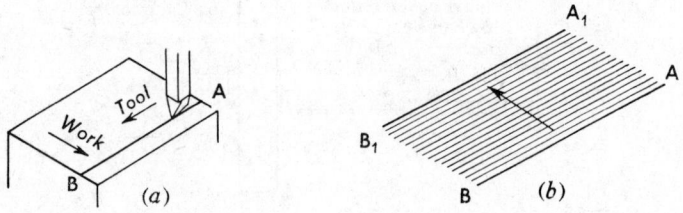

Fig. 282 Action of shaping a flat surface

shop, particularly on account of the quickness with which the work and tool can be set up, and the good standard of accuracy to which a flat surface can be produced.

A diagram of a crank shaper is shown at Fig. 283, the term 'crank' being applied because the motion of its ram is derived from a crank pin (see later). The tool receives its straight-line motion from the ram, which is guided by long, straight ways. Underneath the path of the tool is the work-table, which is supported on ways perpendicular to the ram and travels along these slides when a horizontal surface is being machined. The table ways are mounted on vertical slides so that the whole unit may be moved up and down. Vertical surfaces are machined by employing a vertical feed of the tool or work. On machines not provided with any automatic vertical movements hand operation of the tool slide is used for shaping vertical surfaces. It is now becoming general practice, however, to provide for a power vertical feed for the table, and on some of the larger machines the

The Butler Machine Tool Co. Ltd.

Fig. 283 Crank shaping machine

A. Ram
B. Tool slide
C. Handle for clamping ram
D. Screw for adjusting position of ram
E. Screw for locking tool slide unit
F. Speed change lever
G. Starting lever

H. Vertical/horizontal feed selector
I. Feed directional control
J. Lever for adjustment of feed
K. Horizontal table adjustment screw
L. Vertical table adjustment
M. Feed shaft
N. Toolbox

Specification

	mm
Maximum length of stroke	457
Table, top (length × width)	457 × 438
„ horizontal travel on slides	508
„ vertical travel	305
Tool box, vertical adjustment	152
Number and range of speeds obtainable	8. 15–150 cycles/min.
Number and range of feeds	10. 0·3 to 2·7 mm
Horse power and speed (driving motor)	3·75 kW × 1000

head slide is fed by an arrangement which notches the actuating screw a small amount each time the ram reaches its backward position. Where vertical power feed is absent the table could, of course, be fed by hand, but this is too laborious and its up and down movements are only employed to suit varying heights of work.

The shaper drive

On the machine shown at Fig. 283 the motor is mounted on a platform at the rear and drives the gear-box intake pulley and clutch by an enclosed vee belt drive. The gear-box provides for eight speeds and delivers the power to a pinion which drives the *stroke wheel* (sometimes called the *bull* wheel). Mounted on the face of the stroke wheel is the *crank pin*, which is incorporated with a sliding block working in dovetail slideways running across the face of the wheel. The connection to the ram is arranged by means of a link attached to the underside of the ram and pivoted at the bottom of the machine. This link engages with a *driving block* bored to suit the crank pin and having some arrangement whereby it can slide up and down the link. On the machine we have illustrated the link consists of two rods, but in other designs the link consists of a member with a long slot which accommodates the driving block for sliding up and down whilst the stroke wheel rotates. This slotted link form of design, which has domi-nated the shaping machine for many years, has led to the mechanism being known as the *slotted link* motion. When a link of this form is used it is necessary to employ an additional link for connection to the ram in order to compensate for the distance variation as the link swings through an arc. This is shown as *DE* on Fig. 285, which also shows the slotted link in diagrammatic form. In the design we have illustrated in Fig. 284, this extra link is unnecessary as the length variation is taken up by the rods sliding in the block where they are anchored at the bottom of the machine. The mechanical details of the link motion are all shown on Fig. 284.

A feature of the shaping machine is that for each productive cutting stroke there is an idle return stroke, and the geometry of the slotted link motion is such that this return stroke occupies a shorter time and so mini-mises the time wasted. In Fig. 285(a), which shows a skeleton form of the mechanism, the crank pin B rotates about the centre of the stroke wheel A and at the same time B slides up and down the link CD. This causes the link to oscillate about C and so drive the ram backwards and forwards. The quick return feature is derived from the configuration of the mechanism and will be understood from Fig. 285(b). When the link is at CD_1, tan-

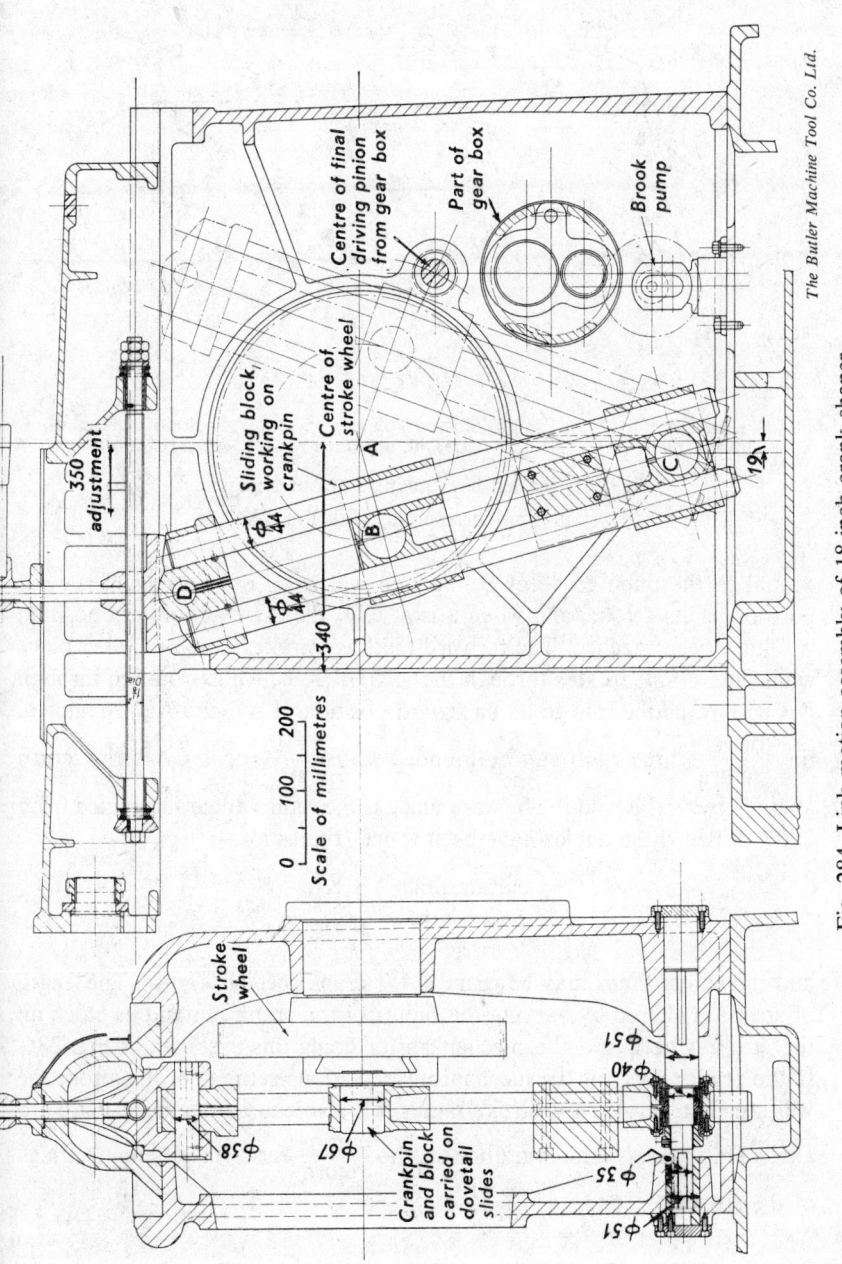

Fig. 284 Link motion assembly of 18-inch crank shaper

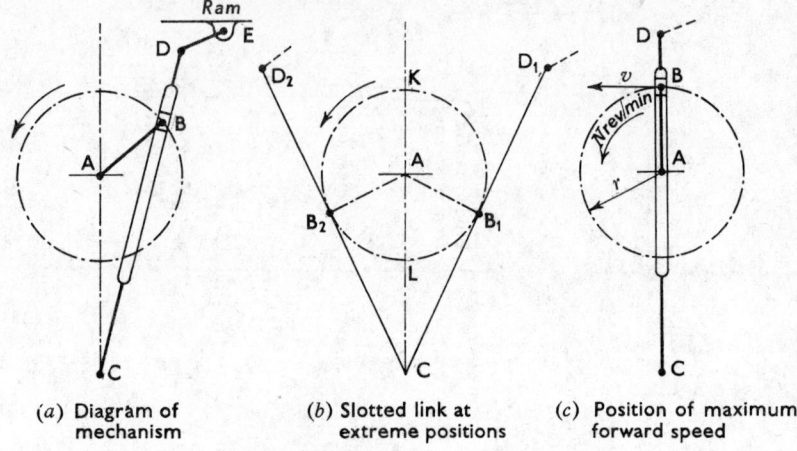

(a) Diagram of mechanism

(b) Slotted link at extreme positions

(c) Position of maximum forward speed

Fig. 285 Slotted link quick return mechanism

(Lettering corresponds with Fig. 284)

gential to the pitch circle of B, the ram will be at the extreme backward position of its stroke, and when it is at CD_2 the extreme forward position will have been reached. The forward (cutting) stroke, therefore, takes place whilst the crank rotates through the angle B_1KB_2, whilst rotation through B_2LB_1 returns the ram to its backward position. If AC and AB are known, the $\dfrac{\text{forward}}{\text{return}}$ time ratio can be found, because $\dfrac{AB}{AC} = \cos \widehat{CAB}$, the return angle is twice this and the forward angle is the return angle subtracted from 360°. When these angles have been found, the ratio

$$\frac{\text{cutting time}}{\text{return time}} = \frac{\widehat{B_1KB_2}}{\widehat{B_2LB_1}}$$

and the actual times may be found if the crank speed is known. The length of stroke is altered by varying the radius of the crank pin and its block on the stroke wheel, and the mechanism for doing this is shown at Fig. 286. If the reader lays out the mechanism for various settings of the stroke, he will see that varying the stroke radius AB also causes the angle B_2LB_1 to alter. This causes some variation in the $\dfrac{\text{cutting}}{\text{return}}$ ratio for different settings of the stroke.

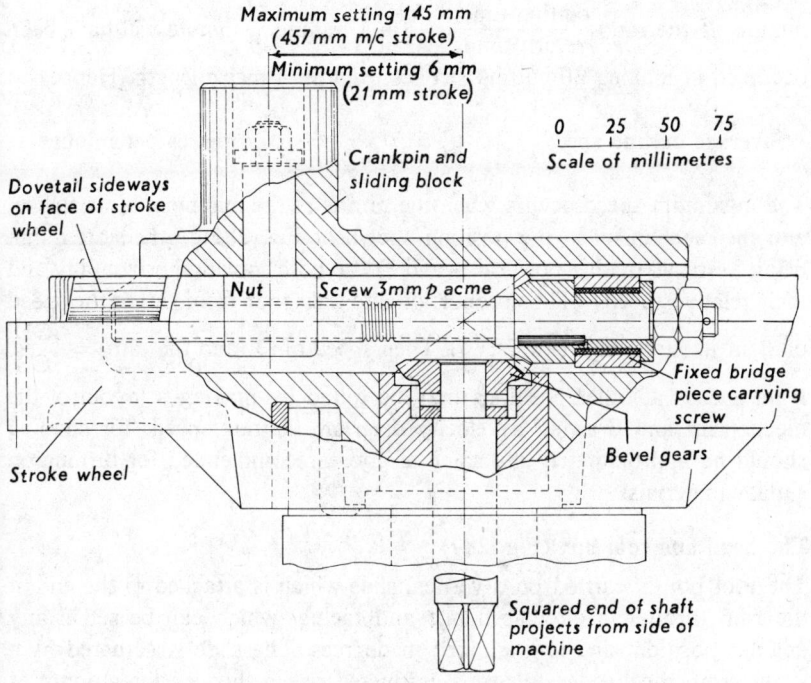

Maximum setting 145 mm
(457 mm m/c stroke)

Minimum setting 6 mm
(21 mm stroke)

Crankpin and
sliding block

Dovetail sideways
on face of stroke
wheel

0   25   50   75
Scale of millimetres

Nut   Screw 3 mm p acme

Fixed bridge
piece carrying
screw

Bevel gears

Stroke wheel

Squared end of shaft
projects from side of
machine

Fig. 286 Details of mechanism for adjusting position of
crank pin block on stroke wheel
(Details from the Butler Machine Tool Co. Ltd.)

Stroke and speed

Shaping machine sizes are designated by the maximum stroke of the ram,
the most usual ranging from 300 mm to 900 mm, with those in the 450 mm
to 600 mm group being the best all-round machines for general purpose
work. The cutting speed of the tool, starting as it does from rest, rising to a
maximum and slowing down to rest again, varies throughout the stroke, but
its average speed can be calculated, and if the dimensions of the mechanism
are known its maximum speed is not difficult to determine. The average
speed, taken over cutting and return strokes, is given by multiplying the
stroke in metres by the number of double strokes per minute. To obtain
the average speed on the cutting stroke we must know the cutting-return
time ratio. Assume a machine working on a $\frac{1}{4}$ metre stroke and making 30
double strokes per minute, then the overall average speed is 15 metres per

minute. If the ratio $\frac{\text{cutting time}}{\text{return time}}$ is $\frac{5}{4}$, in 1 minute, $\frac{5}{9}$ minute will have been occupied in making 30 cutting strokes, each of $\frac{1}{4}$ metre length. Hence

$$\text{average cutting speed} = 7 \cdot 5 \div \frac{5}{9} = 7 \cdot 5 \times \frac{9}{5} = 13 \cdot 5 \text{ metres per minute.}$$

The maximum speed occurs when the ram is at the mid point of its stroke, and the mechanism in the position shown at Fig. 285(c), the slotted link being vertical. In this position B and D are both moving horizontally and their relative speeds will be proportional to their radii from C. Let the speed of B in its circle be v ($v = 2\pi r N$). Then speed of D (and the ram) $= v\dfrac{\text{CD}}{\text{CB}}$.

Ram speeds at other positions may be found by making a layout of the mechanism and drawing a velocity diagram. Cutting speeds for shaping should be approximately the same as those recommended for turning on similar materials.

The head and tool box (Fig. 287)

The tool box is carried on a vertical slide which is attached to the end of the ram through a circular flange and facing, which can be set in any angular position, and is graduated in degrees. The slide is actuated by a screw from the handle on top, a graduated sleeve showing the amount of movement. When the machine is being used for ordinary horizontal machining the slide is clamped in the vertical position and the screw used for putting on the necessary amounts of cut. The slide is also used to traverse the tool up or down for machining vertical surfaces, and by setting the head at some angle for machining angular surfaces. The tool post or clamp is carried in its box on a block, pivoted at the top to allow it to swing forward (sometimes called the 'clapper box'). This allows the tool to lift clear of the path of the cut on the return stroke of the ram. An additional feature is that the tool box may be swung a small amount on either side and clamped by a nut. This is for use when shaping vertical or angular surfaces and, by introducing a horizontal element into the swing, allows the tool, on its return stroke, to clear the face being machined (Fig. 288). The larger and more expensive machines are provided with a swivelling table which allows angular surfaces to be shaped by using the horizontal feed of the table. When the tool slide is being used to traverse cuts it should be used at about the middle of its range of travel. If it is high up the movement is easy, but there may not be sufficient length of engagement

Bowden cable for raising the clapper box on the return stroke.

The Butler Machine Tool Co. Ltd.

Fig. 287 Headslide and tool box

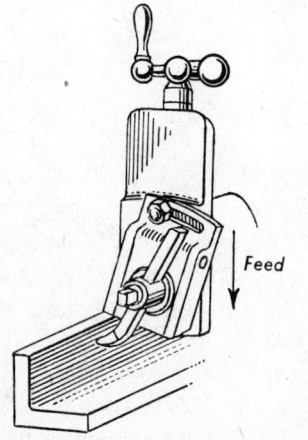

Tool box swung so that tool will clear face on return stroke.

Feed

Fig. 288 Shaping vertical face

to ensure accuracy. When the slide is near the bottom it generally becomes stiff and tiresome to operate by hand. The graduated sleeve on this screw should be calibrated and the value of its markings memorised, since they control the amount of cut being put on when shaping horizontal surfaces. The same applies to the horizontal table lead-screw which controls the cut being put on when shaping vertical surfaces. If there are not graduated sleeves it is a good plan to make and fit them.

Cutting tools

The general patterns of the tools used for shaping are the same as for turning (see Fig. 126, p. 183), but their shanks should be a little more robust to withstand the shock each time a cutting stroke commences. For this reason a good solid shank tool is preferable to a holder into which tool bits are fitted. Slightly more clearance should be ground on the front (about 10° to 12° total), and the other point to consider is the direction of the side rake. When turning, the tool always travels from right to left, so that all straight turning tools have their side rake sloping from left to right when viewed from the shank end. The usual relative tool-work movement in shaping is from left to right, and this requires a side rake sloping in the opposite direction (see Fig. 289(a)). When shaping *down* a vertical face with a left-hand side tool the side rake should slope as shown at Fig. 289(b),

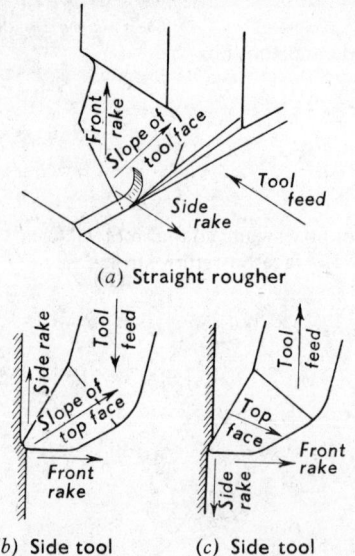

(a) Straight rougher

(b) Side tool cutting down   (c) Side tool cutting up

Fig. 289 Rake directions for shaping tools

but when the bottom has been reached and more cut put on to traverse up again the rake needs to be the opposite way for correct cutting (Fig. 289(c)). It is thus not always possible to work strictly according to theory, but the reader should observe the rule as much as possible, and for the sake of a good finish arrange that the last cut is taken in the correct direction for utilising the advantage of the rake.

Speeds and feeds

When turning in the lathe the material passes over the tool at a uniform rate, but in shaping, as we have seen from our discussion with Fig. 285, the tool starts from rest, attains a maximum speed and then slows down to rest again. It then makes a return stroke of the same character, but in a shorter time. This renders the estimation of a precise cutting speed rather difficult, but the following approximate method may be used as a guide for setting the machine.

Divide twice the length of stroke (in metres) into the lowest allowable cutting speed (for turning) for the material concerned (see Table 15, p. 185). This will give the number of double strokes of the ram per minute.

Thus, for shaping mild steel on a $\frac{1}{4}$ metre stroke, and taking a cutting speed of 20 metres per min.:

Number of machine cycles (double strokes)

$$= \frac{20}{2 \times \frac{1}{4}} = \frac{20}{\frac{1}{2}} = 40 \text{ double strokes per min.}$$

Feed

The feeding mechanism on many shaping machines consists of a ratchet wheel keyed to the table lead-screw, and actuated by a swinging pawl which oscillates under the action of a connecting rod, the end of which is given an eccentric motion from a rotating flange on the side of machine. With such a mechanism the choice of feeds is rather limited, since the greatest movement of the pawl is such as to give up to about four teeth on the ratchet wheel. In general, it will be found that for most jobs one or two teeth of the wheel will prove satisfactory, but the reader should work out for himself the value of the feed given when the ratchet wheel is being indexed by a single tooth. Thus, if the ratchet wheel has, say, 28 teeth and the table lead-screw a pitch of 5 mm:

Feed for 1 tooth $= \frac{1}{28} \times 5 = \frac{5}{28}$ mm = about $0 \cdot 18$ mm per stroke.

It should be remembered that the feed of the tool should occur during the *return* stroke of the ram. On the above arrangement this necessitates arranging the driving end of the connecting rod to one side or other of the driving flange centre, according to whether the table is being traversed to the left or to the right.

Makers are now providing machines with more sophisticated feeding arrangements and the method used on the machine we have shown is explained later.

Holding work—the machine vise

An important method of holding work on the shaper is in the machine vise, and although work can be machined by attaching it to the top or side of the table, the vise is by far the most common method used, except for castings and irregular work. The shaping machine vise is of large, robust construction, designed specially for the machine and attached to the table by a generous, square base. The upper portion is mounted on the base through a graduated turntable, permitting the jaws to be used in line, or perpendicular to the ram, as well as at any other position for angular work. The base is lined up on the machine table by a pair of tenon pieces which fit in a tee slot. Normally the jaw pieces of the shaper vise are not hardened, so that it is possible to clean them up with the tool (Fig. 293(b)) from time to time when they become bruised or damaged. The sliding jaw is moved through a large, square or acme threaded screw by a long, heavy handle, and it is not necessary to use a hammer on the handle to tighten the vise sufficiently for its work. A good, steady increasing pressure applied with both hands and the weight of the body will usually do all that is required. Blows with a hammer mutilate the handle disgracefully as well as being harmful to the vise and distorting the work being held. It is unfortunate that the layout of the machine makes it necessary, in the cross position of the vise jaws, to have the movable jaw take the thrust of the tool, and although the results seem to suffer little in consequence it is always advisable to take a blow or thrust against solid metal whenever possible.

Although, as we have said, the vise is used a great deal for holding work, the most interesting and rewarding jobs are castings and other work which must be set up on the top or side of the table, and it is work of this character which demonstrates the real capabilities of the shaping machine in the hands of an experienced operator. For this reason, in addition to the vise, the machine should be provided with a good selection of bolts, clamps, stops, setting strips, and the usual collection of bits and pieces associated with setting up jobs on a machine.

The table feed

An arrangement for feeding the saddle which provides a more comprehensive range and better control than that employing a ratchet wheel on the lead-screw is shown at Fig. 290, which is the opened up view of the box carrying the feed lever J in Fig. 283. The eccentric cam A rotates with the stroke wheel (centre O) and actuates the picking lever C by means of the roller B. A spring-loaded pawl (not visible) is carried on this lever and this engages with the ratchet wheel F. (The half gear G is attached to the picking lever and this keeps B up against the cam by the action of the spring-loaded rack D.) E is a cam which is keyed to the feed lever J (Fig. 283), and when the box is closed the face of E fits close to the face of A and its edge catches a small length of B projecting from the face of A. When the feed is set at the full, E is in the position allowing B to be subjected to the full throw of A. By rotating E, via J, its cam edge will take B partly away from the full effect of A and so regulate the movement of the picking lever. The intermittent rotation of F is transmitted through bevel

The Butler Machine Tool Co. Ltd.

Fig. 290. Cam feed box.

gears to the feed shaft M (Fig. 283) and thence into the box on the end of the saddle. Here more gearing and clutches with control levers (H and I, Fig. 283) allow selection to be made for direction, and for horizontal or vertical movement.

Setting the machine for accuracy

The shaping machine, probably more than any other, is given much hack work to do, and in some shops never rises above the position of a means of removing metal and doing rough jobs. A good shaper, however, in the right hands, is capable of undertaking a wide range of accurate work and often in less time than would be occupied by alternative methods. Accurate results will be encouraged by a few elementary setting precautions carried out before a job is started and we advise the reader never to go far with any important job (on any machine) without assuring himself that the fundamental conditions leading to the desired result are all that they might be. In addition, he should plan the method and order of his work so as to take advantage of the accuracy built into the machine. He will then be in the position, not of one who is agreeably surprised if the job comes out right, but to whom it will be a rare experience if the final result is wrong.

For parallel horizontal facing, the tool should move parallel to the table, and when the table is traversed across, its surface should be equidistant from the tool over its width. This can be tested by lowering the tool and

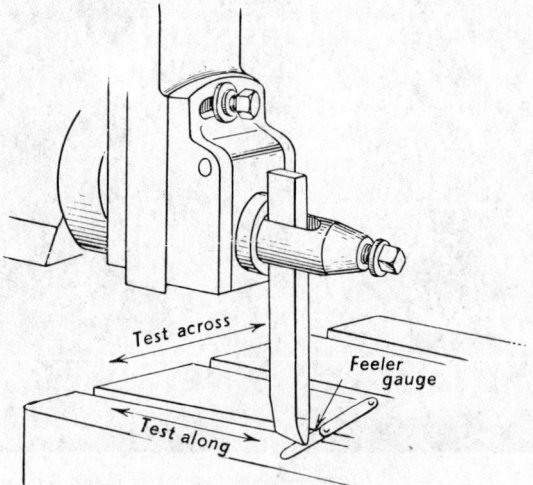

Fig. 291 Testing shaper table for parallelism

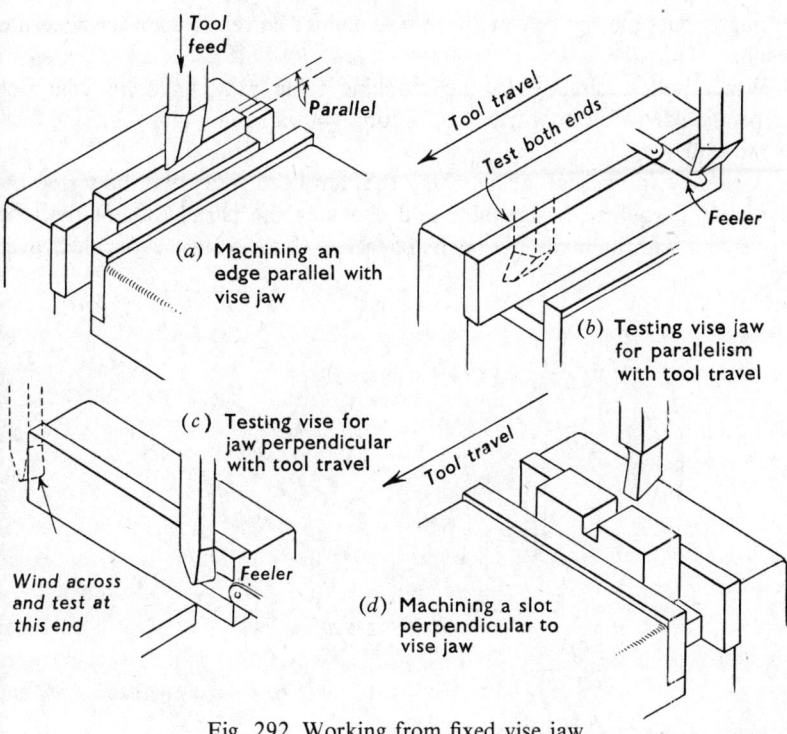

Fig. 292 Working from fixed vise jaw

using a feeler gauge under its point as shown at Fig. 291. The leg and vertical slides should be clamped up before testing. If the table swivels sideways any cross error can be corrected, but otherwise the only way to effect correction is to take a light cut over the table. Before doing this all table and ram slides should be adjusted and the table supporting leg locked. The work-supporting, upper face of the vise should be parallel with its base, so that if the table is correct work held in the vise will also be parallel. When doing ordinary facing the position of the vise jaws is not particularly important, but to machine a slot or shoulder parallel with an edge clamped against the fixed jaw (Fig. 292(a)) the tool should travel in line with that jaw. The test for this is shown at Fig. 292(b). At (c) is shown the test for ensuring that a machined face should be perpendicular with an edge in contact with the jaw (d). An alternative method for this setting is to work from the vise jaw to the face of the table vertical slides with scribing block or height gauge. If the base graduations on the vise are correct a reasonable setting is possible right away, but these cannot be relied upon for accurate results.

Work bolted directly to the machine table may have an edge set perpendicular to the tool travel by setting with a square from the side-face of the table.

Accuracy in vertical facing, using the ram head slide, may be tested by clamping a square to the table and checking the parallelism of the tool movement with its blade by means of feelers (Fig. 293(a)). After the travel

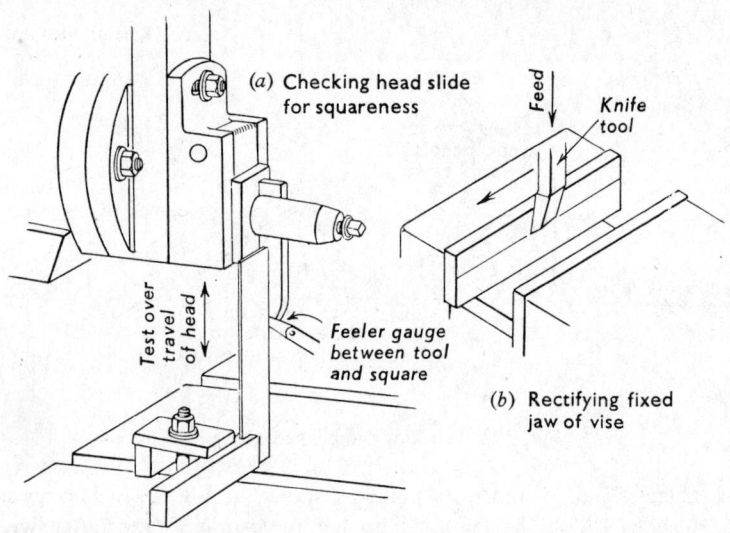

Fig. 293 Accuracy in vertical facing

of the tool has been set correctly it should be used with feelers to check the squareness of the fixed vise jaw with the base, and any error corrected by taking a light cut down with a knife-tool (Fig. 293(b)).

Angular setting

In order to try the machined accuracy of shaped angular work the job must often be removed from the machine and should any correction be necessary, the task of setting it up again to its original position may occupy hours of work. This may be avoided by a careful preliminary setting with vernier protractor or sine bar, and Fig. 294 gives some indications as to how this might be done.

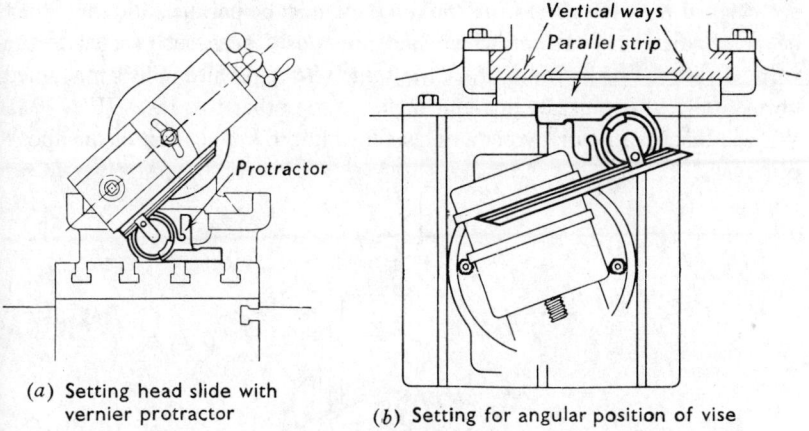

(*a*) Setting head slide with
        vernier protractor

(*b*) Setting for angular position of vise

Fig. 294 Angular setting on shaping machine

Planning and setting work

In the design and construction of machine tools their makers go to a
great deal of trouble to ensure that very accurate geometrical relationships
are maintained between the slides, spindles and faces which control the
surfaces they produce, so that when using these machines it behoves us to
take advantage of such built-in possibilities for accuracy by using the
facilities to best advantage. On the shaping machine the long, accurate
slides of the ram ensure that the point of the tool maintains an unvarying
straight line, and the accuracy of the other machine surfaces ensures that
this line is strictly perpendicular to the vertical and horizontal slides of the
table. Thus, as we have seen from Fig. 282, the combination of the tool
and table movements results in a flat surface, but this can only happen if
other factors which could upset the result are not overlooked. The chief
factors which could be detrimental to the plane accuracy of any single
surface are (a) wear of the edge of the tool as it traverses across a surface,
(b) a loose head slide or a badly adjusted ram slide allowing the tool path
to deviate from a single, constant line, (c) movement of the work due to
faulty clamping or support. Now for the *geometrical relationship* of two or
more surfaces. The line motion of the tool point, when fed downwards
from the head slide accurately set (see Fig. 293), sweeps out a plane which
is perpendicular to the table and perpendicular to the front faces of the
vertical table slides (Fig. 295). If, therefore, two vertical faces of a job

are cleaned up *without moving the job* they must be parallel, and they must be perpendicular to a face which had previously been set parallel to the vertical slides. Furthermore, if, *at the same setting*, a third face is machined horizontally, this must be at right-angles to the other two faces (Fig. 295). We will discuss one or two examples showing the application of the above principles.

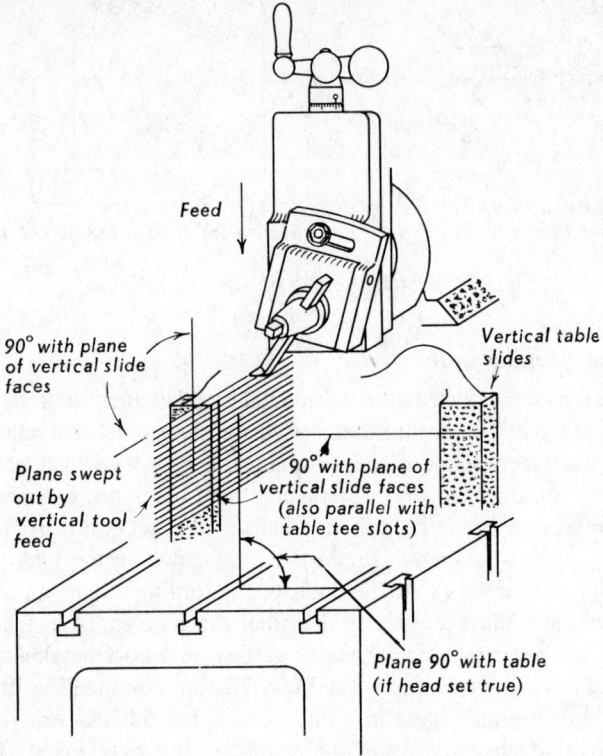

Feed

90° with plane of vertical slide faces

Vertical table slides

Plane swept out by vertical tool feed

90° with plane of vertical slide faces (also parallel with table tee slots)

Plane 90° with table (if head set true)

Fig. 295 Accuracy relationships of a surface shaped by vertical feeding

In Fig. 296, A, B, C and D are the faces to be machined. The horizontal faces must be parallel with the base and both must be *in the same plane*. These faces must be square with the vertical faces, which themselves must be parallel. (The base is machined.)

1. Check the table and head slide (Figs. 291 and 293).

2. Set and clamp the job to the table with a stop at the front end (or hold in the vise).

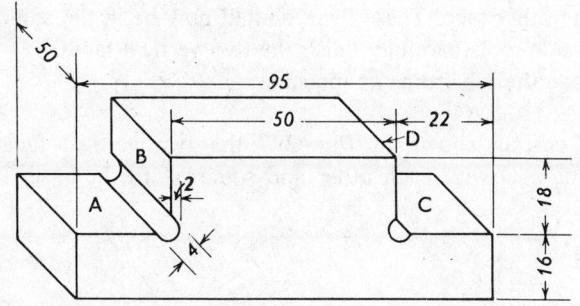

Fig. 296 To be shaped on faces A, B, C and D (see text)

3. Feeding down and across with side tools, rough out the four faces to within about $\frac{1}{2}$ mm of size.

4. With rounded parting tool take out the corner undercuts.

5. Obtain a tool which will cover both horizontal surfaces without being moved after setting in the toolpost (probably the parting tool used for (4) will do this).

6. With this tool machine both horizontal faces to within about 0·10 mm of size, then finish one face, *leave the cut setting alone*, stop the machine, wind the table across, start the machine and take the same

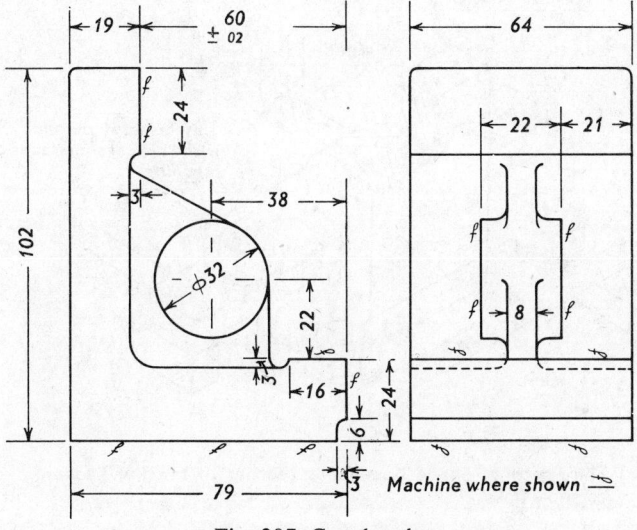

Fig. 297 Cast bracket

cut over the other face. These faces should now be in the same plane.

7. With side tools carefully finish the two vertical faces.

8. Remove the job from the machine.

On the casting shown at Fig. 297 the two vertical faces must be machined parallel with each other and square with the base. The narrow

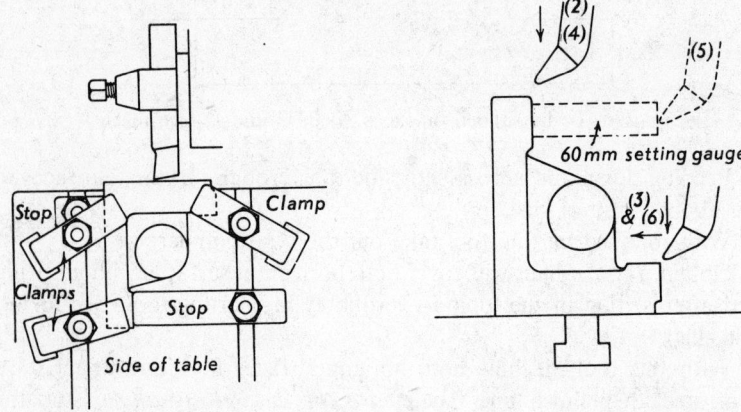

(a) First setting. Shape base (clamped to side of table)

(b) Second setting. Clamped to table (clamps and stop not shown)

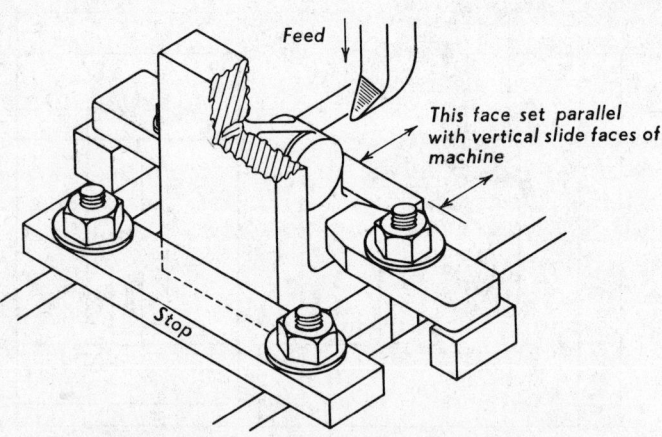

(c) Third setting. Shape boss faces (corner of bracket broken to show boss)

Fig. 298 Shaping operations on bracket (see text)

horizontal facing must be square with the other two, and parallel with the base. The boss faces must be square with the other surfaces.

1st Setting (Fig. 298(a))

1. Shape the base by clamping to an angle plate or to the side of the machine table or by holding in the vise and packing.

2nd Setting (Fig. 298(b))

1. Clamp to the machine table against a stop and line up from a tee slot.

2. By feeding a LH side tool from the head, rough down the upper vertical face until there is about $\frac{1}{2}$ mm left on.

3. Raise the table and with the same tool rough down the other facing as near as possible to the 60 mm dimension. Rough the horizontal facing using the cross feed.

4. Lower the table and finish the upper facing to give drawing thickness.

5. Wind the table over and set the tool to the 60 mm dimension using a gauge or a piece of bar ground to size.

6. Clamp the horizontal slide. Raise the table and feed down from the head to finish the lower facing.

7. Finish the horizontal facing with the cross feed.

The three facings now fulfil the requirements above.

3rd Setting (Fig. 298(c))

1. Clamp across on the table, facings inwards, and set one of the vertical facings parallel with the facings of the machine vertical slides. This is best done by straddling the machine facings with a good parallel strip and working with a height gauge or a setting piece ground parallel on its ends. After setting clamp a stop behind the casting.

2. Mount a side tool in the box and project it out enough for everything to clear the top of the casting. Set the ram on a short stroke and carefully adjust its position so that the tool stops short of the casting face (pull the machine round by hand for this).

3. Rough and finish the sides of the boss by feeding down with R and L hand side tools.

The sides of the boss will now be square with all the other faces.

In Fig. 299 the vee must be accurately parallel with and on the same centre line as the tenon slot. The sides of the vee must make equal angles with the vertical centre line. (The job is a casting with the vee roughly formed.)

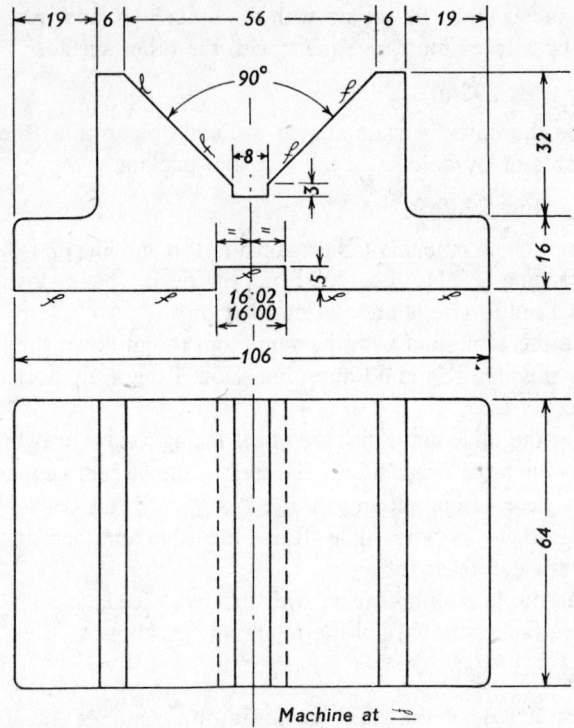

Fig. 299 Vee block

1st Setting (Fig. 300(a))

1. Hold in the vise with the base uppermost. Line up with the stroke of the ram. Rough and finish the base surface to within about ¼ mm, allowing for the opposite (top) surface to clean up.

2. With a parting tool take out the tenon slot to full depth +¼ mm and with ¼ mm to ½ mm each side on the width for opening out. Get this slot as near as possible central with the edges of the base.

3. Finish the tenon slot to width with knife or pointed side tools.

4. Skim up the base edges equidistant from the sides of the tenon slot, measure with vernier calipers to get both sides equal (take the minimum amount off the edges and leave a witness if possible).

5. Finish the surface of the base.

(The tenon slot is now parallel and centrally located with respect to the base edges.)

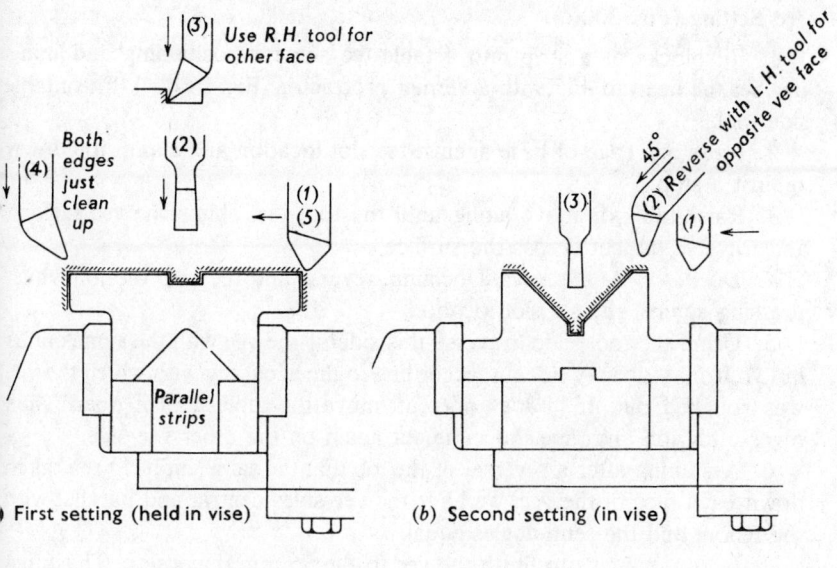

First setting (held in vise)

(b) Second setting (in vise)

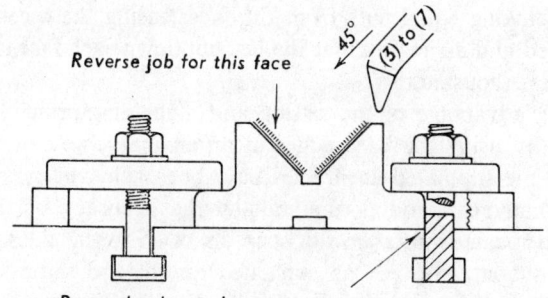

(c) Third setting (clamped to table) stop not shown

Fig. 300 Shaping operations on Vee block
(Numbers on tools refer to operation numbers in text.)

2nd Setting (Fig. 300(b))

Hold in vise or clamp base to table of machine. Line up.

1. Rough and finish top surface.

2. Rough sides of vee leaving about $\frac{1}{2}$ mm on each.

3. Set an 8 mm parting tool central (from base edges) and take out the slot to depth.

3rd Setting (Fig. 300(c))

1. Fit blocks or a strip into a table tee slot for positioning and lining up. Set the head to 45° with a vernier protractor (Fig. 294(a)). Fit suitable side tool.

2. Press one edge of base against tee slot location and clamp job down against a stop.

3. Raise table or move along until the tool just skims the vee surface and traverse the cut across the surface.

4. *Do not move the table.* Unclamp, reverse and reclamp the job whilst pressing against the tee slot location.

5. Using the tool slide, traverse this side of the vee with the same cut as in (3). It may or may not cut, according to the accuracy with which the vee was roughed out. If it does not cut, move the table until it does. Then reverse the job and take the same cut again on the other vee side.

6. Assuming after a reversal of the job that the same cut has been taken down each face of the vee, we have the vee sides central and parallel with the tenon, and the semi-angles equal.

7. It now remains to finish the vee to the 56 mm dimension. This must be done by taking equal cuts from the faces, using the reversal principle just described and arranging that the last cut down each face is a very light skim of a few thousandths.

By taking advantage of the setting and machining principles discussed above, and by using the base edges as an auxiliary, we can now be quite certain that the stipulated accuracies have been achieved.

An alternative method of maintaining the accuracy of the vee sides would be to locate from a tee-slot strip direct on to the sides of the tenon slot whilst shaping the vee as explained for the 3rd setting above. This would avoid cleaning up the edges of the casting for a location. The practice of cleaning up one or both edges, however, as we have described, often provides a more convenient locating face for subsequent shaping and boring operations to the upper side of a casting.

Examples of shaper work

Example 1. To shape up the block shown at Fig. 301(a), all faces to be parallel and square.

1. Obtain a pair of parallel strips of such a height that when the work is rested on them a portion of its thickness will project above the top of the vise jaws. Rest the block on them and press it down whilst tightening

the vise. Set the machine to a stroke about 50 mm longer than the work and move the ram longitudinally so that the tool overruns the job by about 25 mm at each end. Check the thickness to see that there is sufficient on for machining and if so put on sufficient cut to get below the scale. Engage the traverse and let the cut travel across. When the first cut is complete put on a small finishing cut with a finer feed if necessary and take this across (Fig. 301(b)).

2. Turn the work on its edge with a parallel under if necessary, hold in the vise with the machined face against the fixed jaw and clean up one edge square with the face just machined. Obtaining squareness will depend on the squareness of the solid vise jaw and upon the degree of contact between the two surfaces. For example, if the block is not parallel, the pressure of the moving jaw on the other face may cant the work and result in the machined surface contacting the fixed jaw in only one place. This may be avoided by applying the pressure through a small packing block which localises the effect and minimises the tendency (Fig. 301(c)).

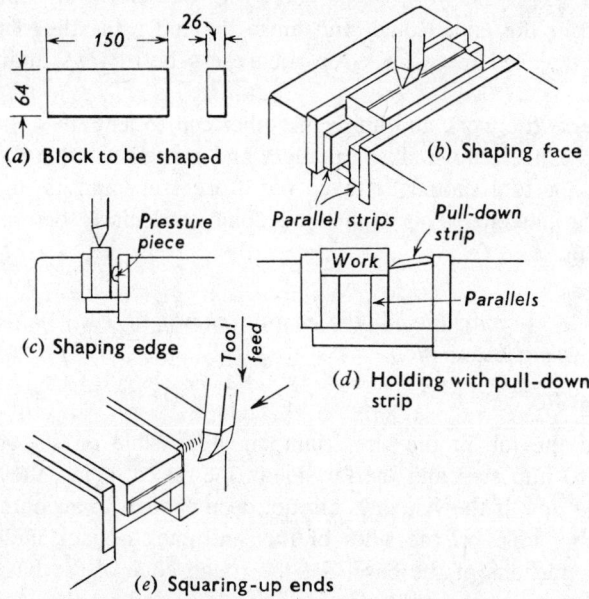

(a) Block to be shaped

(b) Shaping face

(c) Shaping edge

(d) Holding with pull-down strip

(e) Squaring-up ends

Fig. 301 Shaping operations on a block

3. Hold the work on the flat again, as at Fig. 301(b), with the finished face resting on the parallels and the machined edge in contact with the

solid jaw of the vise. After tightening up the vise knock the work down
on to the parallels with a lead, or rawhide hammer, until the strips are
tight at each end. If the unmachined edge in contact with the loose jaw is
not square there may be difficulty in getting contact with the strips due to
the lifting tendency. In the author's experience this used to be overcome
by using what was known as a 'pull down' strip. This is a strip of steel,
bevelled on one edge and placed between the work and the jaw in a slightly
tilted position so that when tightened it exercised a downward force on the
part being held.

Take a cut over the face and check for parallelism with a micrometer.
If the strips are tight this should be in order. Machine the face down to
size allowing that the final cut shall be light with a fine feed.

4. Hold the job by its flat faces, knock the machined edge down on to a
parallel strip or to the vise bottom and finish the other edge to size. Check
for squareness and parallelism.

5. Rotate the vise to the other position with its jaws perpendicular to the
tool travel. Hold the work on parallels with one end protruding a small
amount from the end. Rough and finish the end using the vertical head
travel for the feed (e). Check for squareness both ways and correct if
necessary.

6. Reverse the work and finish the other end to length.

7. With a file remove all sharp edges and corners. During the progress
of the job the tool should be taken out if necessary and its edge touched
up on a grindstone. This should be done particularly before taking a
finishing cut.

*Example 2. To machine up the casting shown at Fig. 302(a). (To be
machined where shown 'f'.)*

1st Setting

1. Hold the job in the vise, clamping if possible on the edge of the
25 mm × 10 mm step and the far side of the raised facing, the step being
at the fixed jaw. If the vise jaws are not deep enough to accommodate the
raised facing hold on the sides of this and pack or jack under the unsupported portions of the base. Set the rough edge of the base approximately level, check the underside for being level and set the edge adjacent
to the 25 mm × 10 mm step parallel with the tool travel.

2. Check to ensure that there is sufficient metal on for machining and
take a cut over the face (put on a cut deep enough to get under the scale)

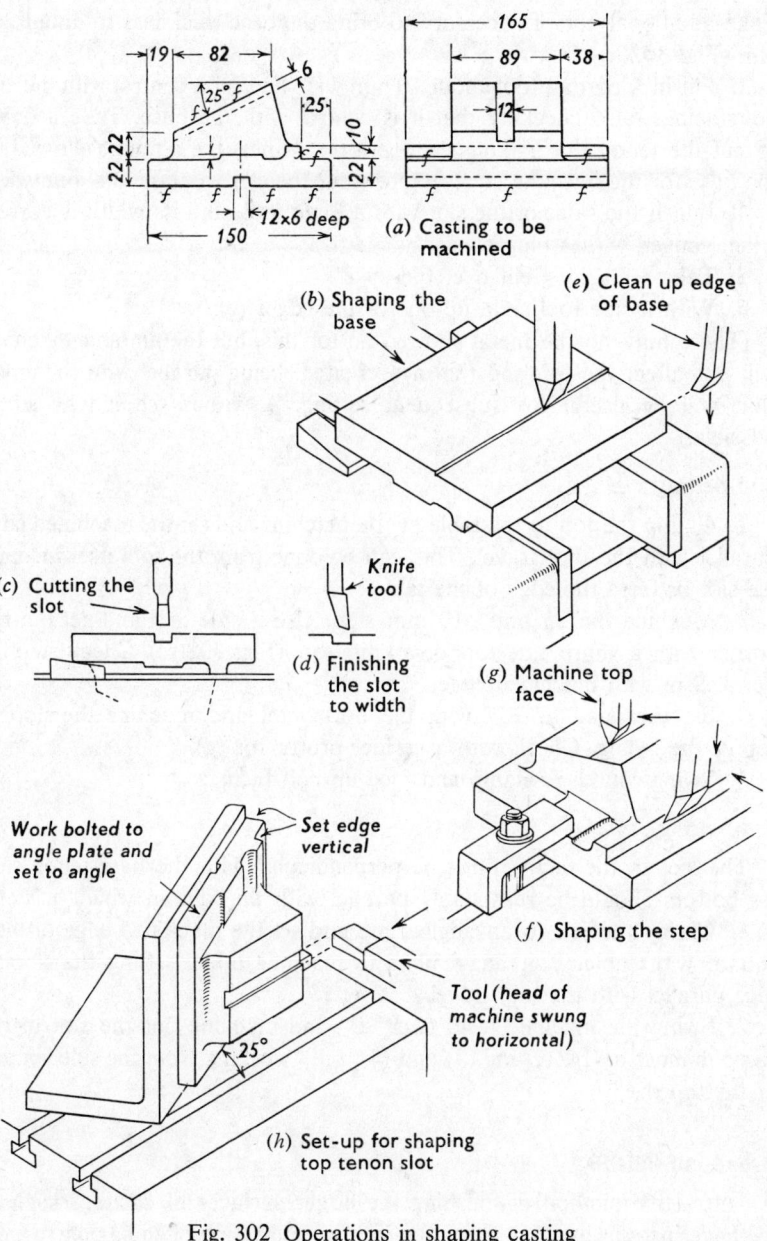

(a) Casting to be machined

(b) Shaping the base

(e) Clean up edge of base

(c) Cutting the slot

(d) Finishing the slot to width

Knife tool

(g) Machine top face

(f) Shaping the step

Work bolted to angle plate and set to angle

Set edge vertical

Tool (head of machine swung to horizontal)

(h) Set-up for shaping top tenon slot

Fig. 302 Operations in shaping casting

Take additional cuts if necessary to bring the base thickness to about $22\frac{1}{2}$ mm (Fig. 302(b)).

3. Put in a parting tool about 10 mm wide and set it central with the end bolt facings, also checking that it is central with the base. Take it down to cut the tenon slot making the depth 6 mm plus the amount allowed on the base for the finishing cut (c). Open out the slot to about $11\frac{1}{2}$ mm wide.

4. Finish the sides of the slot with a knife tool until its width is correct when gauged with a plug or slip gauge (d).

5. Take a finishing cut over the base.

6. With a side tool skim up one of the edges (e).

(There may not be metal allowed on for this, but the amount taken off will not affect the job and the face created, being parallel with the tenon slot, will be useful for subsequent setting. A witness should be left if possible.)

2nd Setting

1. Clamp the job to the table of the machine and set the machined edge parallel with the tool travel. This can be done from the tool itself, from a tee slot or from the edge of the table.

2. Machine the 25 mm $\times$ 10 mm step. Use a side tool and get out the corner with a sharp side tool or a knife tool (Fig. 302(f)). Check step for parallelism with machined edge.

3. Set the head to 25° with the horizontal and machine the sloping top of the facing. Check with a vernier protractor (g).

4. Swing round, re-clamp and face up bolt facings.

3rd Setting

The slot in the facing must be perpendicular with the base tenon slot. Its bottom should be reasonably parallel with the face in which it is cut.

1. Clamp the job to an angle plate and set the machined edge of base square with table. Bolt angle plate to machine table, setting the sloping face parallel with the tool travel.

2. Swing the machine head over 90° and machine out the slot in the same manner as 1st setting (3) and (4), only working from the side instead of the top (h).

Broad cut finishing

An attractive method of finishing the larger surfaces of castings such as we have just discussed is to employ a broad parting tool and a coarse feed.

The tool should be 9 mm to 12 mm wide, its edge ground dead straight and oilstoned to a keen edge (see Fig. 126(e), p. 183). It must be set rigidly in the tool post with the minimum of overhang, and its edge must be parallel with the face being machined. In operation, a one or two-thousandth cut is put on and the tool traversed across with a coarse feed which is best worked by hand, taking about one complete turn of the table feed-screw for each stroke of the ram. The result is a striped finish, which is more attractive than the usual feed marks. The method can only be used on cast-iron and the tool must be kept very keen with an oilstone. Success is helped by lifting the tool clear by hand on the return stroke and by filing a very small chamfer on the leading edge of the face to remove hard scale which might cause the tool to jump. Do not allow any oil or moisture to get on to a surface which is to be so finished.

Example 3. To shape up the casting shown at Fig. 303(a).

This is an example of a job which offers several possibilities for holding and setting up, and when such cases arise in the experience of the reader he must use his own judgment as to which method to employ, bearing in mind time, ease and rigidity of setting, tendency of each method towards accurate results and so on.

One method in this case would be as follows:

1. Clamp the bracket on its side to an angle plate or to the side of the machine table, with the base uppermost. Set the base as near as possible horizontal. Rough and finish the base (Fig. 303(b)).

2. Secure the base to an angle plate with the other surfaces to be machined uppermost. Place jacks or packing under the far end and clamp down on these, holding on the extreme 6 mm of length of the bracket. Set the upper surface horizontal, machine edge of base, 38 mm slot and its lower edge (see later) Fig. 303(c)).

3. Move clamps to hold at bottom of machined slot. Machine portion previously covered by clamps.

A second method, which permits all the machining to be done at one setting, is to hold the bracket partly in the vise and partly by clamps. To do this it might be necessary to move the vise nearer to the edge of the table, and perhaps pack it up higher. Before commencing, the vertical head slide should be tested for its squareness with the table (see Fig. 293(a)).

1. Hold the bracket partly in the vise, pack under the edge of the base and clamp down. Set the base approximately square with the table, and the upper surface approximately horizontal (Fig. 303(d)).

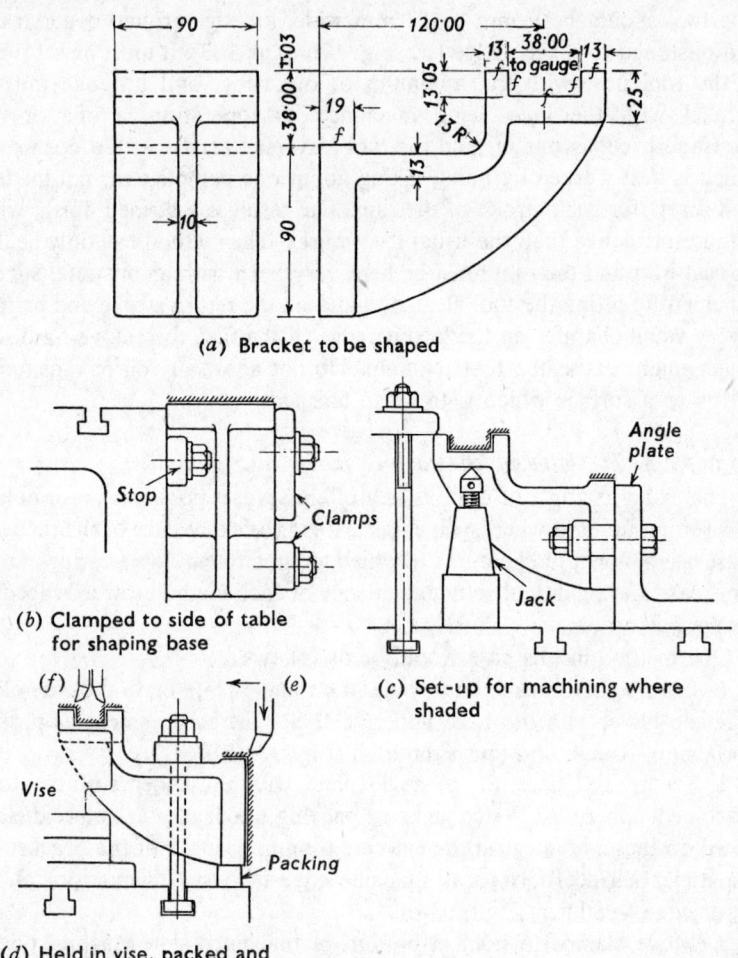

(a) Bracket to be shaped

(b) Clamped to side of table
for shaping base

Stop

Clamps

Angle
plate

Jack

(c) Set-up for machining where
shaded

(f)

(e)

Vise

Packing

(d) Held in vise, packed and
clamped

Fig. 303 Settings for shaping bracket

2. Rough shape surface and edge of base to within about $\frac{1}{2}$ mm of size.
Use side tool and vertical feed for base surface. Before machining check
that the 120·00 and 38·00 mm dimensions may be obtained (Fig. 303(e)).

3. Rough the two facings adjoining the slot, making them 38 mm above
the previously rough machined edge of the base. Rough out the slot to 13
mm deep, 37 mm wide with its bottom side about 102 mm from the base

$(120 - 19 + \frac{1}{2}$ mm to finish on two surfaces). Use side tools to go down the sides and match up the surface in the centre (f).

4. Finish surface and edge of base.

5. Finish the two facings adjoining the slot. Check for the 38·000 mm dimension with a parallel strip across facings, and a vernier between the strip and the base edge.

6. With a knife tool finish the side of the slot nearest the base, making it 101·00 mm from the base. Check over parallel strip pressed on base, and edge being machined, with vernier (101·00 mm + thickness of parallel).

7. Finish width of slot to 38·00 mm (vernier or plug gauge).

8. Finish slot to depth with parting tool. Check with depth gauge.

Safety on the shaping machine

We have already made some reference to safety in Chapter 5, and if by any chance the reader skipped that part of our discussion we hope he will turn back and consider it as seriously as any other part of his work. The shaper is not really any more dangerous than other machines, but as some people will get into trouble in the most innocent situations it may be worth while to issue a few words of advice. We are also encouraged in this where the shaper is concerned because it is often considered, like the drilling machine, to be a kind of general purpose workshop hack which anyone can use for odd jobs without much supervision or advice. Most of the other machine tools have their chief source of danger in members which rotate, and the reader must have heard of lurid examples of hair being torn out by an unguarded drilling spindle or someone being wound up by the carrier of a lathe. We hope he has neither seen nor experienced either, but have no doubt that he has heard accounts which have lost nothing in the telling! The only rotating hazard on the shaper is the squared shaft which projects from the side of the machine for setting the stroke, and this can be dangerous if after setting a stroke we forget to remove the handle from it before starting up. It is also a good precaution not to lean anywhere near this, as the rotating squared end could wind up in loose clothing. The ram is the only other real source of danger, and accidents can occur both by forgetfulness and by knowingly taking chances. On the latter aspect, the writer was once working a machine on a fairly long stroke and thought there was time to micrometer the thickness of the job at its end. Unfortunately the micrometer stuck a little and would not come away in time . . .! On modern, motorised, gear-driven machines the danger from an obstruction to the ram is greater than it was on the earlier belt-driven types, since a belt would often save a serious situation by slipping off. A train of gearing from a high-speed motor, however, is not so obliging. The ram, therefore, should always be considered as a source of danger, and this should never be forgotten, neither should a known risk be entertained. After setting or resetting a job, make sure that neither the tool nor any portion of the ram will foul anything on the table before starting the machine. Better still, pull the machine round by hand to make sure. There is sometimes a temptation to change the stroke or the setting of the ram while the machine is running. This is taking a known risk and should be avoided. Special care is necessary when a fine setting to the forward end of the stroke is necessary for shaping up to an obstruction, and the machine should be pulled round several times before it is started up.

Many mishaps occur through work being insecurely clamped and supported against the force of the cut and it should be remembered that a considerable shock force is exerted by the tool at the beginning of each stroke. Stops should always be used when jobs are clamped direct to the table and clamping should be arranged to come on to solid metal. Faulty clamping and setting up, as well as being dangerous, lead to inaccurate results due to work either moving or distorting. Even if a job is well clamped it may be pushed off the table, or the tool broken, by winding in to a large cut that was not checked or has jumped on due to a loose head slide.

The collection of the cutting chips on a shaping machine is a problem that has never been solved, and when dealing with hard steel particularly it is advisable not to stand at the front of the machine. Also, sweep the floor often if you value the leather on your shoe soles. Finally, since accidents can happen under the best regulated practice, cultivate the ability to stop the machine quickly and always avoid standing in the line of the ram.

Conclusion

We must now, for the time being, leave the reader to develop his knowledge, hoping that he has benefited at least from some of our discussions, and that we have stimulated him to further thought and enquiry. We encourage him to pursue his course, maintaining at all times, and whatever his knowledge, the simple mind of a student, for upon the simplest of foundations are built the most wonderful works of Nature.

Conversion tables

Fractional Sub-divisions of an inch to decimals and to millimetres.

in	in	milli-metres	in	in	milli-metres
1/64	0.015625	0.3969	33/64	0.609375	15.4781
1/32	0.03125	0.7938	5/8	0.625	15.875
3/64	0.046875	1.1906	41/64	0.640625	16.2719
1/16	0.0625	1.5875	21/32	0.65625	16.6688
5/64	0.078125	1.9844	43/64	0.671875	17.0656
3/32	0.09375	2.3812	11/16	0.6875	17.4625
7/64	0.109375	2.7781	45/64	0.703125	17.8594
1/8	0.125	3.175	23/32	0.71875	18.2562
9/64	0.140625	3.5719	47/64	0.734375	18.6531
5/32	0.15625	3.9688	3/4	0.75	19.05
11/64	0.171875	4.3656	49/64	0.765625	19.4469
3/16	0.1875	4.7625	25/32	0.78125	19.8438
13/64	0.203125	5.1594	51/64	0.796875	20.2406
7/32	0.21875	5.5562	13/16	0.8125	20.6375
15/64	0.234375	5.9531	53/64	0.828125	21.0344

Millimetres to inches
Based on 1 inch = 25·4 millimetres

mm	0	1	2	3	4	5	6	7	8	9
	in	in	in	in	in	in	in	in	in	in
—	—	0·03937	0·07874	0·11811	0·15748	0·19685	0·23622	0·27559	0·31496	0·35433
10	0·39370	0·43307	0·47244	0·51181	0·55118	0·59055	0·62992	0·66929	0·70866	0·74803
20	0·78740	0·82677	0·86614	0·90551	0·94488	0·98425	1·02362	1·06299	1·10236	1·14173
30	1·18110	1·22047	1·25984	1·29921	1·33858	1·37795	1·41732	1·45669	1·49606	1·53543
40	1·57480	1·61417	1·65354	1·69291	1·73228	1·77165	1·81102	1·85039	1·88976	1·92913
50	1·96850	2·00787	2·04724	2·08661	2·12598	2·16535	2·20472	2·24409	2·28346	2·32283
60	2·36220	2·40157	2·44094	2·48031	2·51969	2·55906	2·59843	2·63780	2·67717	2·71654
70	2·75591	2·79528	2·83465	2·87402	2·91339	2·95276	2·99213	3·03150	3·07087	3·11024
80	3·14961	3·18898	3·22835	3·26772	3·30709	3·34646	3·38583	3·42520	3·46457	3·50394
90	3·54331	3·58268	3·62205	3·66142	3·70079	3·74016	3·77953	3·81890	3·85827	3·89764
100	3·93701	3·97638	4·01575	4·05512	4·09449	4·13386	4·17323	4·21260	4·25197	4·29134
10	4·33071	4·37008	4·40945	4·44882	4·48819	4·52756	4·56693	4·60630	4·64567	4·68504
20	4·72441	4·76378	4·80315	4·84252	4·88189	4·92126	4·96063	5·0000	5·0394	5·0787
30	5·1181	5·1575	5·1969	5·2362	5·2756	5·3150	5·3543	5·3937	5·4331	5·4724
40	5·5118	5·5512	5·5906	5·6299	5·6693	5·7087	5·7480	5·7874	5·8268	5·8661
50	5·9055	5·9449	5·9843	6·0236	6·0630	6·1024	6·1417	6·1811	6·2205	6·2598
60	6·2992	6·3386	6·3780	6·4173	6·4567	6·4961	6·5354	6·5748	6·6142	6·6535
70	6·6929	6·7323	6·7717	6·8110	6·8504	6·8898	6·9291	6·9685	7·0079	7·0472
80	7·0866	7·1260	7·1654	7·2047	7·2441	7·2835	7·3228	7·3622	7·4016	7·4409
90	7·4803	7·5197	7·5591	7·5984	7·6378	7·6772	7·7165	7·7559	7·7953	7·8346

Fractions of an inch — decimal and millimetre equivalents

Fraction	Inch	mm		Fraction	Inch	mm
1/4	0·25	6·35		27/32	0·84375	21·4312
17/64	0·265625	6·7469		55/64	0·859375	21·8281
9/32	0·28125	7·1438		7/8	0·875	22·225
19/64	0·296875	7·5406		57/64	0·890625	22·6219
5/16	0·3125	7·9375		29/32	0·90625	23·0188
21/64	0·328125	8·3344		59/64	0·921875	23·4156
11/32	0·34375	8·7312		15/16	0·9375	23·8125
23/64	0·359375	9·1281		61/64	0·953125	24·2094
3/8	0·375	9·525		31/32	0·96875	24·6062
25/64	0·390625	9·9219		63/64	0·984375	25·0031
13/32	0·40625	10·3188		1	1	25·4
27/64	0·421875	10·7156		2	2	50·800
7/16	0·4375	11·1125		3	3	76·200
29/64	0·453125	11·5094		4	4	101·600
15/32	0·46875	11·9062		5	5	127·000
31/64	0·484375	12·3031		6	6	152·400
1/2	0·5	12·7		7	7	177·800
33/64	0·515625	13·0969		8	8	203·200
17/32	0·53125	13·4938		9	9	228·600
35/64	0·546875	13·8906		10	10	254·000
9/16	0·5625	14·2875		11	11	279·400
37/64	0·578125	14·6844		12	12	304·800
19/32	0·59375	15·0812				

Millimetres to inches

mm										
200	7·8740	7·9134	7·9528	7·9921	8·0315	8·0709	8·1102	8·1496	8·1890	8·2283
10	8·2677	8·3071	8·3465	8·3858	8·4252	8·4646	8·5039	8·5433	8·5827	8·6220
20	8·6614	8·7008	8·7402	8·7795	8·8189	8·8583	8·8976	8·9370	8·9764	9·0157
30	9·0551	9·0945	9·1339	9·1732	9·2126	9·2520	9·2913	9·3307	9·3701	9·4094
40	9·4468	9·4882	9·5276	9·5669	9·6063	9·6457	9·6850	9·7244	9·7638	9·8031
50	9·8425	9·8819	9·9213	9·9606	10·0000	10·0394	10·0787	10·1181	10·1575	10·1969
60	10·2362	10·2756	10·3150	10·3543	10·3937	10·4331	10·4724	10·5118	10·5512	10·5906
70	10·6299	10·6693	10·7087	10·7480	10·7874	10·8268	10·8661	10·9055	10·9449	10·9843
80	11·0236	11·0630	11·1024	11·1417	11·1811	11·2205	11·2598	11·2992	11·3386	11·3780
90	11·4173	11·4567	11·4961	11·5354	11·5748	11·6142	11·6535	11·6929	11·7323	11·7717
300	11·8110	11·8504	11·8898	11·9291	11·9685	12·0079	12·0472	12·0866	12·1260	12·1654
10	12·2047	12·2441	12·2835	12·3228	12·3622	12·4016	12·4409	12·4803	12·5197	12·5591
20	12·5984	12·6378	12·6772	12·7165	12·7559	12·7953	12·8346	12·8740	12·9134	12·9528
30	12·9921	13·0315	13·0709	13·1102	13·1496	13·1890	13·2283	13·2677	13·3071	13·3465
40	13·3858	13·4252	13·4646	13·5039	13·5433	13·5827	13·6220	13·6614	13·7008	13·7402
50	13·7795	13·8189	13·8583	13·8976	13·9370	13·9764	14·0157	14·0551	14·0945	14·1339
60	14·1732	14·2126	14·2520	14·2913	14·3307	14·3701	14·4094	14·4488	14·4882	14·5276
70	14·5669	14·6063	14·6457	14·6850	14·7244	14·7638	14·8031	14·8425	14·8819	14·9213
80	14·9606	15·0000	15·0394	15·0787	15·1181	15·1575	15·1969	15·2362	15·2756	15·3150
90	15·3543	15·3937	15·4331	15·4724	15·5118	15·5512	15·5906	15·6299	15·6693	15·7087
400	15·7480	15·7874	15·8268	15·8661	15·9055	15·9449	15·9843	16·0236	16·0630	16·1024
10	16·1417	16·1811	16·2205	16·2598	16·2992	16·3386	16·3780	16·4173	16·4567	16·4961
20	16·5354	16·5748	16·6142	16·6535	16·6929	16·7323	16·7717	16·8110	16·8504	16·8898
30	16·9291	16·9685	17·0079	17·0472	17·0866	17·1260	17·1654	17·2047	17·2441	17·2835
40	17·3228	17·3622	17·4016	17·4409	17·4803	17·5197	17·5591	17·5984	17·6378	17·6772
50	17·7165	17·7559	17·7953	17·8346	17·8740	17·9134	17·9528	17·9921	18·0315	18·0709
60	18·1102	18·1496	18·1890	18·2283	18·2677	18·3071	18·3465	18·3858	18·4252	18·4646
70	18·5039	18·5433	18·5827	18·6220	18·6614	18·7008	18·7402	18·7795	18·8189	18·8583
80	18·8976	18·9370	18·9764	19·0157	19·0551	19·0945	19·1339	19·1732	19·2126	19·2520
90	19·2913	19·3307	19·3701	19·4094	19·4488	19·4882	19·5276	19·5669	19·6063	19·6457
500	19·6850	19·7244	19·7638	19·8031	19·8425	19·8819	19·9213	19·9606	20·0000	20·0394

GKN Bolts & Nuts Ltd.

For a full list of conversions refer to BS 350, Parts 1 and 2.

APPENDIX 2

The ISO metric Thread

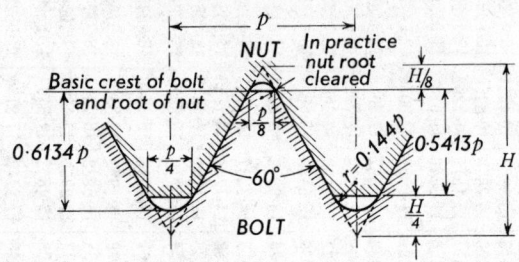

Recommended* Bolt Diameter (mm)	ISO metric Coarse				ISO metric Fine			
	Designation	Pitch (mm)	Bolt Root Dia. (mm)	Nut Core Dia. (mm)	Designation	Pitch (mm)	Bolt Root Dia. (mm)	Nut Co Dia. (m
1·6	M1·6×0·35	0·35	1·17	1·22				
2	M2×0·4	0·4	1·51	1·57				
2·5	M2·5×0·45	0·45	1·95	2·01				
3	M3×0·5	0·5	2·39	2·46				
4	M4×0·7	0·7	3·14	3·24				
5	M5×0·8	0·8	4·02	4·13				
6	M6×1	1	4·77	4·92				
8	M8×1·25	1·25	6·47	6·65	M8×1	1	6·77	6·92
10	M10×1·5	1·5	8·16	8·38	M10×1·25	1·25	8·47	8·65
12	M12×1·75	1·75	9·85	10·11	M12×1·25	1·25	10·47	10·65
16	M16×2	2	13·55	13·84	M16×1·5	1·5	14·16	14·38
20	M20×2·5	2·5	16·93	17·29	M20×1·5	1·5	18·16	18·38
24	M24×3	3	20·32	20·75	M24×2	2	19·55	21·84
30	M30×3·5	3·5	25·71	26·21	M30×2	2	25·55	27·84
36	M36×4	4	31·09	31·67	M36×3	3	32·32	32·75
42	M42×4·5	4·5	36·48	37·13	M42×4	4	37·09	37·67
48	M48×5	5	41·87	42·59	M48×4	4	43·09	43·67
56	M56×5·5	5·5	49·25	50·05	M56×4	4	51·09	51·67
64	M64×6	6	56·64	57·50	M64×4	4	59·09	59·67

Note method of designating thread: M(Dia)×(Pitch), e.g. M10×1·5.

* BS 3643 also specifies second choice and third choice.

APPENDIX 3

The Unified Thread

Thread form profile identical with the ISO metric thread on Appendix 2.
Particulars of Diameters and Pitches

Bolt Diameter (in)	Unified Coarse				Unified Fine			
	Designation	Thread per in	Bolt Root Dia. (in)	Nut Core Dia. (in)	Designation	Thread per in	Bolt Root Dia. (in)	Nut Core Dia. (in)
$\frac{1}{4}$ (0·250)	$\frac{1}{4}$–20. UNC	20	0·1887	0·1959	$\frac{1}{4}$–28. UNF	28	0·2062	0·2113
$\frac{5}{16}$ (0·3125)	$\frac{5}{16}$–18. UNC	18	0·2443	0·2524	$\frac{5}{16}$–24. UNF	24	0·2614	0·2674
$\frac{3}{8}$ (0·375)	$\frac{3}{8}$–16. UNC	16	0·2983	0·3073	$\frac{3}{8}$–24. UNF	24	0·3239	0·3299
$\frac{7}{16}$ (0·4375)	$\frac{7}{16}$–14. UNC	14	0·3499	0·3602	$\frac{7}{16}$–20. UNF	20	0·3762	0·3834
$\frac{1}{2}$ (0·50)	$\frac{1}{2}$–13. UNC	13	0·4056	0·4167	$\frac{1}{2}$–20. UNF	20	0·4387	0·4459
$\frac{9}{16}$ (0·5625)	$\frac{9}{16}$–12. UNC	12	0·4603	0·4723	$\frac{9}{19}$–18. UNF	18	0·4943	0·5024
$\frac{5}{8}$ (0·625)	$\frac{5}{8}$–11. UNC	11	0·5135	0·5266	$\frac{5}{8}$–18. UNF	18	0·5568	0·5649
$\frac{3}{4}$ (0·750)	$\frac{3}{4}$–10. UNC	10	0·6273	0·6417	$\frac{3}{4}$–16. UNF	16	0·6733	0·6823
$\frac{7}{8}$ (0·875)	$\frac{7}{8}$–9. UNC	9	0·7387	0·7547	$\frac{7}{8}$–14. UNF	14	0·7874	0·7977
1 (1·00)	1–8. UNC	8	0·8466	0·8647	1–12. UNF	12	0·8978	0·9098
$1\frac{1}{8}$ (1·125)	$1\frac{1}{8}$–7. UNC	7	0·9497	0·9704	$1\frac{1}{8}$–12. UNF	12	1·0228	1·0348
$1\frac{1}{4}$ (1·250)	$1\frac{1}{4}$–7. UNC	7	1·0747	1·0954	$1\frac{1}{4}$–12. UNF	12	1·1478	1·1598
$1\frac{3}{8}$ (1·375)	$1\frac{3}{8}$–6. UNC	6	1·1705	1·1946	$1\frac{3}{8}$–12. UNF	12	1·2728	1·2848
$1\frac{1}{2}$ (1·50)	$1\frac{1}{2}$–6. UNC	6	1·2955	1·3196	$1\frac{1}{2}$–12. UNF	12	1·3978	1·4098
$1\frac{3}{4}$ (1·750)	$1\frac{3}{4}$–5. UNC	5	1·5046	1·5335				
2 (2·00)	2–4$\frac{1}{2}$. UNC	4$\frac{1}{2}$	1·7274	1·7594				
thence by steps of $\frac{1}{4}$ in to 4 in								

(The clearing of the root of the nut by a radius as shown causes this
diameter to be increased by $0·072p$.)

APPENDIX 4

Whitworth and British Standard Fine Threads

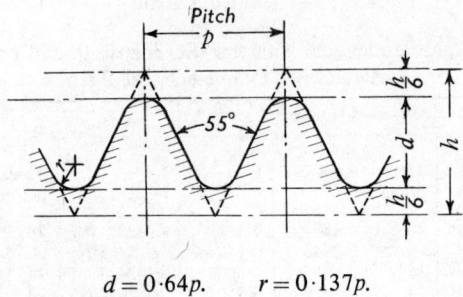

$$d = 0 \cdot 64p. \qquad r = 0 \cdot 137p.$$

Bolt Dia. (in)	Whitworth				British Standard Fine		
	Threads per inch	Pitch of Thread (in)	Thread Core Dia. (in)	Tapping Size	Threads per inch	Thread Core Dia. (in)	Tapping Size (in)
$\frac{1}{4}$	20	0·050	0·186	5 mm	26	0·201	$5\frac{1}{4}$ mm
$\frac{5}{16}$	18	0·0556	0·241	$\frac{1}{4}$ in	22	0·254	0·261 in
$\frac{3}{8}$	16	0·0625	0·295	$\frac{5}{16}$,,	20	0·311	8 mm
$\frac{7}{16}$	14	0·0714	0·346	$\frac{23}{64}$,,	18	0·366	$\frac{3}{8}$ in
$\frac{1}{2}$	12	0·083	0·393	$\frac{13}{32}$,,	16	0·420	$\frac{27}{64}$ in
$\frac{9}{16}$	12	0·083	0·456	$\frac{15}{32}$,,	16	0·483	$12\frac{1}{2}$ mm
$\frac{5}{8}$	11	0·091	0·509	$\frac{17}{32}$,,	14	0·534	$\frac{35}{64}$ in
$\frac{3}{4}$	10	0·100	0·622	$\frac{41}{64}$,,	12	0·643	$\frac{21}{32}$,,
$\frac{7}{8}$	9	0·111	0·733	$\frac{3}{4}$,,	11	0·759	$\frac{25}{32}$,,
1	8	0·125	0·840	$\frac{55}{64}$,,	10	0·872	$\frac{57}{64}$,,
$1\frac{1}{4}$	7	0·143	1·067	$1\frac{3}{32}$,,	9	1·108	$1\frac{1}{8}$,,
$1\frac{1}{2}$	6	0·167	1·287	$1\frac{5}{16}$,,	8	1·340	$34\frac{1}{2}$ mm
$1\frac{3}{4}$	5	0·200	1·494	$1\frac{33}{64}$,,	7	1·567	$1\frac{19}{32}$ in
2	$4\frac{1}{2}$	0·222	1·715	$1\frac{3}{4}$,,	7	1·817	$1\frac{27}{32}$,,

Note. The tapping sizes given above for Whitworth and BSF threads are based on the formula $T = D - 1 \cdot 1328p$, and give a thread about $88 \cdot 5\%$ of full form (see page 290).

APPENDIX 5

British Association (BA) Threads

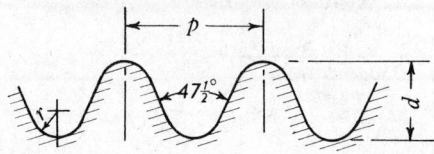

$$d = 0.6p. \qquad r = \frac{2p}{11}.$$

BA No.	0	1	2	3	4	5
Diameter (mm)	6	5·3	4·7	4·1	3·6	3·2
Pitch (mm). . .	. 1	0·9	0·81	0·73	0·66	0·59
Core Diameter (mm)	. 4·8	4·22	3·73	3·22	2·81	2·49
Core Diameter (in).	. 0·189	0·166	0·147	0·127	0·111	0·098

BA No.	6	7	8	9	10	11
Diameter (mm). .	. 2·8	2·5	2·2	1·9	1·7	1·5
Pitch (mm). . .	. 0·53	0·48	0·43	0·39	0·35	0·31
Core Diameter (mm)	. 2·16	1·92	1·68	1·43	1·28	1·13
Core Diameter (in).	. 0·085	0·076	0·066	0·056	0·05	0·044

BA No.	12	13	14	15	16	17
Diameter (mm). .	. 1·3	1·2	1·0	0·9	0·79	0·7
Pitch (mm). . .	. 0·28	0·25	0·23	0·21	0·19	0·17
Core Diameter (mm)	. 0·96	0·90	0·72	0·65	0·56	0·50
Core Diameter (in).	. 0·038	0·035	0·028	0·025	0·022	0·020

BA No.	18	19	20	21	22	23
Diameter (mm). .	. 0·62	0·54	0·48	0·42	0·37	0·29
Pitch (mm). . .	. 0·15	0·14	0·12	0·11	0·098	0·09
Core Diameter (mm)	. 0·44	0·37	0·34	0·29	0·25	0·22
Core Diameter (in).	. 0·017	0·015	0·013	0·012	0·011	0·009

APPENDIX 6

British Standard Pipe Threads (BSP)

(Thread Form—Whitworth)

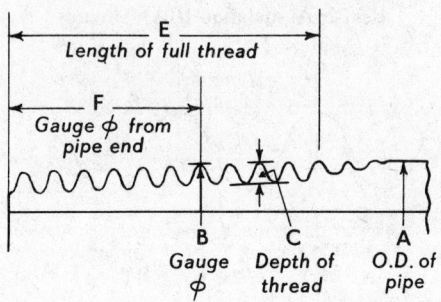

Thread on end of pipe

Size = Bore of Pipe	A	B	C	Core dia.	E	F Standard	Max	Min	Thread per in
in	in	in	in	in	in	in	in	in	
$\frac{1}{8}$	$\frac{15}{32}$	0·383	0·0230	0·337	$\frac{3}{8}$	$\frac{5}{32}$ (0·1563)	0·18	0·13	28
$\frac{1}{4}$	$\frac{17}{32}$	0·518	0·0335	0·451	$\frac{7}{16}$	$\frac{3}{16}$ (0·1875)	0·22	0·16	19
$\frac{3}{8}$	$\frac{11}{16}$	0·656	0·0335	0·589	$\frac{1}{2}$	$\frac{1}{4}$ (0·2500)	0·29	0·21	19
$\frac{1}{2}$	$\frac{27}{32}$	0·825	0·0455	0·734	$\frac{5}{8}$	$\frac{1}{4}$ (0·2500)	0·29	0·21	14
$\frac{3}{4}$	$1\frac{1}{16}$	1·041	0·0455	0·950	$\frac{3}{4}$	$\frac{3}{8}$ (0·3750)	0·44	0·31	14
1	$1\frac{11}{32}$	1·309	0·0580	1·193	$\frac{7}{8}$	$\frac{3}{8}$ (0·3750)	0·44	0·31	11
$1\frac{1}{4}$	$1\frac{11}{16}$	1·650	0·0580	1·534	1	$\frac{1}{2}$ (0·5000)	0·58	0·42	11
$1\frac{1}{2}$	$1\frac{29}{32}$	1·882	0·0580	1·766	1	$\frac{1}{2}$ (0·5000)	0·58	0·42	11
2	$2\frac{3}{8}$	2·347	0·0580	2·231	$1\frac{1}{8}$	$\frac{5}{8}$ (0·6250)	0·73	0·52	11
$2\frac{1}{2}$	3	2·960	0·0580	2·844	$1\frac{1}{4}$	$\frac{11}{16}$ (0·6875)	0·80	0·57	11
3	$3\frac{1}{2}$	3·460	0·0580	3·344	$1\frac{3}{8}$	$\frac{13}{16}$ (0·8125)	0·95	0·68	11
$3\frac{1}{2}$	4	3·950	0·0580	3·834	$1\frac{1}{2}$	$\frac{7}{8}$ (0·8750)	1·02	0·73	11
4	$4\frac{1}{2}$	4·450	0·0580	4·334	$1\frac{5}{8}$	1 (1·0000)	1·17	0·83	11
$4\frac{1}{2}$	5	4·950	0·0580	4·834	$1\frac{5}{8}$	1 (1·0000)	1·17	0·83	11
5	$5\frac{1}{2}$	5·450	0·0580	5·334	$1\frac{3}{4}$	$1\frac{1}{8}$ (1·1250)	1·31	0·94	11
6	$6\frac{1}{2}$	6·450	0·0580	6·334	2	$1\frac{3}{8}$ (1·3750)	1·60	1·15	11
7	$7\frac{1}{2}$	7·450	0·0640	7·322	$2\frac{1}{8}$	$1\frac{3}{8}$ (1·3750)	1·60	1·15	10
8	$8\frac{1}{2}$	8·450	0·0640	8·322	$2\frac{1}{4}$	$1\frac{1}{2}$ (1·5000)	1·75	1·25	10
9	$9\frac{1}{2}$	9·450	0·0640	9·322	$2\frac{1}{4}$	$1\frac{1}{2}$ (1·500)	1·75	1·25	10
10	$10\frac{1}{2}$	10·450	0·0640	10·322	$2\frac{3}{8}$	$1\frac{1}{8}$ (1·6250)	1·90	1·35	10

APPENDIX 7

Morse Tapers

The Morse taper is an international standard, and as originally conceived, its elements were set out in inch units. It is possible that in certain parts of the world it will continue this way, so we reproduce in the tabulated information the original inch unit dimensions, together with a table of their metric conversions

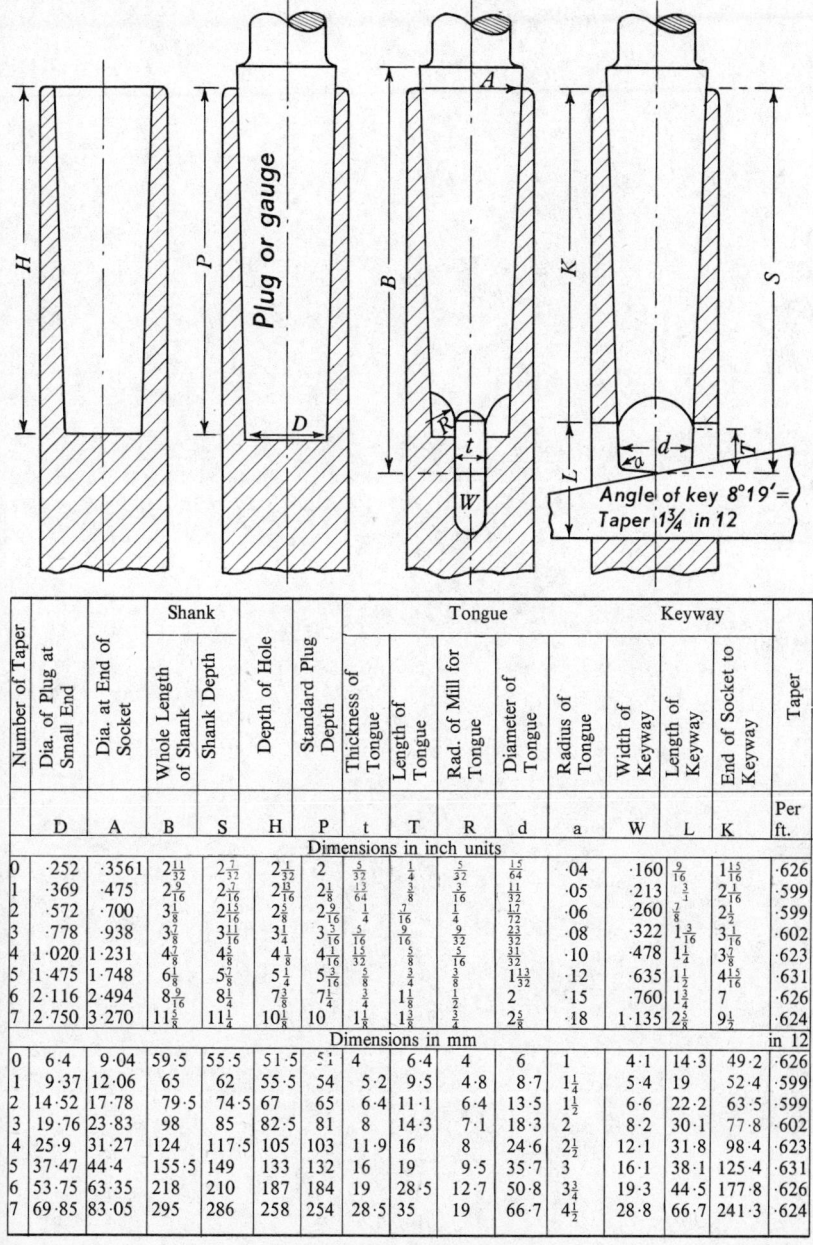

Angle of key 8°19′ = Taper 1¾ in 12

Number of Taper	Dia. of Plug at Small End	Dia. at End of Socket	Shank: Whole Length of Shank	Shank Depth	Depth of Hole	Standard Plug Depth	Tongue: Thickness of Tongue	Length of Tongue	Rad. of Mill for Tongue	Diameter of Tongue	Radius of Tongue	Keyway: Width of Keyway	Length of Keyway	End of Socket to Keyway	Taper
	D	A	B	S	H	P	t	T	R	d	a	W	L	K	Per ft.
Dimensions in inch units															
0	·252	·3561	$2\frac{11}{32}$	$2\frac{7}{32}$	$2\frac{1}{32}$	2	$\frac{5}{32}$	$\frac{1}{4}$	$\frac{5}{32}$	$\frac{15}{64}$	·04	·160	$\frac{9}{16}$	$1\frac{5}{16}$	·626
1	·369	·475	$2\frac{9}{16}$	$2\frac{7}{16}$	$2\frac{13}{16}$	$2\frac{1}{8}$	$\frac{13}{64}$	$\frac{3}{8}$	$\frac{3}{16}$	$\frac{11}{32}$	·05	·213	$\frac{3}{4}$	$2\frac{1}{16}$	·599
2	·572	·700	$3\frac{1}{8}$	$2\frac{15}{16}$	$2\frac{5}{8}$	$2\frac{9}{16}$	$\frac{1}{4}$	$\frac{7}{16}$	$\frac{1}{4}$	$\frac{17}{32}$	·06	·260	$\frac{7}{8}$	$2\frac{1}{2}$	·599
3	·778	·938	$3\frac{7}{8}$	$3\frac{11}{16}$	$3\frac{1}{4}$	$3\frac{3}{16}$	$\frac{5}{16}$	$\frac{9}{16}$	$\frac{9}{32}$	$\frac{23}{32}$	·08	·322	$1\frac{3}{16}$	$3\frac{1}{16}$	·602
4	1·020	1·231	$4\frac{7}{8}$	$4\frac{5}{8}$	$4\frac{1}{8}$	$4\frac{1}{16}$	$\frac{15}{32}$	$\frac{5}{8}$	$\frac{5}{16}$	$\frac{31}{32}$	·10	·478	$1\frac{1}{4}$	$3\frac{3}{8}$	·623
5	1·475	1·748	$6\frac{1}{8}$	$5\frac{5}{8}$	$5\frac{1}{4}$	$5\frac{3}{16}$	$\frac{5}{8}$	$\frac{3}{4}$	$\frac{3}{8}$	$1\frac{3}{32}$	·12	·635	$1\frac{1}{2}$	$4\frac{5}{16}$	·631
6	2·116	2·494	$8\frac{9}{16}$	$8\frac{1}{4}$	$7\frac{7}{8}$	$7\frac{1}{4}$	$\frac{3}{4}$	$1\frac{1}{8}$	$\frac{1}{2}$	2	·15	·760	$1\frac{3}{4}$	7	·626
7	2·750	3·270	$11\frac{1}{8}$	$11\frac{1}{4}$	$10\frac{1}{8}$	10	$1\frac{1}{8}$	$1\frac{3}{8}$	$\frac{3}{4}$	$2\frac{5}{8}$	·18	1·135	$2\frac{5}{8}$	$9\frac{1}{2}$	·624
Dimensions in mm															in 12
0	6·4	9·04	59·5	55·5	51·5	51	4	6·4	4	6	1	4·1	14·3	49·2	·626
1	9·37	12·06	65	62	55·5	54	5·2	9·5	4·8	8·7	$1\frac{1}{4}$	5·4	19	52·4	·599
2	14·52	17·78	79·5	74·5	67	65	6·4	11·1	6·4	13·5	$1\frac{1}{2}$	6·6	22·2	63·5	·599
3	19·76	23·83	98	85	82·5	81	8	14·3	7·1	18·3	2	8·2	30·1	77·8	·602
4	25·9	31·27	124	117·5	105	103	11·9	16	8	24·6	$2\frac{1}{2}$	12·1	31·8	98·4	·623
5	37·47	44·4	155·5	149	133	132	16	19	9·5	35·7	3	16·1	38·1	125·4	·631
6	53·75	63·35	218	210	187	184	19	28·5	12·7	50·8	$3\frac{3}{4}$	19·3	44·5	177·8	·626
7	69·85	83·05	295	286	258	254	28·5	35	19	66·7	$4\frac{1}{2}$	28·8	66·7	241·3	·624

Index